T0235019

Postharvest Oxidative Stress
in Horticultural Crops

FOOD PRODUCTS PRESS®
Crop Science
Amarjit S. Basra, PhD
Senior Editor

Mineral Nutrition of Crops: Fundamental Mechanisms and Implications by Zdenko Rengel

Conservation Tillage in U.S. Agriculture: Environmental, Economic, and Policy Issues by Noel D. Uri

Cotton Fibers: Developmental Biology, Quality Improvement, and Textile Processing edited by Amarjit S. Basra

Heterosis and Hybrid Seed Production in Agronomic Crops edited by Amarjit S. Basra

Intensive Cropping: Efficient Use of Water, Nutrients, and Tillage by S. S. Prihar, P. R. Gajri, D. K. Benbi, and V. K. Arora

Physiological Bases for Maize Improvement edited by María E. Otegui and Gustavo A. Slafer

Plant Growth Regulators in Agriculture and Horticulture: Their Role and Commercial Uses edited by Amarjit S. Basra

Crop Responses and Adaptations to Temperature Stress edited by Amarjit S. Basra

Plant Viruses As Molecular Pathogens by Jawaid A. Khan and Jeanne Dijkstra

In Vitro Plant Breeding by Acram Taji, Prakash P. Kumar, and Prakash Lakshmanan

Crop Improvement: Challenges in the Twenty-First Century edited by Manjit S. Kang

Barley Science: Recent Advances from Molecular Biology to Agronomy of Yield and Quality edited by Gustavo A. Slafer, José Luis Molina-Cano, Roxana Savin, José Luis Araus, and Ignacio Romagosa

Tillage for Sustainable Cropping by P. R. Gajri, V. K. Arora, and S. S. Prihar

Bacterial Disease Resistance in Plants: Molecular Biology and Biotechnological Applications by P. Vidhyasekaran

Handbook of Formulas and Software for Plant Geneticists and Breeders edited by Manjit S. Kang

Postharvest Oxidative Stress in Horticultural Crops edited by D. Mark Hodges

Encyclopedic Dictionary of Plant Breeding and Related Subjects by Rolf H. G. Schlegel

Handbook of Processes and Modeling in the Soil-Plant System edited by D. K. Benbi and R. Nieder

The Lowland Maya Area: Three Millennia at the Human-Wildland Interface edited by A. Gomez-Pompa, M.F. Allen, S. Fedick, and J.J. Jiménez-Osornio

Biodiversity and Pest Management in Agroecosystems, Second Edition by Miguel A. Altieri and Clara I. Nicholls

Postharvest Oxidative Stress in Horticultural Crops

D. Mark Hodges, PhD
Editor

Food Products Press
An Imprint of The Haworth Press, Inc.
New York • London • Oxford

First published 2003 by Haworth Press, Inc

Published 2019 by Routledge
52 Vanderbilt Avenue, New York, NY 10017
2 Park Square Milton Park, Abingdon Oxon OX14 4RN

Routledge is an imprint of the Taylor & Francis Group, an informa business

© 2019 by Taylor & Francis.

Cover design by Lora Wiggins.

Library of Congress Cataloging-in-Publication Data

Postharvest oxidative stress in horticultural crops / D. Mark Hodges, editor.
    p. cm.
    Includes bibliographical references and index.
    ISBN 1-56022-962-4 (alk. paper)—ISBN 1-56022-963-2
    1. Horticultural crops—Postharvest diseases and injuries. 2. Horticultural crops—Effect of oxygen on. 3. Oxidation, Physiological. I. Hodges, D. Mark.

SB319.5 .P67 2003
635'.0468—dc21

                                             2002068607

CONTENTS

ABOUT THE EDITOR

D. Mark Hodges, PhD, is a Research Scientist specializing in plant stress physiology at the Atlantic Food and Horticulture Research Centre of Agriculture and Agri-Food–Canada. His primary research interests involve plant oxidative stress. Early in his career, Dr. Hodges examined the role that antioxidants played in limiting chilling-induced oxidative stress in differentially chilling-sensitive maize plants. Currently, he is investigating the role of oxidative stress in senescence of horticultural crops. He is also focusing on identifying and developing technologies to enhance stress tolerance, postharvest storage quality, and functional food/nutraceutical value in horticultural crops.

Dr. Hodges is an authority on plant oxidative stress and has published numerous articles in various prestigious journals. He is also Adjunct Professor at Acadia University and trains graduate students in addition to his research duties. He provides leadership in organizing and fostering collaboration in international horticultural research and research dissemination.

CONTRIBUTORS

John M. DeLong, PhD, is Research Scientist, Agriculture and Agri-Food Canada, Atlantic Food and Horticulture Research Centre, Kentville, Nova Scotia.

Charles F. Forney, PhD, is Postharvest Physiologist at the Atlantic Food and Horticulture Research Centre, Kentville, Nova Scotia.

Gene Edward Lester, PhD, is Research Postharvest Plant Physiologist, USDA-ARS, Kika de la Garza Subtropical Agricultural Research Center, Weslaco, Texas.

Susan Lurie, PhD, is Professor in the Department of Postharvest Science, Agricultural Research Organization, Volcani Center, Bet Dagan, Israel.

Andrea Masia, PhD, is Biochemist/Plant Physiologist in the Dipartimento di Colture Arboree, Bologna, Italy.

Robert K. Prange, PhD, is Research Scientist, Agriculture and Agri-Food Canada, Atlantic Food and Horticulture Research Centre, Kentville, Nova Scotia.

Albert C. Purvis, PhD, is Professor of Postharvest Physiology in the Horticulture Department, The University of Georgia Coastal Plain Experiment Station, Tifton, Georgia.

Mulpuri V. Rao is Group Leader, Plant Transformation for Paradigm Genetics, Inc., RTP, North Carolina.

Peter M. A. Toivonen, PhD, is Postharvest Physiologist in the Food Research Program, Agriculture and Agri-Food Canada, Pacific Agri-Food Research Centre, Summerland, British Columbia.

Christopher B. Watkins is Associate Professor in the Department of Horticulture, Cornell University, Ithaca, New York.

Wendy V. Wismer, PhD, is Assistant Professor of Agricultural, Food and Nutritional Sciences at the University of Alberta, Edmonton, Alberta.

Preface

Oxidative stress occurs when the generation of active oxygen species (e.g., H_2O_2, $\cdot OH$) exceeds the capacity of the organism to maintain cellular redox homeostasis; peroxidized lipids and denatured proteins and nucleic acids are often the result. Oxidative stress not only has detrimental effects on plant function but is also associated with animal diseases such as atherosclerosis, cancers, various neurological disorders, and advanced macular degeneration. Only relatively recently has the induction of oxidative stress been associated with such stresses to growing plants (preharvest) as chilling, drought, ambient pollutants, and high light levels. In the past few years, the impact of oxidative stress on postharvest quality has also gained attention. Whether arising from preharvest practices, storage conditions, endogenous or exogenous triggers, or general senescent processes, oxidative stress affects the appearance, quality, shelf life, and nutritional value of harvested horticultural crops.

This compilation brings together internationally respected authorities in postharvest oxidative stress who provide new insights and approaches for individuals interested in this dynamic area. Topics in this book cover factors inducing oxidative stress in postharvest crops (e.g., storage temperatures and atmospheres), effects of oxidative stress on postharvest produce (scald, senescence dynamics), regulation of oxidative stress in postharvest commodities (active oxygen species production, antioxidants), and novel technologies to enhance resistance of postharvest crops to oxidative stress (genetic engineering, physical and/or chemical treatments). This book will be extremely useful to advanced students, teachers, and researchers in plant physiology, biochemistry, molecular biology, biotechnology, breeding, and horticulture, and especially to those who work in the field of plant stress.

This book would not have been possible without the cooperation of the contributing authors who have shared their particular knowledge for the benefit of the international scientific community. I would also like to thank Dr. Amarjit Basra, Punjab Agricultural University, for the original invitation to edit this book and for his continued support during the process.

Chapter 1

Overview: Oxidative Stress and Postharvest Produce

D. Mark Hodges

ACTIVE OXYGEN SPECIES

All aerobic organisms require oxygen as an essential component of their metabolism. However, life with oxygen also involves potential danger. The paramagnetism of molecular oxygen (O_2) denotes unpaired electrons; a triplet molecule of O_2 in the ground state contains two electrons with parallel spins. The addition of a pair of electrons would require a spin inversion in order to conform to the Pauli Exclusion Principle by which a pair of spin-opposed electrons cannot be transferred from a reductant to O_2 without inversion of an electronic spin. The length of time required to invert the spin of an electron, although a very rapid process, is relatively long in comparison to the half-life of collisional complexes. Because it is kinetically limited by this relatively slow spin inversion process, O_2 has a much greater tendency to react with radicals and single unpaired electrons than it does with substrates that donate pairs of electrons. Although O_2 itself is not particularly toxic, it forms the basis of by-products that are highly reactive and that pose a potential for severe cellular damage and/or lethality. These active oxygen species (AOS) arise from ground state O_2 as products of its reduction or excitation to the singlet state. The spin restriction of O_2 means that it is most readily reduced one electron at a time, allowing for spin inversion in the interval between productive collisions (Fridovich, 1991). A complete reduction of O_2 requires four electrons, with H_2O being the end product. When the reduction of oxygen proceeds in univalent steps, reactive intermediates are produced (Halliwell and Gutteridge, 1985). In order, these are the superoxide anion (O_2^-), hydrogen peroxide (H_2O_2), and the hydroxyl radical ($\cdot OH$). Physical excitation of O_2 can lead to the production of singlet oxygen ($^1O_2$).

The generation of O_2^- occurs in such locales as the mitochondria, chloroplasts, glyoxysomes, peroxisomes, and nuclei of many types of living organisms. Autooxidation of substrates such as hydroquinones (e.g., ubiquinone), ferredoxins, and thiols in the presence of O_2 can generate O_2^- as a catalytic by-product, providing a partial explanation for O_2^- production during normal metabolism involving mitochondrial and/or chloroplastic electron transport chains (Foyer et al., 1994; Purvis, 1997). Superoxide can also evolve as a by-product of normal metabolism involving enzymes, such as lipoxygenase, aldehyde oxidase, galactose oxidase, and xanthine oxidase (Halliwell, 1981; Chamulitrat et al., 1991; Scandalios, 1993; del Rio et al., 1998; Distefano et al., 1999). Although O_2^- is not as highly toxic as some of the other AOS (e.g., $^1O_2$), it can act within biological membranes as a powerful nucleophile or electrophile (Alscher and Amthor, 1988). Superoxide also appears to convert the metal ion Mn^{+2} into a more reactive species. Most important, O_2^- will react nonenzymatically with H_2O_2 in the metal catalyzed Haber-Weiss reaction to form $\cdot OH$, which, as will be discussed later in this section, is one of the more potent forms of AOS.

Hydrogen peroxide production has been observed in such organelles as chloroplasts (Cakmak and Marscher, 1992), peroxisomes (Badger, 1985), and mitochondria (Prasad et al., 1994). Hydrogen peroxide can arise from a multitude of reactions, including dismutation of O_2^-, β-oxidation of fatty acids, and peroxisomal reactions involving glyoxylate oxidation. Hydrogen peroxide is the most stable member of the AOS because its two outer orbitals are completely filled. Similar to O_2^-, H_2O_2 acts as both a mild reductant and an oxidant, although it is a much stronger oxidizing agent. As such, H_2O_2 is relatively unreactive; it does not react efficiently with organic substrates, although it does tend to form complexes with transition metals. However, H_2O_2 has been observed to play a significant role in DNA breakage and in the oxidation of thiol-containing proteins, such as the thioredoxin-modulated enzymes of the chloroplast stroma, although most toxic effects directly attributable to H_2O_2 seem to occur only at nonphysiological concentrations unless in the presence of metal catalysts (Fridovich, 1976). Nevertheless, as mentioned above, H_2O_2 can interact with O_2^- in the Haber-Weiss reaction to form the very toxic $\cdot OH$ radical. As H_2O_2 can readily diffuse across membranes, the presence of this AOS can thus potentially induce indirect damage. Some biochemical pathways do require H_2O_2 as an oxidizer. Examples of these include the polymerization of aromatic alcohols (e.g., *p*-coumaryl, coniferyl, and sinapyl alcohols) during lignin formation, and cross-linking of tyrosine residues in structural proteins (Liu et al., 1995) and in biosynthetic reactions involving peroxidases in hydroxylating and halogenating reactions in the peroxisome. Both H_2O_2 and O_2^- have also been suggested to be involved in cell signaling (Prasad et al., 1994; Foyer et al., 1994).

The hydroxyl radical is the product of the univalent reduction of H_2O_2 and is one of the strongest biochemical oxidizing agents known (Salin, 1988). Such a reduction process, known as the Haber-Weiss reaction, involves an interaction between H_2O_2 and O_2^- catalyzed by certain

$$O_2^- + H_2O_2 \rightarrow \cdot OH + OH^- + O_2$$

transition metal ions or chelates, such as those of copper or iron. The reductive cleavage of H_2O_2 by a reduced metal, such as Fe(II), to form $\cdot OH$ is known as the Fenton reaction. The hydroxyl radical is an extremely unspecific species that is highly reactive toward proteins, nucleic acids, and/or lipids. In particular, $\cdot OH$ is very reactive toward proteins and may cause modification of almost all amino acid residues, covalent cross-linking, and fragmentation.

When electronically excited species of oxygen are formed due to the elevation of one of the outer shell electrons to a higher orbital, which inverts its spin, the resulting antiparallel spin is referred to as the singlet state. There are two excited states of $^1O_2$, the second (^{1} g) being extremely short-lived and rapidly inactivated by collisional quenching to form the first singlet state ($^1\Delta g$). The stability of $^1\Delta g$ is greater than that of (^{1} g) and, as a result, many of the singlet reactions involve the $^1\Delta g$ species. Although $^1O_2$ can be produced from various sources, such as from a by-product of lipoxygenase activity (Chamulitrat et al., 1991), the conditions which most favor $^1O_2$ generation are found within the actively photosynthesizing chloroplast (Knox and Dodge, 1985). The major mechanism of $^1O_2$ formation is by energy or electron transfer from photoexcited compounds, such as chlorophyll, when they are in the singlet state. However, singlet state chlorophyll may also convert to the triplet state, which can then interact with ground state O_2 to form $^1O_2$. Alternatively, chlorophyll can undergo radiative decay (fluorescence) (Larson, 1988). Singlet oxygen can take part in addition reactions to enes and dienes to form hydroperoxides and endoperoxides, which may propagate AOS and other free radicals in chain-type reactions and lead to loss of membrane integrity through lipid peroxidation. Specific amino acids such as histidine, methionine, and tryptophan are also quite susceptible to attack by $^1O_2$ (Knox and Dodge, 1985).

GENERAL AOS ACTIVITY

With the notable exception of the highly unspecific $\cdot OH$, AOS have particular specificities for various cellular components and, dependent upon the availability and concentration of cellular targets, an average lifetime measured in nanoseconds (e.g., $\cdot OH$) to milliseconds (e.g., O_2^-, H_2O_2)

(Saran et al., 1988). AOS play significant roles in lipid peroxidation (Zheng and Yang, 1991; DeLong and Steffen, 1998; van Breusegem et al., 1998), polysaccharide cleavage (Monk et al., 1989), and in both nucleic acid (Floyd et al., 1989; Becana et al., 1998) and protein (Casano and Trippi, 1992; Iturbe-Ormaetxe et al., 1998) degradation. For plants, typical symptoms of the presence of excess amounts of AOS include inhibition of chloroplast development (Poskuta et al., 1974), bleaching of pigments (Elstner and Osswald, 1994), disruption in membrane integrity (Biedinger et al., 1990), inactivation of many types of enzymes resulting from damage, along with subsequent attacks by proteases (Landry and Pell, 1993; Casano et al., 1994), and lesions and mutations of DNA. AOS have also been implicated in the regulation, properties, and/or dynamics of induced or natural senescent processes (Droillard et al., 1987; Thompson et al., 1991; Philosoph-Hadas et al., 1994). Indeed, one of the major characteristics of senescence in plant tissues is increased lipid peroxidation (Kunert and Ederer, 1985; Lacan and Baccou, 1998). Lipid peroxidation is also thought to play an important role in the biosynthesis of ethylene (Paulin et al., 1986), a hormone involved in senescence regulation. Although the mechanism of ethylene induction is unclear, it may arise as a lipid-fragmentation product through reactions of lipoxygenase (Gardner and Newton, 1987), an enzyme catalyzing the production of hydroperoxy-conjugated dienes from polyunsaturated fatty acids, or perhaps through increased accessibility of its precursor 1-aminocyclopropane-1-carboxylic acid (ACC) to ACC oxidase as a result of membrane deterioration (Pell et al., 1997).

AOS GENERATION

AOS can often be generated as by-products of normal cellular metabolism in mitochondria, chloroplasts, peroxisomes, and nuclei of many types of living organisms (del Rio et al., 1989; Prasad et al., 1994). For example, O_2^- can arise from electron transport chains and reactions catalyzed by enzymes such as aldehyde, glucose, and xanthine oxidases (Scandalios, 1993). Production of H_2O_2 can occur through β-oxidation of fatty acids, peroxisomal photorespiration reactions involving glyoxylate oxidation, and cell wall peroxidase activities during lignin formation. Singlet oxygen can be generated from by-products of lipoxygenase activity (Thompson et al., 1991), electron transport chains, and transfer of energy from photoexcited compounds such as chlorophyll.

The production and destruction of AOS is clearly a regulated cellular phenomenon. However, the term *oxidative stress* has been coined to describe situations in which the generation of AOS exceeds the capacity of the organism to maintain cellular redox homeostasis. For example, exposure of plant tissue to stresses such as chilling, freezing, ultraviolet (UV) irradiation, ozone, drought, high light levels, herbicides (e.g., paraquat, diquat), pollutants, heavy metals, salts, pathogens (e.g., *Cercospora*), mechanical damage, and hypoxia/anoxia often results in induction and/or significant enhancement of AOS production (Foyer et al., 1994; Gossett et al., 1994; Yoshida et al., 1994; Fadzillah et al., 1996; Rao et al., 1996; Fryer et al., 1998; Iturbe-Ormaetxe et al., 1998; Li et al., 1998; May et al., 1998; Shalata and Tal, 1998; Hodges et al., 1999; Hodges, 2001). Although the magnitude and cellular locations of AOS generation may differ dependent upon the species, tissue, and/or developmental age of the plant, and with the nature, length, and degree of the stress applied, proliferation of AOS in plant systems appears to be a common denominator of many types of stresses. It has been surmised that cross-tolerance to several forms of environmental stress could occur if oxidative stress experienced by the plant could be minimized (Malan et al., 1990; Willekens et al., 1994).

Transfer of electrons from electron transport chains to compounds or molecules other than intended is a major route of AOS production under many types of stresses (Elstner, 1991; Foyer et al., 1994). This transfer can occur through overenergization of the electron transport chains, a situation which promotes electron leakage to O_2, thereby reducing it. Moreover, an overenergized chloroplastic electron transport chain can prolong or enhance the photoexcited state of chlorophyll, potentially inducing $^1O_2$ generation (Wise and Naylor, 1987; Somersalo and Krause, 1989). Often stresses can affect protein, lipid, and lipid-protein interactions and integrity (Graham and Patterson, 1982; Nishida and Murata, 1996), resulting in serious consequences to metabolism involving redox reactions and potentially facilitating AOS generation. Other stresses, such as exposure to ozone, can generate AOS directly by reacting with cellular constituents (Ranieri et al., 1996).

Biotic or abiotic stresses are not the only factors associated with the induction or enhanced production of AOS; these species have also been implicated in induced or natural senescent processes and their dynamics (Philosoph-Hadas et al., 1994; Hodges and Forney, 2000). Senescence has been generally defined as a genetically regulated process that leads to the death of cells, organs, or whole organisms (Noodén and Guiamét, 1989). In plants, senescence is often accompanied by morphological changes and/or alterations in biochemical and biophysical properties of metabolism. Philosoph-Hadas et al. (1994) and Meir et al. (1995) demonstrated that general leaf reductant properties were negatively correlated with the rate of senescence for a number of herbaceous species. Hodges and Forney (2000) provided evidence that H_2O_2 regulation plays an important role in both the dynamics and severity of postharvest senescence of spinach. Increases in superoxide dismutase activities (which produces H_2O_2 as a product) and decreases in catalase activities (which scavenges H_2O_2) have been observed in peroxisomes of senescing pea leaves (Pastori and del Rio, 1994, 1997). Peroxisome catalase has a considerably higher affinity for H_2O_2 than those of other peroxidases (del Rio et al., 1998). It has been hypothesized that imbalanced H_2O_2 production/scavenging leads to H_2O_2 accumulation in peroxisomes, and that some of this H_2O_2, which can readily cross membranes, may diffuse out into the cytosol, inducing or accelerating senescence (Kanazawa et al., 2000).

The generation of AOS in senescing material can be enhanced through a variety of mechanisms. For example, autooxidation of compounds such as flavins and quinones, the production of which increases during senescence, can give rise to O_2^- (Droillard et al., 1987). More important, however, senescence and fruit ripening are characterized by lipid peroxidation and membrane deterioration (Thompson, 1988). Subsequent loss of membrane and organellar integrity can lead to increased AOS production through unregulated electron transfer reactions (Purvis, 1997) and/or production of polyunsaturated fatty acids, the substrate for lipoxygenase enzymes. Lipoxygenases have been postulated to be involved in plant growth, development, defense against wounding and pathogens, senescence, and ripening (for review, see Paliyath and Droillard, 1992). Lipoxygenase catalyzes the incorporation of O_2 by the peroxidative modification of polyunsaturated fatty acids, producing hydroperoxy conjugated dienes (9- or 13-hydroperoxyoctadeca(e)noate) (Dörnenburg and Davies, 1999). One method by which lipoxygenase can be regenerated is through the reduction of fatty acid hydroperoxides, producing fatty acid alkoxyl radicals which can react with O_2 to form O_2^- (Chamulitrat et al., 1991). Recombination of two peroxy radicals, proposed as intermediates in lipoxygenase-catalyzed oxidations of polyunsaturated fatty acids, may also lead to the production of a single $^1O_2$ molecule (Kanofsky and Axelrod, 1986; Thompson et al., 1991). Chain reactions of hydroperoxides can also promote AOS production; for example, conjugated dienes can react with Fe^{+2} to form $\cdot OH$. AOS can de-esterfy phospholipids in cellular membranes, inducing further release of polyunsaturated fatty acids, and conjugated dienes can exacerbate further loss in membrane integrity (Thompson et al., 1991). A cycle of membrane degradation, AOS generation, and enhanced senescence results. Lipoxygenase is also implicated in the synthesis of jasmonic acid, a known promoter of plant cell senescence (Hildebrand, 1989), and ethylene (Ievinsh, 1992). Furthermore, there is evidence that lipoxygenase and AOS play a role in the biosynthesis of ethylene (Lynch et al., 1985; Wise and Naylor, 1988; Kacperska and Kubacka-Zêbalska, 1989), which has been demonstrated to enhance the rate of membrane lipid breakdown and accelerate senescence and ripening (Tucker, 1993;

Kim and Wills, 1995). Paulin et al. (1986) hypothesized that membrane breakdown generates AOS, and these AOS then promote ethylene biosynthesis. Lipoxygenase activities have been shown to increase during early stages of senescence and fruit ripening (Cheour et al., 1992; Ealing, 1994; Riley et al., 1996), although elevated lipoxygenase activities do not necessarily represent a universal characteristic of senescent plant tissue (Peterman and Siedow, 1985).

Senescence-related genetic modifications (e.g., alterations in specific gene expression, translation, transcription, and/or posttranscription), associated directly and/or indirectly with increased levels of AOS, may also play a significant role in AOS accumulation during senescence and/or ripening. Although the sources, severity, and modes of AOS generation may differ, once excess levels of AOS are being produced, they can autocatalyze their own production by damaging lipids and proteins. For example, peroxidation of lipids initiates a chain reaction that leads to the accumulation of lipid hydroperoxides.

CONCLUSION

There is little doubt that AOS have the capacity to limit the establishment, growth, yield, and quality of horticultural crops. Moreover, AOS have been implicated in postharvest storage disorders, limitations in shelf life, product quality, and nutritional decline. AOS can accumulate when horticultural material is exposed to stresses such as chilling temperatures, ozone, poor postharvest storage protocols, mechanical damage during production, harvesting, processing, transport, and/or marketing. Levels of AOS are regulated by their rates of generation, their rate of reaction with target proteins, lipids, and/or nucleic acids, their potential rate of degradation, and their rate of scavenging/neutralizing by enzymatic and/or nonenzymatic antioxidants. The mediation of AOS levels and oxidative stress in postharvest material is, therefore, multifactorial, and the following contributions discuss a selection of these factors. It is hoped that this compilation will serve to stimulate further research in the relatively nascent field of oxidative stress in postharvest commodities.

REFERENCES

Alscher, R.G., and J.S. Amthor (1988). The physiology of free-radical scavenging: Maintenance and repair processes. In *Air Pollution and Plant Metabolism*, S. Schulte-Hostede, N.M. Darrall, L.W. Blank, and A.R. Wellburn (Eds.). London, United Kingdom: Elsevier Applied Science, pp. 94-115.

Badger, M.R. (1985). Photosynthetic oxygen exchange. *Annual Review of Plant Physiology* 36:27-53.

Becana, M., J.F. Moran, and I. Iturbe-Ormaexte (1998). Iron-dependent oxygen free radical generation in plants subjected to environmental stress: Toxicity and antioxidant protection. *Plant and Soil* 201:137-147.

Biedinger, U., R.J. Youngman, and H. Schnabl (1990). Differential effects of electrofusion and electropermeabilization parameters on the membrane integrity of plant protoplasts. *Planta* 180:598-602.

Cakmak, I., and H. Marscher (1992). Magnesium deficiency and high light intensity enhance activities of superoxide dismutase, ascorbate peroxidase, and glutathione reductase in bean leaves. *Plant Physiology* 98:1222-1227.

Casano, L.M., H.R. Lacano, and V.S. Trippi (1994). Hydroxyl radicals and a thylakoid-bound endopeptidase are involved in light and oxygen-induced proteolysis in oat chloroplasts. *Plant Cell Physiology* 108:145-152.

Casano, L.M., and V.S. Trippi (1992). The effect of oxygen radicals on proteolysis in isolated oat chloroplasts. *Plant Cell Physiology* 33:329-332.

Chamulitrat, W., M.F. Hughes, T.E. Eling, and R.P. Mason (1991). Superoxide and peroxyl radical generation from the reduction of polyunsaturated fatty acid hydroperoxides by soybean lipoxygenase. *Archives of Biochemistry and Biophysics* 290:153-159.

Cheour, F., J. Arul, J. Makhlouf, and C. Willemot (1992). Delay of membrane lipid degradation by calcium treatment during cabbage leaf senescence. *Plant Physiology* 100:1656-1660.

del Rio, L.A., V.M. Fernandez, F.L. Ruperez, L.M. Sandalio, and J.M. Palma (1989). NADH induces the generation of superoxide radicals in leaf peroxisomes. *Plant Physiology* 89:728-731.

del Rio, L.A., L.M. Sandalio, F.J. Corpas, H.E. Huertas, J.M. Palma, and G.M. Pastori (1998). Activated oxygen-mediated metabolic functions of leaf peroxisomes. *Physiologia Plantarum* 104:673-680.

DeLong, J.M., and K.L. Steffen (1998). Lipid peroxidation and α-tocopherol content in α-tocopherol-supplemented thylakoid membranes during UV-B exposure. *Environmental and Experimental Botany* 39:177-185.

Distefano, S., M.J. Palma, I. McCarthy, and L.A. del Rio (1999). Proteolytic cleavage of plant proteins by peroxisomal endoproteases from senescent pea leaves. *Planta* 209:308-313.

Dörnenburg, H., and C. Davies (1999). The relationship between lipid oxidation and antioxidant content in postharvest vegetables. *Food Reviews International* 15:435-453.

Droillard, M.J., A. Paulin, and J.C. Massot (1987). Free radical production, catalase, and superoxide dismutase activities and membrane integrity during senescence of petals of cut carnations *(Dianthus caryophyllus)*. *Physiologia Plantarum* 71:197-202.

Ealing, P.M. (1994). Lipoxygenase activity in ripening tomato fruit pericarp tissue. *Phytochemistry* 36:547-552.

Elstner, E.F. (1991). Mechanisms of oxygen activation in different compartments of plant cells. In *Active Oxygen/Oxidative Stress and Plant Metabolism, Current*

Topics in Plant Physiology, E.J. Pell and K.L. Steffen (Eds.). Rockville, MD: American Society of Plant Physiologists, pp. 13-25.

Elstner, E.F., and W. Osswald (1994). Mechanisms of oxygen activation during plant stress. In *Oxygen and Environmental Stress in Plants: Proceedings of the Royal Society of Edinburgh Vol 102,* R.M.N. Crawford, G.A.F. Hendry, and B.A. Goodman (Eds.). Edinburgh, United Kingdom: Royal Society of Edinburgh, pp. 131-154.

Fadzillah, N.M., V. Gill, R. Finch, and R.H. Burdon (1996). Chilling, oxidative stress and antioxidant responses in shoot cultures of rice. *Planta* 199:552-556.

Floyd, R.A., M.S. West, W.E. Hogsett, and D.T. Tingey (1989). Increased 8-hydroxyguanine content of chloroplast DNA from ozone-treated plants. *Plant Physiology* 91:644-647.

Foyer, C.H., P. Descourvières, and K.J. Kunert (1994). Protection against oxygen radicals: An important defence mechanism studied in transgenic plants. *Plant, Cell and Environment* 17:507-523.

Fridovich, I. (1976). Superoxide dismutases. *Annual Review of Biochemistry* 44: 146-159.

Fridovich, I. (1991). Molecular oxygen: Friend and foe. In *Active Oxygen/Oxidative Stress and Plant Metabolism, Current Topics in Plant Physiology,* E.J. Pell and K.L. Steffen (Eds.). Rockville, MD: American Society of Plant Physiologists, pp. 1-5.

Fryer, M.J., J.R. Andrews, K. Oxborough, D.A. Blowers, and N.R. Baker (1998). Relationship between CO_2 assimilation, photosynthetic electron transport, and active O_2 metabolism in leaves of maize in the field during periods of low temperature. *Plant Physiology* 116:571-580.

Gardner, H.W., and J.W. Newton (1987). Lipid hydroperoxides in the conversion of 1-aminocyclopropane-1-carboxylic acid to ethylene. *Phytochemistry* 26:621-626.

Gossett, D.R., E.P. Millhollon, and M.C. Lucas (1994). Antioxidant response to NaCl stress in salt-tolerant and salt-sensitive cultivars of cotton. *Crop Science* 34:706-714.

Graham, D., and B.D. Patterson (1982). Responses of plants to low, non-freezing temperatures: Proteins, metabolism, and acclimation. *Annual Review of Plant Physiology* 33:347-372.

Halliwell, B. (1981). *Chloroplast Metabolism: The Structure and Function of Chloroplasts in Green Leaf Cells.* Oxford, England: Clarendon Press.

Halliwell, B., and J.M.C. Gutteridge (1985). *Free Radicals in Biology and Medicine.* Oxford, England: Clarendon Press.

Hildebrand, D.F. (1989). Lipoxygenases. *Physiologia Plantarum* 76:249-253.

Hodges, D.M. (2001). Chilling effects on active oxygen species and their scavenging systems in plants. In *Crop Responses and Adaptations to Temperature Stress,* A.S. Basra (Ed.). Binghamton, NY: The Haworth Press, pp. 53-76.

Hodges, D.M., J.M. Delong, C.F. Forney, and R.K. Prange (1999). Improving the thiobarbituric acid-reactive assay for estimating lipid peroxidation in plant tissues containing anthocyanin and other interfering compounds. *Planta* 207:604-611.

Hodges, D.M., and C.F. Forney (2000). The effects of ethylene, depressed oxygen and elevated carbon dioxide on antioxidant profiles of senescing spinach leaves. *Journal of Experimental Botany* 51:645-655.

Ievinsh, G. (1992). Soluble lipoxygenase activity in rye seedlings as related to endogenous and exogenous ethylene and wounding. *Plant Science* 82:155-159.

Iturbe-Ormaetxe, I., P.R. Escuredo, C. Arrese-Igor, and M. Becana (1998). Oxidative damage in pea plants exposed to water deficit or paraquat. *Plant Physiology* 116:173-181.

Kacperska, A., and M. Kubacka-Zêbalska (1989). Formation of stress ethylene depends both on ACC synthesis and on the activity of free radical generating system. *Physiologia Plantarum* 77:231-237.

Kanazawa, S., S. Sano, T. Koshiba, and T. Ushimaru (2000). Changes in antioxidative enzymes in cucumber cotyledons during natural senescence: Comparison with those during dark-induced senescence. *Physiologia Plantarum* 109:211-216.

Kanofsky, J.R., and B. Axelrod (1986). Singlet oxygen production by soybean lipoxygenase isozymes. *Journal of Biological Chemistry* 261:1099-1104.

Kim, G.H., and R.B.H. Wills (1995). Effect of ethylene on storage life of lettuce. *Journal of the Science of Food and Agriculture* 69:197-201.

Knox, J.P., and A.D. Dodge (1985). Singlet oxygen and plants. *Phytochemistry* 24:889-896.

Kunert, K.J., and M. Ederer (1985). Leaf aging and lipid peroxidation: The role of the antioxidants vitamin C and E. *Physiologia Plantarum* 65:85-88.

Lacan, D., and J.-C. Baccou (1998). High levels of antioxidant enzymes correlate with delayed senescence in nonnetted muskmelon fruits. *Planta* 204:377-382.

Landry, L.G., and E.J. Pell (1993). Modification of rubisco and altered proteolytic activity in O_3-stressed hybrid poplar *(Populus maximowizii × trichocarpa). Plant Physiology* 101:1355-1362.

Larson, R.A. (1988). The antioxidants of higher plants. *Phytochemistry* 27:969-978.

Li, L., J. Van Staden, and A.K. Jäger (1998). Effects of plant growth regulators on the antioxidant system in seedlings of two maize cultivars subjected to water stress. *Plant Growth Regulation* 25:81-87.

Liu, L., K.L. Eriksson, and J.F.D. Dean (1995). Localization of hydrogen peroxide production in *Psium sativum* L. using epi-polarization microscopy to follow cerium perhydroxide deposition. *Plant Physiology* 107:501-506.

Lynch, D.V., S. Sridhara, and J.E. Thompson (1985). Lipoxygenase-generated hydroperoxides account for the non-physiological features of ethylene formation from 1-aminocyclopropane-1-carboxylic acid by microsomal membranes of carnations. *Planta* 164:121-125.

Malan, C., M.M. Greyling, and J. Gressel (1990). Correlation between CuZn superoxide dismutase and glutathione reductase, and environmental and xenobiotic stress tolerance in maize inbreds. *Plant Science* 157-166.

May, M.J., T. Vernoux, R. Sánchez-Fernádez, M. Van Montagu, and D. Inzé (1998). Evidence for postranscriptional activation of γ-glutamylcysteine synthetase during plant stress responses. *Proceedings of the National Academy of Sciences of the United States of America* 95:12049-12054.

Meir, S., J. Kanner, B. Akiri, and S. Philosoph-Hadas (1995). Determination and involvement of aqueous reducing compounds in oxidative defense systems of various senescing leaves. *Journal of Agriculture and Food Chemistry* 43:1813-1819.

Monk, L.S., K.V. Fagerstedt, and R.M.M. Crawford (1989). Oxygen toxicity and superoxide dismutase as an antioxidant in physiological stress. *Physiologia Plantarum* 76:456-459.

Nishida, I., and N. Murata (1996). Chilling sensitivity in plants and cyanonbacteria: The crucial contribution of membrane lipids. *Annual Review of Plant Physiology and Plant Molecular Biology* 47:541-568.

Noodén, L.D., and J.J. Guiamét (1989). Regulation of assimilation and senescence by the fruit in monocarpic plants. *Physiologia Plantarum* 77:267-274.

Paliyath, G., and M.J. Droillard (1992). The mechanisms of membrane deterioration and disassembly during senescence. *Plant Physiology and Biochemistry* 30:789-812.

Pastori, G.M., and L.A. del Rio (1994). An activated-oxygen-mediated role for peroxisomes in the mechanism of senescence of *Psium sativum* L. leaves. *Planta* 193:385-391.

Pastori, G.M., and L.A. del Rio (1997). Natural senescence of pea leaves: An activated oxygen-mediated function for peroxisomes. *Plant Physiology* 89:159-164.

Paulin, A., M.J. Droillard, and J.M. Bureau (1986). Effect of a free radical scavenger 3,4,5-trichlorophenol, on ethylene production and on changes in lipids and membrane integrity during senescence of petals of cut carnations *(Dianthus caryophyllus)*. *Physiologia Plantarum* 67:465-471.

Pell, E.J., C.D. Schlagnhaufer, and R.N. Arteca (1997). Ozone-induced oxidative stress: Mechanisms of action and reaction. *Physiologia Plantarum* 100:264-273.

Peterman, T.K., and J.N. Siedow (1985). Behaviour of lipoxygenase during establishment, senescence, and rejuvenation of soybean cotyledons. *Plant Physiology* 78:690-695.

Philosoph-Hadas, S., S. Meir, B. Akiri, and J. Kanner (1994). Oxidative defense systems in leaves of three edible herb species in relation to their senescence rate. *Journal of Agriculture and Food Chemistry* 42:2376-2381.

Poskuta, J., M. Mikulska, M. Faltynowicz, B. Bielak, and B. Wroblewska (1974). Chloroplast development. *Zeitschrift Pflanzenphysiol* 73:387-393.

Prasad, T.K., M.D. Anderson, B.A. Martin, and C.R. Stewart (1994). Evidence for chilling-induced oxidative stress in maize seedlings and a regulatory role for hydrogen peroxide. *Plant Cell* 6:65-74.

Purvis, A.C. (1997). The role of adaptive enzymes in carbohydrate oxidation by stresses and senescing plant tissues. *HortScience* 32:1165-1168.

Ranieri, A., G. D'Urso, G. Lorenzini, and G.F. Soldatini (1996). Ozone stimulates apoplastic antioxidant systems in pumpkin leaves. *Physiologia Plantarum* 97:381-387.

Rao, M.V., G. Paliyath, and D.P. Ormrod (1996). Ultaviolet-B- and ozone-induced biochemical changes in antioxidant enzymes of *Arabidopsis thaliana*. *Plant Physiology* 110:125-136.

Riley, J.C.M., C. Willemot, and J.E. Thompson (1996). Lipoxygenase and hydroperoxide lyase activities in ripening tomato fruit. *Postharvest Biology and Technology* 7:97-107.

Salin, M.L. (1988). Toxic oxygen species and protective systems of the chloroplast. *Physiologia Plantarum* 72:681-689.

Saran, M., C. Michel, and W. Bors (1988). Reactivities of free radicals. In *Air Pollution and Plant Metabolism*, S. Schulte-Hostede, N.M. Darrall, L.W. Blank, and A.R. Wellburn (Eds.). London, United Kingdom: Elsevier Applied Science, pp. 76-92.

Scandalios, J.G. (1993). Oxygen stress and superoxide dismutases. *Plant Physiology* 101:7-12.

Shalata, A., and M. Tal (1998). The effect of salt stress on lipid peroxidation and antioxidants in the leaf of the cultivated tomato and its wild salt-tolerant relative *Lycopersicon pennellii*. *Physiologia Plantarum* 104:169-174.

Somersalo, S., and G.H. Krause (1989). Photoinhibition at chilling temperatures. Fluorescence characteristics of unhardened and cold-acclimated spinach leaves. *Planta* 177:409-416.

Thompson, J.E. (1988). The molecular basis for membrane deterioration during senescence. In *Senescence and Aging in Plants*, L.D. Noodén and A.C. Leopold (Eds.). San Diego, CA: Academic Press, pp. 51-83.

Thompson, J.E., J.H. Brown, G. Paliyath, J.F. Todd, and K. Yao (1991). Membrane phospholipid catabolism primes the production of activated oxygen in senescing tissues. In *Active Oxygen/Oxidative Stress and Plant Metabolism, Current Topics in Plant Physiology*, E.J. Pell and K.L. Steffen (Eds.). Rockville, MD: American Society of Plant Physiologists, pp. 57-66.

Tucker, G.A. (1993). Introduction. In *Biochemistry of Fruit Ripening*, G.B. Seymour, J.E. Taylor, and G.A. Tucker (Eds.). London, United Kingdom: Chapman and Hall, pp. 1-51.

van Breusegem, F., M. Van Montagu, and D. Inzé (1998). Engineering stress tolerance in maize. *Outlook on Agriculture* 27:115-124.

Willekens, H., W. Van Camp, M. Van Monatgu, D. Inzé, C. Langebartels, and H. Sandermann Jr. (1994). Ozone, sulfur dioxide, and ultraviolet B have similar effects on mRNA accumulation of antioxidant genes in *Nicotiana plumbaginifolia* L. *Plant Physiology* 106:1007-1014.

Wise, R.R., and A.W. Naylor (1987). Chilling-enhanced photooxidation: Evidence for the role of singlet oxygen and superoxide in the breakdown of pigments and endogenous antioxidants. *Plant Physiology* 83:278-282.

Wise, R.R., and A.W. Naylor (1988). Stress ethylene does not originate directly from lipid peroxidation during chilling-enhanced photooxidation. *Journal of Plant Physiology* 133:62-66.

Yoshida, M., I. Nouchi, and S. Toyama (1994). Studies on the role of active oxygen in ozone injury to plant cells. II. Effect of antioxidants on rice protoplasts exposed to ozone. *Plant Science* 95:207-212.

Zheng, R., and Z. Yang (1991). Lipid peroxidation and antioxidative defense systems in early leaf growth. *Journal of Plant Growth Regulation* 10:187-199.

Chapter 2

Postharvest Response
of Horticultural Products to Ozone

Charles F. Forney

INTRODUCTION

The use of ozone for postharvest sanitation and decay control of fresh fruits, vegetables, and cut flowers during handling and storage has been investigated for commercial application. Ozone is an effective substitute for chlorine to reduce microbial contamination in water used for cooling and washing produce. However, the addition of ozone to storage room air has shown mixed effectiveness in the reduction of decay and maintenance of produce quality. Many ozone concentrations and exposure durations have been tested on a range of different fresh commodities, and ozone was found to have a variety of effects on microbial organisms as well as on the fresh commodities.

Ozone is a reactive compound that imposes oxidative stress on exposed fresh commodities. This stress can result in a variety of responses ranging from tissue damage to induction of defense mechanisms that may reduce postharvest deterioration. The interaction of ozone with the volatile ethylene has also been an area of interest; the destruction of ethylene, which can induce fruit ripening and plant senescence, has been suggested to reduce decay and prolong storage life of fresh commodities.

In this chapter, the chemical nature of ozone, its effects on microorganisms, its physiological effects on plant tissues, and its use in postharvest handling of fresh commodities are discussed.

OZONE CHEMISTRY

Chemical Properties

Ozone (O_3), also known as triatomic oxygen, is a naturally occurring, highly reactive form of oxygen. Ozone is a gas at ambient and refrigerated temperatures and has a boiling point of $-112°C$. It has a pungent and characteristic odor detectable in air by humans at a concentration of about 0.010 μL/L (Nebel, 1981). It is a potent oxidant with an oxidation potential of −2.09 volts (V) and is rated fifth in its thermodynamic oxidation potential after fluorine, chlorine trifluoride, atomic oxygen, and hydroxyl free radicals (Graham, 1997). As an oxidizer, ozone is 1.5 times stronger than chlorine and is effective against a much wider spectrum of microorganisms than chlorine and other disinfectants (Xu, 1999). Ozone decomposes to form numerous free-radical species (Bablon et al., 1991). The most prominent free radical produced is the hydroxyl radical (·OH), which has a half-life of only microseconds (Graham, 1997). It is an important transient species and acts to propagate additional radicals that may contribute to the overall oxidizing power of ozone (Kim et al., 1999b).

Ozone's solubility in water is only about 0.2 mg/L at 20°C, but increases as temperature decreases to 0.54 mg/L at 0°C (Bablon et al., 1991). Grimes et al. (1983) have shown that, in aqueous solutions, ozone decomposes primarily to the hydroxyl radical and that the presence of phenolic compounds enhances this decomposition. In fact, many different compounds act to initiate, promote, or inhibit free-radical formation from ozone (Bablon et al., 1991). The half-life of ozone in room-temperature water is only twenty minutes (Xu, 1999). In water with suspended soil and organic matter, the half-life of ozone is reduced to less than one minute (Suslow, 1998). The rate of ozone decomposition increases concurrent with the pH of the solution containing ozone; at a pH of about 10, ozone decomposes instantaneously. Ozone decomposes into oxygen, and thus there are no safety concerns about residual ozone.

The half-life of ozone in air depends on temperature, humidity, and the presence of reactable substrates. The addition of carrots to storage chambers caused ozone to be lost at a faster rate than when chambers were empty (Liew, 1992). Similar observations were made in citrus storage rooms (Palou et al., 2001). Nagy (1959) reported that the half-life of ozone at 20°C is about three minutes, but increases to six minutes at 4°C. This increased half-life of ozone may be due to the decrease in volatile compounds in the air at low temperatures.

Reactions

Ozone is most reactive with compounds having nucleophilic sites including oxygen (O), nitrogen (N), sulfur (S), phosphorus (P), and carbon-carbon double bonds (Bablon et al., 1991; Nebel, 1981). Ozone reacts with many biologically important compounds including unsaturated fatty acids such as those found in membrane lipids, ring-containing compounds including NAD(P)H, and aromatic and sulfur-containing amino acids (Mudd et al., 1969, 1974).

Ozone also oxidizes sulfhydryl groups to disulfides and sulphonic acids, potentially inhibiting the activity of various metabolically important enzymes including acyl-CoA-thioesterase, thiokinases, acyltransferases, and glycerinealdehyde-3-phosphate dehydrogenase (Hippeli and Elstner, 1996). In addition, ozone oxidizes glutathione and thus changes the cellular redox balance (Hippeli and Elstner, 1996; Miller, 1987). Reaction of ozone or the free radicals it generates is often restricted by its penetration into the cell or tissue, and many reactions are limited to the cell wall and plasma membrane (Bablon et al., 1991).

There has been much interest in the reaction of ozone with the simple hydrocarbon ethylene (C_2H_4). Ethylene is produced by fruit and other plant material and can stimulate fruit ripening and senescence. The theoretical mechanism by which ozone destroys ethylene is summarized by Anglada et al. (1999). Studying the energetics of possible reactions, they suggest that ozone combines with ethylene to form 1,2,3-trioxolane (POZ), an energy-rich ozonide. The unstable POZ molecule then breaks down through several possible mechanisms to form various radicals, formaldehyde, carbon dioxide, carbon monoxide, hydrogen, and water. When equal quantities of ozone and ethylene react, decomposition products average 66 ± 11 percent as formaldehyde (HCHO), 28 ± 4 percent carbon monoxide (CO), 19 ± 3 percent carbon dioxide (CO_2), and 3 ± 1 percent formic acid (HCOOH) (Horie and Moortgat, 1991). Atkinson (1990) reported that the reaction of ozone with ethylene at room temperature and ambient atmospheric pressure yields 1.0 HCHO $+ 0.37$ $CH_2OO + 0.12$ $HO_2 + 0.13$ $H_2 + 0.19$ $CO_2 + 0.44$ CO $+ 0.44$ H_2O.

Production of Ozone

Natural Occurrence

Ozone is generated naturally from O_2 by ultraviolet irradiation from the sun and from the high energy discharge of lightning (Xu, 1999). Ozone is also formed in the lower levels of the atmosphere as a result of photochemical oxidation of hydrocarbons from automobile and industrial emissions, as well as from photocopiers, electrical transformers, and other electrical devices.

Commercial Generation

Most commonly, commercial ozone is generated by the use of corona discharge. A corona is formed when high voltage at high frequency is applied to a specialized electrode (Nebel, 1981). Many different types of electrodes have been developed using flat plates as well as tubes (White, 1999).

A simple plate electrode consists of two parallel conductive plates separated by a dielectric material, such as glass or ceramic, and a narrow discharge gap (Figure 2.1) (Bablon et al., 1991; Graham, 1997; Kim et al., 1999b). When high voltage is applied across the gap, a continuous corona is produced. Molecular oxygen (O_2) is passed through the gap and the corona excites the oxygen electrons causing the molecules to split. The resulting atomic oxygen reacts with other oxygen molecules to form ozone. A corona discharge-type generator fed with air yields 1 to 3 percent (w/w) ozone. This can be increased to 2 to 12 percent (w/w) ozone if pure oxygen is used as the reaction gas. Humidity or hydrocarbons in the reaction gas reduces the efficiency of ozone production by inhibiting the reaction or reacting with the generated ozone (Bablon et al., 1991).

Ozone is also generated using ultraviolet light at wavelengths lower than 200 nm (Bablon et al., 1991). Radiation from ultraviolet bulbs can create concentrations of ozone in the range of 1,200 µL/L or 0.1 percent (w/w) (Graham, 1997). Ozone generators can be equipped with ultraviolet (UV) bulbs, or bulbs can be placed in refrigerated cold rooms. Depending on the type of bulbs and their distribution in a room, ozone concentrations in room air can be maintained at trace levels to several µL/L (Ewell, 1942; Nagy, 1959). Ozone concentrations in rooms irradiated with UV lamps are reduced as temperature and humidity are increased (Ewell, 1942).

PHYSIOLOGICAL EFFECTS ON PLANTS

Ozone, being highly reactive, causes many physiological changes in living plants, including harvested fresh fruits and vegetables. Ozone-induced physiological changes affect rates of decay, ripening, and overall quality. Although little research has been done on physiological changes in fresh produce resulting from postharvest ozone exposure, there is a significant

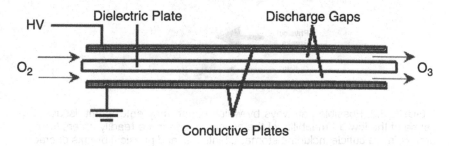

FIGURE 2.1. Components of a high-voltage electrode used to generate ozone by corona discharge.

body of literature on the physiological effects of ozone on growing plants relative to their exposure to ozone as a pollutant. Therefore, much of the following discussion will rely on this literature for insights to explain postharvest responses to ozone exposure.

Pathway of Entry

The physiological effects on plant tissues are dependent on contact by ozone or its reactive products with cellular sites. Tingey and Taylor (1982) described a conceptual overview of the pathway by which ozone enters plant tissues. Gas phase ozone must first pass through the plant cuticle, which is fairly resistant to ozone penetration. Kerstiens and Lendzian (1989) reported that ozone uptake through the most permeable cuticles is 10,000-fold less than through an open stomate. Therefore, most ozone enters plant tissues through openings or breaks in the cuticle. In growing plants, this normally occurs through the stomata. In harvested fruits and vegetables, open stomata may not be available, and paths of entry include lenticels and cuts or cracks in the cuticle (Figure 2.2), although the resistance to mass flow through these paths is often much greater than through open stomata. The rate of ozone flux through

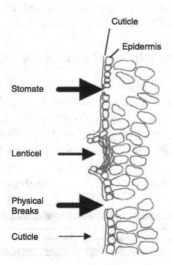

FIGURE 2.2. Possible pathways by which ozone may enter plant tissues. Because of the low permeability of the cuticle, ozone more readily enters through breaks in the cuticle including stomata, lenticels, and physical breaks or cracks. Once in the tissue, ozone diffuses in the free space and is dissolved in the cell wall water. Arrow size indicates relative flux of ozone through different pathways of entry.

these openings is determined by the concentration gradient between the ambient air and cellular surface and the resistance to mass transfer along the diffusion path. Morphological features such as trichomes and pubescence may contribute to this resistance. Concentrations of ozone reaching the cellular reaction sites are also reduced by reactions of ozone with volatile compounds released by the plant tissue, and nonoxidized organic surfaces on the plant tissue. Monoterpenes released by some plants react with ozone in and around the plant tissue, reducing this concentration gradient and the net flux of ozone into the plant tissue (Grimsrud et al., 1975).

Once in the tissue, gas-phase ozone must diffuse into the liquid phase on the wet cell surface (apoplast). The low water solubility of ozone slows this process. Once in the liquid phase, ozone decomposes to produce hydrogen peroxide, hydroxyl, hydroperoxyl, superoxide anion, and other free radicals (Wellburn and Wellburn, 1996). Ozone and its reactive products react with molecules in the apoplastic solution as well as components of the cell wall and plasma membrane. As ozone and its products are highly reactive, when it does reach the cell wall and plasma membrane it rapidly finds reactive sites and very little ozone is able to reach the cytoplasm. Heath (1988) suggests that the average distance reactive species produced by ozone can travel before reacting with the solvent is only a few nanometers, making it unlikely that these reactive products penetrate past the plasma membrane. The direct disruption of normal membrane function, which is expressed as increased leakiness and inhibition of active pumps and transporters, may alter normal metabolism of the whole cell. However, ozone induces many reactions normally elicited by viral and microbial pathogens, suggesting an ozone receptor may exist that initiates a signal chain, which results in additional metabolic effects (Sandermann, 1996). Kangasjärvi et al. (1994) suggested that plasma membrane lipid peroxidation caused by ozone-derived oxygen radicals may contribute to the signal transduction pathway, inducing defense responses in the cells. Peroxidation products of the plasma membrane, jasmonic acid and methyl jasmonate, are suggested to play a role in this pathway. Ethylene and salicylic acid may also play a role in signal transduction (Kangasjärvi et al., 1994; Sandermann et al., 1998).

Defense Mechanisms

The prevention of ozone entry into plant tissue is a plant's first line of defense. This is generally dependent on the physical state of the surface of the plant or plant part at the time of ozone exposure. As mentioned, open stomata are the major pathways by which ozone enters plant tissue. Any factor that results in the closure of stomata generally results in an increased tolerance to ozone. Increased ozone tolerance is associated with water stress and nutrient deficiency and can be explained by the associated reduction in stomatal conductance (Chappelka and Samuelson, 1998). Tingey and Hogsett (1985) demonstrated that water-stressed bean plants sprayed with fusicoccin, which stimulates stomatal opening, results in ozone damage similar to that in well-watered plants. Similarly, if well-watered plants are sprayed with abscisic acid, which induces stomatal closure, they become tolerant to ozone. Plants appear to respond to ozone stress by reducing stomatal conductance. However, ozone-induced stomatal closure may be a secondary effect, as stomatal conductance decreases slowly in most plants during ozone exposure and could be explained by changes in mesophyll metabolic activity (Chevone et al., 1990). In harvested fruits and vegetables, which are normally stored in the dark, stomata are generally closed, reducing the significance of this pathway of entrance.

After ozone has penetrated into the plant tissue, the next line of defense is in the liquid phase of the apoplast (Schraudner et al., 1997). Free-radical scavengers and antioxidants in this liquid phase can reduce or eliminate ozone and its reactive products, including hydrogen peroxide, hydroxyl radical, peroxyl radical, singlet oxygen, and superoxide, before they cause cellular damage. Many detoxifying systems are present in the aqueous phase of the apoplast.

Ascorbic Acid

Ascorbic acid (AA) is present in the apoplast and is an important antioxidant that removes ozone and other active forms of oxygen before they damage plant cells. Ascorbic acid reacts rapidly with ozone and many of its reactive products, including superoxide and peroxy radicals, to prevent cell damage (Figure 2.3) (Heath, 1988; Luwe et al., 1993; Luwe and Heber, 1995; Smirnoff, 1996). Concentrations of AA in the apoplast of leaves range from 0.5 to 3.0 µmol/mL and increase in response to ozone exposure (Luwe and Heber, 1995). Ozone-induced damage of spinach leaves decreases as AA content and its redox state increase (Luwe et al., 1993). Rabotti and Ballarin-Denti (1998) associate high concentrations of AA in beech leaves early in the growing season with resistance to ozone injury. Ozone sensitivity also relates to AA synthesis in the plant. Conklin et al. (1996) demonstrated that an *Arabidopsis* mutant, which produces only 30 percent of the normal concentration of AA, is more susceptible to ozone injury.

During exposure to ozone, plants attempt to maintain a constant redox potential in the apoplast. In many cases, this results in an increase in AA concentration in the plant tissue. When spinach plants are exposed to 0.06 or 0.12 µL/L ozone, a transient threefold increase in apoplastic AA is observed after six hours, which returns to initial concentrations after 24 hours (Luwe

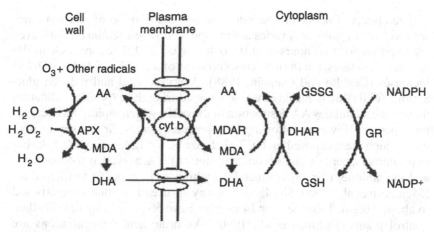

FIGURE 2.3. Proposed role of ascorbic acid (AA) in defending plant cells against ozone-induced oxidative stress. AA from the cytoplasm is translocated into the apoplast where it reacts with ozone (O_3) and other radicals, as well as hydrogen peroxide (H_2O_2) with the enzyme ascorbate peroxidase (APX). Oxidation of AA results in the formation of monodehydroascorbate (MDA) and dehydroascorbate (DHA). DHA is translocated back into the cytoplasm where it is reduced back to AA through the ascorbate-glutathione cycle by the enzyme dehydroascorbate reductase (DHAR). Reduced glutathione (GSH) is oxidized to GSSG through this reaction. GSSG is reduced back to GSH through the action of glutathione reductase (GR) and the reducing power of NADPH. A membrane-bound cytochrome b (cyt b) may also reduce MDA to AA in the apoplast. Cyt b is reduced by AA in the cytoplasm and MDA is reduced back to AA through monodehydroascorbate reductase (MDAR) or through the ascorbate-glutathione cycle.

and Heber, 1995). This increase in AA coincides with an increase in dehydro-ascorbate (DHA), the oxidized form of AA, resulting in the redox state of AA in the apoplast remaining fairly constant, at about 90 percent. In beech leaves exposed to 0.06 µL/L ozone for one week, apoplastic AA is fivefold and DHA 8.5-fold greater than in the controls. This difference is lost after four weeks, but the redox potential of AA again remains constant. During long-term treatment of beech plants with 0.15 to 0.20 µL/L ozone, visible leaf damage occurs even though the apoplastic AA redox potential is maintained, indicating that at this high concentration a significant quantity of ozone or its reactive products is not reduced by AA. In broad bean, which has lower levels of AA and is less tolerant to ozone, 0.06 µL/L ozone causes a decrease in apoplastic and cellular AA redox potential, indicating that plants differ in their ability to adapt to and tolerate ozone stress.

The effects of ozone on the induction and distribution of AA and other antioxidants in plant cells varies among species. When *Sedum album* leaves are exposed for two hours to 0.10, 0.40, or 0.60 µL/L ozone, AA in the apoplast decreases, but the total concentration of ascorbic acid (AA + DHA) increases (Castillo and Greppin, 1988). Ascorbic acid and reduced glutathione (GSH) are depleted in whole leaves during the exposure, indicating that ozone stimulates AA destruction in the cells, which rapidly recovers after exposure. However, following long-term exposures, fir and spruce needles, which were exposed to 0.037 µL/L ozone for 12 hours per day during two summers, contain higher concentrations of AA, as well as α-tocopherol and glutathione, than control needles, which were exposed to filtered air (Mehlhorn et al., 1986). Similarly, Norway spruce plants that were exposed to about 0.05 µL/L ozone over 14 months have 50 percent more GSH than control plants (Dohmen et al., 1990). Ascorbic acid concentrations are threefold greater in strawberry fruit treated with 0.35 µL/L ozone for three days at 2°C than in control fruit (Pérez et al., 1999). However, this difference is not maintained during a four-day shelf-life treatment at 20°C in air as AA levels increase in the control fruit and decrease in the ozone-treated fruit.

At high concentrations of ozone, this mechanism to reduce ozone and its reactive products is not able to fully defend the plant from ozone-induced injury. In spinach and beech leaves exposed to 0.06 to 0.12 µL/L ozone, no changes in concentrations of AA, DHA, or glutathione occur in the whole tissue (Luwe and Heber, 1995). However, after six weeks of fumigation with 0.15 µL/L ozone, beech leaves have elevated concentrations of DHA and show a loss of chlorophyll, indicating that they are unable to fully detoxify ozone at this level and duration. Beech leaves exposed to 0.80 µL/L ozone show a rapid loss of reduced AA and GSH in the cell, but a twofold increase in AA in the apoplast after 24 hours.

In conjunction with AA, ascorbate peroxidase (APX) protects plants from ozone by removing ozone-generated H_2O_2 from the apoplast by catalyzing the reaction between AA and H_2O_2 to form monodehydroascorbate (MDA) and water (Figure 2.3) (Willekens et al., 1997). Örvar and Ellis (1997) demonstrated that transgenic tobacco plants expressing reduced APX mRNA levels and APX catalytic activity are injured more by ozone exposure than are control plants. In *Arabidopsis* plants exposed to ozone, cytosolic APX mRNA increases (Conklin and Last, 1995). Ozone also increases apoplastic APX activity in cell lines from grape leaves (Sgarbi et al., 1999), *Sedum album* leaves (Castillo and Greppin, 1986), and beech leaves (Rabotti and Ballarin-Denti, 1998), which is associated with greater ozone tolerance. In Norway spruce needles, APX activity increases threefold in the apoplast but only 0.3-fold in the cells following two days of a seven-hour 0.16 µL/L ozone exposure (Castillo et al., 1987).

Ascorbate-Glutathione Cycle

In order to maintain reduced AA in the apoplast, the plant utilizes the ascorbate-glutathione cycle in the cytoplasm (Figure 2.3) (Smirnoff, 1996). Oxidation of AA in the apoplast results in the formation of DHA, which easily passes into the cytoplasm along concentration gradients (Rautenkranz et al., 1994). In the cytoplasm, DHA is reduced to AA by the enzyme dehydroascorbate reductase (DHAR), which utilizes GSH. Reduced AA is then translocated back across the plasma membrane to the apoplast along the concentration gradient (Rautenkranz et al., 1994; Smirnoff, 1996). Oxidized glutathione (GSSG) is reduced to GSH by glutathione reductase (GR) utilizing NADPH. To regulate apoplastic AA concentrations, active transport is needed to move AA back into the cytoplasm from the apoplast (Rautenkranz et al., 1994).

Plants subjected to ozone fumigation increase AA and activity of the ascorbate-glutathione cycle enzymes purportedly in order to maintain the AA redox potential (Castillo and Greppin, 1988). In pumpkin leaves, a five-day, five hours/day exposure to 0.15 µL/L ozone increases apoplastic concentrations of AA and DHA in mature leaves (Ranieri et al., 1996). Intracellular concentrations of glutathione increase slightly and no glutathione is detected in the apoplast. Similarly, exposure of *Populus* leaves to 0.18 µL/L ozone increases total glutathione levels, which is maintained for 21 hours after ozone exposure (Sen Gupta et al., 1991). At higher ozone concentrations, spinach leaves exposed to 0.30 µL/L ozone have reduced redox potentials of AA in the apoplast, while in the cells AA oxidation state remains constant and GSH becomes oxidized (Luwe and Heber, 1995).

Activity of GR increases in barley leaves exposed to 0.20 µL/L ozone for seven hours/day for five days (Price et al., 1990). Similar increases in GR activity are reported in spinach (Tanaka et al., 1988) and in pea (Mehlhorn and Wellburn, 1987) as a result of ozone exposure. However, attempts to enhance GR activity to increase ozone tolerance in transgenic tobacco has met with mixed results (Broadbent et al., 1995). Some tobacco lines that have enhanced GR activity also have reduced ozone sensitivity, measured as reduced stress ethylene production and maintained chlorophyll fluorescence (Fv/Fm). However, the lines showing the greatest increase in GR activity had no decrease in ozone sensitivity. There is also no increase in ozone tolerance in transformed tobacco plants that express a GR gene from *Escherichia coli* in the cytosol, which resulted in 1 to 3.5 times higher GR activity (Aono et al., 1991).

Activity of DHAR has not been detected in the apoplast (Castillo and Greppin, 1988). However, ozone exposure increases DHAR activity (Castillo and Greppin, 1988), as well as total levels of glutathione in the cytoplasm (Schraudner et al., 1997), emphasizing its important role in regenerating AA.

In addition to the regeneration of AA in the cytoplast through the AA-glutathione cycle, other regeneration mechanisms are proposed. Smirnoff (1996) suggested that the MDA radical produced from the oxidation of AA may be reduced, in part, back to AA by plasma membrane-bound cytochrome *b* with cytoplasmic AA acting as an electron donor (Figure 2.3). It is also suggested that apoplastic α-tocopherol may reduce oxidized forms of AA.

Other Antioxidants

In addition to AA and the components of the ascorbate-glutathione cycle, the antioxidant enzymes superoxide dismutase (SOD) and catalase (CAT) as well as polyamines and phenolic compounds play a major role in the defense system of plants against oxidative stress (Bors et al., 1989; Heath, 1988).

Superoxide dismutase. Superoxide dismutase is an enzyme that converts the radical anion superoxide (O_2^-) to H_2O_2 and O_2 (Larson, 1988). Following this conversion, H_2O_2 must be removed by catalases or peroxidases to prevent the conversion of H_2O_2 to hydroxyl radicals through the Haber-Weiss reaction ($H_2O_2 + O_2^- \rightarrow OH^- + O_2 + \cdot OH$). Many different isozymes of SOD are present in the plant including FeSOD and Cu/ZnSOD in the chloroplasts, MnSOD in the mitochondria, and Cu/ZnSOD in the cytosol. In addition, significant quantities of SOD are reported to be active in the apoplast where it converts ozone-produced superoxides to H_2O_2, which is then converted to H_2O by APX (Castillo et al., 1987).

The role that SOD plays in the plant's defense system has, in part, been studied through molecular manipulation of SOD genes. When transgenic tobacco plants are engineered to produce enhanced levels of MnSOD, ozone injury is reduced three- to fourfold following seven days of a 14-hour exposure to ozone averaging 0.06 µL/L but peaking at 0.09 to 0.13 µL/L (Van Camp et al., 1994). This improved ozone tolerance is observed when activity is enhanced in the chloroplast, but to a lesser extent when activity is enhanced in the mitochondria. Van Camp et al. suggest that the observed protection is a result of SOD scavenging toxic reaction products of ozone that may reach intracellular sites. However, when tobacco plants are transformed to overproduce Cu/ZnSOD in the chloroplast, there is no increase in ozone tolerance (Pitcher et al., 1991). The lack of protection in this second study may be due to the high concentration of ozone exposure (0.3 µL/L for five hours) or differences in the level of SOD production. Van Camp et al.

(1994) suggested that for increased SOD activity to be effective in protecting a plant from ozone injury, adequate H_2O_2 scavenging must also be present.

Ozone is also reported to enhance SOD production. A fumigation of 0.50 μL/L ozone for five hours induces large increases in cytosolic Cu/ZnSOD and mitochondrial MnSOD gene expression, but reduces chloroplastic FeSOD expression in tobacco (Hérouart et al., 1993). Similar results are observed in barley (Azevedo et al., 1998) and *Populus* leaves (Sen Gupta et al., 1991). Badiani et al. (1993) reported diurnal changes in SOD, APX, and CAT activity in bean leaves, with increases in activity corresponding to increases in ambient ozone concentrations. However, Heath (1988), citing several studies, reported that ozone exposure has little effect on the activity of SOD.

Catalase. The enzyme catalase (CAT) may play an indirect role in reducing ozone injury in plant tissues. Catalase is a H_2O_2-scavenging enzyme that is present in the cytoplasm and organelles of plant cells. Willekens et al. (1997) demonstrated that when catalase gene expression is inhibited in tobacco, the plants are damaged by ozone exposure, while noninhibited plants are not. Exposure of barley leaves to 0.12 μL/L ozone increases CAT activity in normal plants, while a CAT-deficient mutant does not increase CAT activity and develops injury (Azevedo et al., 1998). Ozone induction of CAT, however, is variable and depends on species and ozone exposure conditions (Heath, 1988).

Polyamines. Polyamines may play a role in ozone tolerance. These compounds are implicated in the inhibition of lipid peroxidation of membranes (Schraudner et al., 1997; Tadolini, 1988), activation of membrane-bound ATPases (Heath, 1988), and reduction of ethylene formation (Evans and Malmberg, 1989). Significant quantities of polyamines occur in the apoplast of tobacco, but they are poor scavengers of radicals (Bors et al., 1989).

Ozone stimulates the accumulation of polyamines in a variety of plants including barley (Rowland-Bamford et al., 1989) and Norway spruce (Dohmen et al., 1990). Arginine decarboxylase (ACD), a rate-limiting enzyme in polyamine synthesis, increases in activity as a result of ozone exposure in barley (Rowland-Bamford et al., 1989) and tobacco (Langebartels et al., 1991). When ACD is inhibited by the specific inhibitor difluoromethylarginine, foliar damage to barley leaves caused by ozone increases, suggesting that polyamines play a protective role in preventing ozone injury (Rowland-Bamford et al., 1989). In addition, when tomato (Ormrod and Beckerson, 1986) or tobacco (Bors et al., 1989) plants are fed polyamines prior to ozone exposure, injury is reduced. Short exposures to ozone stimulate the production of free and conjugated putrescine in an ozone-tolerant tobacco cultivar, while in an ozone-sensitive cultivar ozone-induced putrescine production is weak and ethylene production is induced (Langebartels et al., 1991). Monocaffeoyl-putrescine, which is an effective scavenger of oxyradicals, increases in the apoplastic fluid of the tolerant cultivar to concentrations of 50 to 400 μmol/L and may scavenge ozone-derived oxyradicals at their site of generation, thus protecting the leaf from injury. Short ozone exposures also alter polyamine concentrations in potato leaves (Reddy et al., 1991). After a four-hour exposure to 0.20 μL/L ozone, putrescine levels are higher than in control plants up to 20 hours after treatment. Spermidine levels also increase as a result of the ozone treatment.

Phenolic compounds. Phenolic compounds possess several properties that make them effective protectants against oxidative damage (Larson, 1995). When phenolic compounds react with oxidizing agents, phenoxyl radicals are formed which are relatively stable due to the resonance delocalization of the unpaired electron in the ring structure. Therefore, phenoxyl radicals tend not to initiate further free-radical reactions. Many phenols are partly ionized at physiological pHs, making them more effective antioxidants. The head group of α-tocopherol is an example of a phenolic compound with strong antioxidant properties that effectively reacts with superoxide, singlet oxygen, and peroxy-radical to prevent cellular damage (Heath, 1988). Membrane-bound α-tocopherol in the apoplast can be regenerated to its reduced form by AA (Rautenkranz et al., 1994; Schraudner et al., 1997). Flavonoids are also very effective antioxidants, especially those with ortho-hydroxylation, which stabilizes phenoxyl radicals.

Ozone tolerance is associated with the presence of phenolic compounds, and ozone induces or enhances the production of phenolics in plants. In soybean, ozone tolerance is associated with the presence of kaempferol glycosides, a class of flavanols (Foy et al., 1995), and injurious treatments of 0.70 μL/L for two or three hours result in the accumulation of the isoflavonoids daidzein, coumestrol, and sojagol (Keen and Taylor, 1975). In parsley, the flavone glycoside, 6"-*O*-malonylapiin, as well as apiin, increases two- to threefold immediately following a 10-hour exposure to 0.20 μL/L ozone (Eckey-Kaltenbach et al., 1993). However, in the apoplast, concentrations of apiin and AA increase ten- and twofold, respectively, as a result of the ozone treatment, indicating that phenolics are located where they can effectively protect the plant from injury by ozone and its reactive products. In pumpkin leaves, a five-day, five hours/day exposure to 0.15 μL/L ozone causes phenols to decrease in the apoplast but increase in the cells, suggesting that under these conditions phenolics are destroyed faster in the apoplast than they can be replaced (Ranieri et al., 1996). In tobacco, the flavonoid, rutin (quercetin-3-rhamnosyl-glucoside), which may contribute to ozone tolerance, accumulates in ozone-tolerant tobacco but not in an ozone-sensitive line as a result of ozone and other stresses (Steger-Hartmann et al., 1994).

Cellular Damage

When the natural defenses in the plant cells cannot neutralize the oxidative stress imposed by ozone, cellular damage occurs. This can be expressed as changes in physiology as well as in visual damage.

Ozone can degrade the plant cuticle both during growth and in storage. Plums grown under 0.094 µL/L ozone develop thinner cuticles with reduced wax deposition and subsequently lose more weight in storage than do control plums (Crisosto et al., 1993). Cranberries held in 0.60 µL/L ozone for five weeks at 15°C have 25 percent less surface lipids on the fruit compared to control fruit, resulting in three times more weight loss than air-stored fruit (Norton et al., 1968). Elstner et al. (1985) indicated that ozone may degrade the wax cuticle of spruce needles resulting in reduced resistance to physical and biological stresses. Exposure of grapes to ~ 4,000 µL/L ozone for > 40 minutes also results in the degradation of the cuticle seen as microscopic veinlike cracks on the epidermis (Sarig et al., 1996).

Degradation of the cuticle leads to greater penetration of ozone into the plant cells. Ozone contact with plant cells causes damage in a manner similar to many other oxidative stresses and affects the general metabolism of plants. Ozone entering leaves through the stomata results in localized cell death, usually in the abaxial mesophyll layer in leaves of sensitive plants (Chevone et al., 1990). This damage results in lost photosynthetic capacity and reduces plant growth, although no visual damage may be apparent. In forest species, ozone stress reduces carbon fixation, increases foliar and root respiration, and alters allocation and distribution of carbon and nutrients (Chappelka and Samuelson, 1998). During vegetative growth, ozone exposure causes more of a reduction in root growth than of shoot growth, which may be the result of an accelerated senescence of older leaves that feed the roots (Chevone et al., 1990). In fruiting plants, ozone exposure reduces the amount of assimilates partitioned to the fruit (Bennett et al., 1979). In general, ozone is more damaging to older leaves than to younger ones, inhibiting partitioning of assimilated carbon from the leaf (Chevone et al., 1990).

When injury from ozone is severe, visual symptoms of damage become apparent. Injury is characterized by the development of water-soaked spots and whitish-tan lesions, which are often associated with the loss of chlorophyll and carotenoids (Khan et al., 1996; Sakaki et al., 1983). Epinasty can be induced by ozone and is believed to be the result of induced ethylene production (Reddy et al., 1991). Premature senescence of leaves in a variety of species is also reported to result from ozone exposure (Chappelka and Samuelson, 1998; Miller et al., 1999; Simini et al., 1992; Tingey and Taylor, 1982). A more detailed description of phytotoxic responses of different fresh fruits and vegetables is presented later in this chapter.

Mechanisms of Damage

The mechanism of ozone damage in plant tissue is not well established. Some suggested mechanisms include damage to biomembranes by reaction with polyunsaturated fatty acids or sulfhydryl groups on membrane proteins, destruction of AA and peroxidase enzymes in the apoplast, or disruption of the superoxide dismutase-ascorbate-glutathione antioxidant system in the chloroplast (Chevone et al., 1990). Of these suggested mechanisms, the effects of ozone on the plasma membrane are the focus of the majority of studies. After ozone enters the plant tissue, it rapidly reacts with components of the cell walls and the apoplastic fluid forming singlet oxygen, H_2O_2, hydroxyl radicals, and superoxide anion radicals (Heath, 1988). These active oxygen species may induce secondary signals that induce various metabolic changes (Kangasjärvi et al., 1994; Sandermann, 1996). Regardless of the mechanism, it appears that if ozone or the production of these reactive intermediates exceeds the detoxification capacity of the cell, damage occurs.

A common indicator of ozone injury in plants is increased membrane permeability (Tingey and Taylor, 1982). Increases in membrane permeability are measured as increased leakage of a variety of solutes including amino acids, sugars, and potassium. Eckey-Kaltenbach et al. (1993) reported that the conductivity of apoplast fluids doubles in parsley leaves following an eight-hour exposure to 0.20 µL/L ozone. Increases in potassium leakage are believed to be the result of ozone directly affecting specific transport sites, and are reversible if ozone exposure is removed (Heath and Frederick, 1979). However, with prolonged exposure, irreparable membrane damage occurs. At low ozone doses, membrane proteins are oxidized, while at higher doses lipids undergo peroxidation, both of which can alter membrane permeability (Tingey and Taylor, 1982). Changes in membrane permeability ultimately alters the cellular environment and leads to altered metabolism throughout the cell.

Lipid peroxidation by ozone or ozone-induced free radicals also may play an important role in cellular damage (Schraudner et al., 1997). This process can result in the formation of lipid hydroperoxides (LOOHs) from unsaturated fatty acids in the plasma membrane. LOOHs can form other LOOHs through chain-type reactions or can decompose in the presence of transition metals or enzymes including glutathione peroxidase, glutathione transferase, or lyases to form short-chain aldehydes, pentane, and ethane. The formation of LOOHs results in reduced membrane fluidity and function.

In addition to the direct action of ozone on membrane lipids, ozone can alter lipid metabolism (Sakaki et al., 1990). Spinach leaves fumigated for two hours with 0.50 μL/L ozone have higher levels of free fatty acids, which alter lipid synthesis when compared with control plants.

The primary target of ozone attack appears to be the plasma membrane, therefore membranes of other organelles are less affected. The ozone-induced reduction of photosynthetic carbon fixation in wheat has been correlated with a reduction in ribulose biphosphate and an increase in the ATP/ADP ratio and the triose phosphate/3-phosphoglycerate ratio, but not to an increase in triose phosphate (Lehnherr et al., 1988). This suggests that ozone exposure limits photosynthesis by reducing carboxylation or the regeneration of ribulose bisphosphate, rather than by affecting electron transport in the chloroplast membranes. Similarly, in hybrid poplar leaves, light-dependent reactions involving electron transport are less affected by long-term ozone exposure than metabolism associated with the pentose-phosphate cycle in the cytoplasm (Reich, 1983).

Ethylene Production

Exposure of plants to ozone can induce ethylene production. Induced ethylene production is often transient and dependent upon ozone concentration and exposure time. In potato leaves exposed to 0.2 μL/L ozone for four hours, a transient fivefold increase in ethylene production occurs immediately following exposure but is no longer observed after 20 hours (Reddy et al., 1991). Similarly in pea seedlings, a seven-hour exposure to 0.05 to 0.15 μL/L ozone more than doubles ethylene production (Mehlhorn and Wellburn, 1987). However, when pea seedlings are grown under a daily seven-hour exposure to these same concentrations of ozone, ethylene production is reduced below that of untreated seedlings, indicating that pea seedlings are able to develop ozone tolerance. Leaf injury develops in three-week-old seedlings exposed once to the seven-hour treatment, but no injury occurs in those treated daily.

Ethylene production is an indicator of ozone sensitivity. Ozone-sensitive species produce more ethylene as a result of ozone exposure than do tolerant species (Schraudner et al., 1997). Tingey et al. (1976) found that ethylene evolution following a four-hour exposure to ozone concentrations ranging from 0 to 0.75 μL/L is a reliable measure of a plant's ozone sensitivity. Following this short ozone exposure, ethylene production peaks after two to four hours and returns to pretreatment rates after 48 hours. When the ozone-sensitive and -tolerant tobacco cultivars Bel W3 and Bel B are exposed to 0.15 μL/L ozone, ethylene production peaks after one to two hours of exposure (Langebartels et al., 1991). The ozone stress causes an 11-fold increase in ethylene production in 'Bel W3', which continues to be enhanced 14 hours after treatment, but only a twofold increase in 'Bel B', which drops back to control levels. Craker (1971) found a two-hour exposure of 0.25 μL/L ozone causes a two- to threefold increase in ethylene production in tomato, tobacco, and bean plants; increased concentrations enhance ethylene production, and sensitive tomato cultivars produce more ethylene than do tolerant cultivars. Wellburn and Wellburn (1996) compared the acclimation responses of ozone-sensitive and -tolerant selections of six different species of plants after two days of eight-hour exposures to ozone, peaking each day at 0.12 μL/L. No visual damage occurred in any of the plants. However, ethylene increased in all six sensitive selections while remaining constant or decreasing in the tolerant plants. The amount of ozone-induced ethylene and the duration of its production is proportional to ozone dose and reflects the ability of the plant tissue to recover following exposure (Craker, 1971; Tingey et al., 1976).

Some of the physiological effects of ozone are mediated through ethylene. Stress ethylene production correlates with injury but precedes its development (Tingey, 1980). The induction of ethylene by ozone in pea seedlings appears to play a role in the development of leaf damage; when ozone-induced ethylene production is inhibited with the ethylene biosynthesis inhibitor aminoethoxyvinylglycine (AVG), leaf injury caused by ozone exposure is almost eliminated (Mehlhorn and Wellburn, 1987).

Respiration

Respiration is reported to increase when plants are exposed to ozone and is often greater when visual injury results (Miller, 1987). In harvested blueberry fruit, a one- or two-day exposure to 0.2 μL/L ozone induces a transient increase in respiration (Song et al., 2001). In broccoli and carrots, 0.7 μL/L ozone also induces a transient increase in respiration, which is accompanied by a peak of ethylene production (unpublished data). No difference in respiration rates of strawberry fruit or spinach leaves is detected between ozone-treated and control samples during a seven-day exposure to 0.05 or 0.50 μL/L ozone at 5°C (Ikeda et al., 1998), although timing of the respiration measurements may have been such that a transient increase was not detected.

Other Physiological Effects

The oxidative stress from relatively mild ozone treatments induces physiological changes that impart resistance in plant organs to infection by pathogens or to degradation caused by normal senescence or ripening. These changes may be in the form of induction of defense genes, which includes the induction of phytoalexins that impart resistance to pathogen infection, or formation of cellular barriers.

Defense Genes

Koch et al. (1998) reported that ozone induces activity of a variety of defense genes in hybrid poplar plants including phenylalanine ammonia-lyase (PAL), the first enzyme in the phenyl-propanoid biosynthesis pathway; *O*-methyltransferase, a pathogenesis-related protein involved in lignin synthesis; PR-1, a pathogenesis-related protein characteristic of salicylic acid-mediated responses; and WIN3.7, a wound-inducible proteinase inhibitor that can also be induced by methyl jasmonate. Induction of these genes is associated with ozone tolerance when comparing tolerant and susceptible poplar clones. In parsley leaves, 13 genes are induced while 11 are repressed by a 10-hour exposure to 0.20 µL/L ozone (Eckey-Kaltenbach et al., 1997). Expression of the defense genes for PAL, chalcone synthase, chitinase, and β-1,3-glucanase increases in tobacco leaves following an ozone exposure (Ernst et al., 1992).

Ozone is also reported to induce the production of phytoalexins in plants. Exposure of grapes to ~ 4,000 µL/L ozone induces the production of the stilbenes, resveratrol, and pterostilbene, and concentrations of these compounds peak 24 hours after a five- to ten-minute ozone treatment (Sarig et al., 1996). These compounds are also induced when fruit are inoculated with *Rhizopus* or exposed to ultraviolet light. In soybean plants, injurious treatments of 0.70 µL/L ozone for two or three hours result in an increased accumulation of the isoflavonoids daidzein, coumestrol, and sojagol, which are also induced by fungal pathogens (Keen and Taylor, 1975). Another phytoalexin induced by ozone is chitinase, an enzyme produced by many plants that imparts resistance to fungal invasion (Punja and Zhang, 1993). This enzyme degrades chitin, which is a structural component of the cell wall of many phytopathogenic fungi. Schraudner et al. (1992) reported that a five-hour exposure to 0.15 µL/L ozone induces chitinase activity in tobacco and that most of this induced activity is intracellular. Ozone also stimulates mRNA production for this enzyme (Ernst et al., 1992).

Cellular Barriers

Ozone triggers plant responses similar to those triggered by fungal and viral pathogens, including the formation of cellular barriers to pathogens by the formation of lignins, extensins, and callose (Sandermann, 1998). Ozone-induced β-1,3-glucanase activity in tobacco is associated with increased cell wall callose formation (Ernst et al., 1992; Schraudner et al., 1992). In soybean leaves, ozone-induced changes are believed to stabilize cell walls against microbial attack (Booker et al., 1991). These include increases in peroxidase activity and changes in the cell walls, including impregnation with phenolic esters, suberization, and lignification.

POSTHARVEST USE ON FRESH PRODUCE

Over the years, attempts have been made to use the reactive properties of ozone to preserve the quality of fresh fruits and vegetables. Most of these attempts have focused on the antimicrobial effects of ozone and its usefulness to inhibit or prevent decay, although, as discussed earlier, ozone may induce many physiological changes in horticultural products, both positive and negative. Two main applications of ozone have been to treat water used in the handling of fresh produce and to add ozone to storage-room air. The success of these treatments to maintain quality and extend storage life of fresh produce has been variable.

Water Treatment

Water used to cool, transport, or wash fresh produce following harvest must be sanitized to prevent the buildup of microorganisms. This is important in preventing infections that can result in decay during storage or marketing, as well as in reducing the potential contamination of produce with organisms that are harmful to humans. This is often done through the addition of chlorine to water (Boyette et al., 1993). However, ozone can be used as a substitute for chlorine in treating and sanitizing water (Smilanick et al., 1999; Xu, 1999). Due to its high oxidation potential of -2.07 V, less ozone is required to treat water compared with chlorine, which has an oxidation potential of -1.36 V (Kim et al., 1999b). Concentrations of 1 to 1.5 mg/L ozone give similar effectiveness for microbial control as 100 mg/L chlorine. Ozone can kill many microorganisms including plant pathogens, human pathogens, spore-forming protozoa, and viruses (Nebel, 1981; Rice, 1999). In addition to reducing microbial contamination, ozone effectively oxidizes chemical contaminants in water, including those responsible for odors and discoloration, as well as iron, manganese, and sulfur, without producing any by-products of human health concern (Kim et al., 1999b; Nebel, 1981). Effective water processing treatments consist of 0.5 to 5 mg/L ozone in the water for up to five minutes (Nebel, 1981). Ozone persists for only minutes in clean water, while chlorine persists for hours. For effective sanitation, water must not contain excessive organic material. Broadwater et al. (1973) showed that at low concentrations, ozone loses its effectiveness to kill bacteria when organic matter is present in the water. However, at higher concentrations of 3.8 µg/mL, Ogawa et al. (1990) found that the addition of 0.5 g of loamy soil/L does not affect ozone's ability to inactivate fungal spores.

Bacteria

Ozone is effective in inactivating both gram-negative and gram-positive bacteria in either vegetative or spore form (Kim et al., 1999b). However, inactivation rates vary depending on species, state of growth, ozone concentration, and treatment conditions. The viability of the bacteria *Escherichia coli, Mycobacterium fortuitum,* and *Salmonella typhimurium* in clean water are reduced by 4.0, 1.0, and 4.3 log, respectively, when treated with about 0.25 mg/L ozone at 24° C for 1.7 minutes (Farooq and Akhlaque, 1983). Vegetative cells are killed more easily by ozone than spores (Kondo et al., 1989). Vegetative cells of *Bacillus cereus* require only 0.12 mg/L ozone for a > 2.0 log reduction in five minutes compared with 2.29 mg/L for spores for a similar reduction (Broadwater et al., 1973). Proteins on the spore cell wall provide a protective barrier against ozone. When these proteins are removed, the effectiveness of ozone to kill spores increases about 100-fold (Foegeding, 1985). Additional factors such as acidic pH (Foegeding, 1985) and the addition of metallozeolites, AA, and isoascorbic acid (Naitoh, 1992a,b) may improve the effectiveness of ozone.

Fungi

Spores of various postharvest pathogenic fungi also are controlled in water used for the postharvest handling of fresh produce. Spores of *Monolinia fructicola, Geotrichum citri-aurantii, Penicillium italicum,* and *Botrytis cinerea* are killed in about one minute when exposed to 16.5°C water containing 1.5 mg/L ozone, while *Penicillium digitatum, Penicillium expansum,* and *Rhizopus stolonifer* require an exposure of two to three minutes (Smilanick et al., 1999). Similarly, Spotts and Cervantes (1992) demonstrated that inhibition of spore germination increases with ozone concentration and exposure time. The LD_{95} for *Botrytis cinerea, Mucor piriformis,* and *Penicillium expansum* is an ozone concentration of less than 1 mg/L in a five-minute exposure or between 1.5 and 2.5 mg/L in a one-minute exposure.

Produce Quality

In addition to its use for sanitation, ozone dissolved in water is used to control decay in fresh produce as an alternative to other chemical treatments. Dissolved ozone in water used in either a spray or flume system can reduce populations of microbes on the surface of fresh produce (Xu, 1999). Immersing produce for two or more minutes reduces surface microbial populations by 90 to 99 percent (USFDA, 1997). When strawberries are immersed in water containing 4 mg/L ozone for two minutes, aerobic mesophilic bacteria, yeast, and molds are reduced by over 90 percent (Smilanick et al., 1999). Ogawa et al. (1990) found that *Botrytis cinerea* spores on the surface of uninjured tomatoes are inactivated when exposed to a solution of 3.8 mg/L ozone for 10 minutes. However, these surface treatments have a limited effect in reducing decay. Grey mold of grapes is reduced about 50 percent when immersed for one to four minutes in 10 mg/L ozone-containing water (Smilanick et al., 1999). However, effectiveness is irregular and the researchers believe that cracks on the berry surface or around the pedicel may protect some spores from the ozone treatment. In addition, latent infection in the fruit may be present and responsible for some of the fruit decay. When ozone (0.31 mg/L) is compared with sodium hypochlorite (54 mg/L) in a flotation tank used for pears, the number of viable propagules of *Alternaria* spp. is 10-fold greater in the ozone-treated tank than in the chlorine-treated tank, but there are no differences in populations of *Cladosporium* spp. or *Penicillium* spp. (Spotts and Cervantes, 1992). Similar rates of decay develop in fruit floated in either the ozone- or the chlorine-treated tanks following five months of storage.

The use of ozone-containing wash treatments also reduces microbial populations on fresh-cut produce. Ozone is particularly effective in killing *Escherichia coli,* which is of concern to the fresh-cut produce industry (Xu, 1999). Bacterial counts are reduced by about 2 log colony forming units/g in fresh-cut lettuce washed with water injected with 1.3 mmol/L ozone at a rate of 0.5 L/minute for three minutes (Kim et al., 1999b). Kim et al. (1999a) found that the most effective treatment to reduce microbial contamination of shredded lettuce is a combination of ozone bubbling and high-speed stirring. Using a similar ozone wash, Kondo et al. (1989) obtained a > 90 percent reduction in bacterial counts on the surface of Chinese cabbage.

Because ozone washes are primarily a surface treatment, ozone-containing water is not effective in reducing decay of wound-inoculated fruit. Several studies with citrus (Smilanick et al., 1999), tomatoes (Ogawa et al., 1990), and pears (Spotts and Cervantes, 1992) have shown the ineffectiveness of these treatments to reduce the development of decay during storage. These treatments fail even when concentrations of ozone as high as 12 mg/L and durations of up to 20 minutes are used. However, Spotts and Cervantes (1992) noted that lesion size is slightly reduced in wound-inoculated pears that are treated with ozone-containing water. Pathogens present in a wound appear to be well protected from oxidation by ozone, probably as a result of limited penetration due to the reaction of ozone with reactive compounds present in the wound.

Air Treatments

Treatments of produce in water are short in duration, normally lasting only minutes, and air treatments are normally longer and are often continuous while the product is in storage. As a result, in addition to effects on microorganisms, there are often physiological responses of the product to ozone as described previously. For over 70 years, research has been conducted on the use of gaseous ozone in storage-room air to control product decay. These studies have been conducted on a variety of microorganisms and commodities including apples, potatoes, tomatoes, strawberries, broccoli, pears, cranberries, citrus, peaches, and grapes (Rice et al., 1982). Treatment concentrations and durations vary substantially among these studies, resulting in a wide range of results from control of decay to injury and reduced storage life of the product.

Bacteria

Ozone in air is also effective in inhibiting bacteria. The viability of *Erwinia carotovora* pv *carotovora* inoculated onto agar is reduced by exposure to 0.05 µL/L ozone (Forney et al., 2001). In unpublished work, we have also seen a reduction in the viability of *Escherichia coli* and *Pseudomonas fluorescens* as a result of exposure to 0.5 µL/L ozone. In meat tenderizing rooms that contained about 0.10 µL/L ozone, bacteria on agar are killed after 48 to 72 hours of continuous exposure (Nagy, 1959). Bacteria on the surface of spinach, lettuce, and strawberry fruit are reduced following exposure to 0.05 or 0.50 µL/L ozone at 5°C (Ikeda et al., 1998). At 0.05 µL/L ozone, bacterial contamination is decreased to about one-half to one-third of that of the nontreated controls, while bacteria are nearly eliminated by treatment with 0.50 µL/L.

Sensitivity of bacteria to ozone is affected by many factors including media, growth stage, temperature, humidity, and air ions. The rate of kill by ozone in air is affected by the type of agar the bacterial cells are plated onto and the delay in the application of ozone following plating (Fan et al., 2002). Similarly, bacteria on different surfaces have different responses to ozone exposure, possibly due to differences in reactive sites on the surfaces (Elford and van den Ende, 1942). Natural substances secreted by bacterial cells may react with ozone and thus protect the cells. The stage of growth may also affect its sensitivity to ozone indicating why low concentrations of ozone for long durations are effective in killing bacteria as the bacteria go through several growth cycles (Nagy, 1959). Da Silva et al. (1998) demonstrated that cultured bacteria of four different genera change their sensitivity to ozone during growth. Immediately after plating, cells die during the first 20 minutes of ozone treatment but then become resistant. When cultured for four hours, the bacteria again become sensitive to ozone treatment, but after 10 hours of culture the bacteria are resistant and not affected by ozone. High levels of humidity result in higher rates of kill compared with low humidity

(Nagy, 1959; Naitoh, 1992b). Ishizaki et al. (1986) found that preconditioning *Bacillus* spores to high humidities reduces the time required to kill spores with ozone. This may indicate the need for water to facilitate ozone penetration of the cell wall (Elford and van den Ende, 1942) or that humidity affects the physiology of the bacteria resulting in increased sensitivity to ozone.

Fungi

Air treatments with ozone also inhibit or kill fungal organisms. Heagle (1973) reported that ozone exposure can inhibit the growth, germination, and sporulation of various fungi, but the effectiveness is dependent on the species, growth stage, and ozone concentration and exposure time. Hibben and Stotzky (1969) found that species of fungi vary in their sensitivity to ozone when spores on agar are exposed to ozone concentrations ranging from 0.1 to 1.0 µL/L for up to six hours. Spores of *Chaetomium* spp., *Stemphylium sarcinaeforme, Stemphylium loti,* and *Alternaria* spp. are relatively insensitive; *Trichoderma viride, Aspergillus terreus, Aspergillus niger, Penicillium egyptiacum, Botrytis allii,* and *Rhizopus stolonifer* are moderately sensitive, with reductions in germination following exposure to 1.0 and 0.5 µL/L ozone for four to six hours; and *Fusarium oxysporum, Colletotrichum lagenaruim, Verticillium albo-atrum,* and *Verticillium dahliae* are most sensitive with no germination following a two-hour exposure to 0.5 µL/L ozone. A low concentration of ozone (0.1 µL/L) stimulates spore germination of *Trichoderma viride, Aspergillus terreus, Penicillium egyptiacum, Rhizopus stolonifer,* and *Verticillium dahliae.* In other studies, exposure of 1.95 µL/L ozone for one to several hours kills *Botrytis cinerea* and *Penicillium expansum* spores on glass plates (Schomer, 1948). Spores of *Penicillium expansum* are not affected by a 24-hour exposure to 0.11 µL/L ozone but are nearly completely killed by an exposure to 0.53 µL/L, while spores of *Botrytis cinerea* are not affected (Hildebrand et al., 2001). There is a good correlation between the inhibition of spore germination and the product of the ozone concentration and the exposure time (C × T), indicating that the toxic effect of ozone is related to the amount of ozone contacting the spores.

Additional factors that alter the effectiveness of ozone in killing fungal spores include spore morphology, moisture, and substrate (Heagle, 1973). Multicelled, pigmented spores and spores with thick cell walls are more resistant than single-celled spores or those with thin walls (Hibben and Stotzky, 1969). Single spores are killed more readily than clumped spores (Schomer, 1948). Hibben and Stotzky (1969) suggested that water may stimulate metabolism in spores, making them more sensitive to ozone than dry spores are. They also reported that the type of medium influences the tolerance of spores to ozone. Spores on water agar are more resistant to ozone than spores on richer media. Viable airborne spores were also reduced by ozone treatments of 0.05 to 0.25 µL/L ozone in a commercial onion storage room (Song et al., 2000).

Mycelial growth is also affected by exposure to ozone. Hibben and Stotzky (1969) observed that fungi growth into agar media and aerial mycelium is repressed in the presence of ozone. We have made similar observations that mycelial growth of *Penicillium expansum* and *Botrytis cinerea* is progressively reduced with increasing ozone concentrations of 0.10 to 0.50 µL/L (Hildebrand et al., 2001). Rich and Tomlinson (1968) found that exposure of conidiophores of *Alternaria solani* to 0.1 µL/L ozone for two hours or 1.0 µL/L ozone for four hours stops elongation, and causes swelling and collapse of the cell wall at the apical tips. Conidiophores begin to clongate again when removed from the ozone treatment. If sporulating cultures of *Alternaria solani* are exposed to 1.0 µL/L ozone for 30 minutes, the conidia germinate while still attached to the conidiophores (Rich and Tomlinson, 1968). Sporulation is suppressed in some fungi but is stimulated in *Colletotrichum lagenarium* (Hibben and Stotzky, 1969) and *Alternaria oleraceae* (Treshow et al., 1969).

Nonlethal ozone exposures affect the physiology of microorganisms in a manner similar to that reported in plants (Heagle, 1973). Treshow et al. (1969) reported a loss of pigmentation and neutral lipids in fungi exposed to ozone. However, how these potential changes affect the ability of bacteria or fungi to infect and decay fresh horticultural commodities is not clearly identified.

Produce Quality

The antimicrobial effect of ozone, as well as its reported destruction of ethylene, has resulted in a significant amount of interest in adding ozone to the storage environment for fresh produce. The direct effects of ozone on microorganisms has resulted in applications to decontaminate storage rooms and handling facilities, as well as reduce decay of produce. The use of ozone to reduce ethylene is also purported to delay senescence and ripening of many fresh products. In addition to these effects, ozone can elicit physiological changes that can affect product quality. The nature of the packaging and air circulation in the room may determine the effectiveness of gaseous ozone (Harding, 1968). Due to the high reactivity of ozone, its penetration into cartons and packages is critical for effective treatments to be achieved.

Decay. In efforts to control postharvest decay, a wide range of ozone concentrations have been tested in storage environments, as continuous or intermittent treatments. In general, fruit are more tolerant to ozone than vegetables and therefore higher concentrations of ozone can be used on fruit without risk of damage. Ozone has been added to the storage air of a wide variety of fruits and vegetables with mixed results. In general, ozone is effective in inhibiting mycelium growth and the resulting nesting, as well as sporulation, which may reduce the spread of decay during storage. However, its effectiveness to slow or prevent decay is variable, depending on crop, cultivar, decay organism, and storage conditions.

Although apples are tolerant to high concentrations of ozone, it has little effect on reducing decay. The decay of fruit inoculated with *Penicillium expansum* is not prevented by continuous exposure to ozone concentrations ranging from 0.1 to 3 μL/L (Baker, 1933; Schomer, 1948; Hildebrand et al., 2001). Similar results are reported for fruit inoculated with *Botrytis cinerea* (Hildebrand et al., 2001) and *Phialophora malorum* (Schomer, 1948). Under some conditions, the expansion of decay lesions is delayed (Schomer, 1948) and spore production of infected fruit is reduced (Hildebrand et al., 2001). Under high concentrations of ozone and after long-term storage, fruit decay increases, indicating potential damage to the fruit (Baker, 1933; Schomer, 1948).

Ozone has caused some reduction of stone fruit decay. Peaches dip-inoculated in spore suspensions of *Monilinia fructicola* and *Rhizopus* spp. and exposed to ozone ranging from 0 to 1.0 μL/L for 24 hours at 1.5°C, followed by 24 hours at 22°C, have reduced incidence and severity of decay (Ridley and Sims, 1967). However, results vary depending on cultivar and harvest, and in some cases no benefit is observed. Spalding (1968) found that 0.50 μL/L ozone does not reduce decay of peaches inoculated with *Rhizopus* or *Monolinia,* but the spread of decay, nesting, and mycelial growth is reduced.

Postharvest decay of citrus is also reduced by ozone treatments but with varying effectiveness. Ozone (1.0 μL/L) inhibits decay of scratch-inoculated oranges (33 percent) and lemons (42 percent) held 15 days at 15.5°C and 85 percent relative humidity (RH) in open storage boxes (Harding, 1968). Sporulation of infected fruit is completely inhibited by ozone. However, when fruit are held in vented or nonvented cardboard cartons, 1.0 μL/L ozone has no effect on decay and reduces sporulation only slightly. In another trial, Harding (1968) found that decay of Valencia oranges in open storage boxes held in 1.0 μL/L ozone at 15.5°C and 85 percent RH is not reduced, but sporulation of *Penicillium italicum* and *Penicillium digitatum* is inhibited. Palou et al. (2001) found that the incidence of decay on scratch-inoculated orange and lemon fruit held in 0.3 to 1.0 μL/L ozone at 10°C is delayed by about one week and sporulation was prevented or reduced. Hildebrand et al. (2001) reported that surface colonization of lemons by *Penicillium* spp. is reduced by 75 percent when fruit are held in ozone concentrations of 0.10 to 0.50 μL/L.

The effectiveness of ozone on berry crops, whose short storage life is usually limited by decay, has also been explored. Spalding (1966) found that ozone concentrations ranging from 0.1 to 10 µL/L do not reduce *Botrytis* rot of strawberries held at 13 to 15.5°C. Ozone treatments of 0.35 µL/L for three days at 2°C result in 15 percent less decay of fresh strawberry fruit after two additional days at 20°C, but this reduction in decay is lost after four days, when the ozone-treated fruit have as much or more decay than the controls (Pérez et al., 1999). Spalding (1968) found that 0.50 µL/L ozone does not reduce decay of strawberries inoculated with *Rhizopus* or *Botrytis*, but nesting and mycelium growth are reduced. Ozone concentrations of 0.50 µL/L suppress fungi growth on strawberry fruit when held at 5°C for seven days but cause discoloration (Ikeda et al., 1998). In highbush blueberries, we found that continuous ozone treatments of 0.20 and 0.50 µL/L ozone at 10°C delays decay development by about 1 week (unpublished data), while Spalding (1968) reported that a two-day treatment of 0.50 µL/L ozone at 1.5°C followed by four days in air at 21°C has no effect on decay. Treatment of blackberries with 0.10 and 0.30 µL/L ozone inhibits fruit decay during storage in clamshell containers for 12 days at 2°C and 90 percent RH (Barth et al., 1995). Ozone is not very effective in reducing decay of fresh cranberries during storage. Decay is not reduced in 'Early Black' fruit stored for two months at 4.4°C in 0.27 µL/L ozone (Norton et al., 1968). Concentrations of 0.60 µL/L ozone cause physiological damage in 'Howes' fruit stored for five weeks at 15.6°C, resulting in twofold more rot than in the controls.

Gray mold, caused by *Botrytis cinerea*, is reduced in table grapes by ozone treatments. Short fumigations of grapes with ~ 4,000 µL/L ozone reduce the microflora on the surface of the berries after a 20- to 40-minute exposure and reduce subsequent decay of grape berries (Sarig et al., 1996). Ozone treatments of 30 to 40 minutes are effective in eliminating all decay of berries held for six days at 20°C, even when fruit are heavily inoculated with *Rhizopus* spores. Sarig et al. (1996) found that ozone treatment induces the production of the phytoalexins resveratrol and pterostilbene, which may help to explain the reduction in decay. Hildebrand et al. (2001) also found an 87 percent reduction in the incidence of decayed berries after three weeks as a result of a continuous treatment of 0.20 µL/L ozone at 10°C.

Packaged tomatoes held in ripening rooms with or without 0.02 to 0.05 µL/L ozone ripen at similar rates and have the same amount of watery sour rot caused by *Geotrichum candidum* (Mallison and Spalding, 1966). Ozone concentrations inside the tomato cartons, however, reach only 0.003 µL/L, and ethylene levels in the ozone-treated rooms average one-half to one-third of the levels in the control rooms.

Fumigation of cantaloupes for eight hours a day with 20 μL/L ozone does not reduce decay following 14 days of storage at 13°C (Barger et al., 1948). Spalding (1968) found that a continuous exposure of 0.50 μL/L ozone for seven days at 7°C plus an additional five days at 16°C without treatment also does not reduce decay of cantaloupes, but ozone-treated fruit have less surface mold.

Growth of *Botrytis cinerea* and *Sclerotinia sclerotiorum* in carrot roots is progressively slowed with increasing ozone concentration ranging from 1 to 22 μL/L for eight hours a day (Liew and Prange, 1994). These treatments inhibit the development of mycelium, spores, and sclerotia on lesions, but also cause physiological damage to the carrots. Decay on carrots inoculated with *Botrytis cinerea* is reduced by about 70 percent under continuous exposure of 0.12 μL/L ozone for three weeks at 10°C (Hildebrand et al., 2001). Similar results occur with *Sclerotinia sclerotiorum* on freshly harvested carrots, but not on stored carrots. No control of *Botrytis cinerea* or *Sclerotinia sclerotiorum* on celery occurs when treated with 0.12 to 0.50 μL/L ozone at 10°C for three weeks (Hildebrand et al., 2001). Similarly, Spalding (1968) found that 0.50 μL/L ozone does not reduce *Alternaria* rot of green beans, but nesting and mycelium growth are reduced. Surface mold on onions caused by *Penicillium* spp. is reduced by ozone treatments of 0.05 (day) and 0.25 (night) μL/L, causing no physiological effects on sprouting, rooting, or decay (Song et al., 2000).

Gaseous ozone treatments reduce disease development on flowers. Low concentrations of ozone inhibit disease development on gladiolus flowers inoculated with *Botrytis gladiolorum* and chrysanthemums inoculated with *Botrytis* spp. (Magie, 1963). Ozone treatments decrease the severity of gray mold development in geranium flowers when inoculated flowers are exposed to 0.35 or 0.55 μL/L ozone for four hours (Manning et al., 1970). However, a four-hour treatment of 0.15 to 0.45 μL/L ozone does not prevent or reduce the infection of poinsettia bracts by *Botrytis cinerea* (Manning et al., 1972).

Interaction with negative air ions. Air ions are charged molecules found in the air that can be generated through high voltage ion generators. Some types of corona discharge also produce air ions (Forney et al., 2001). A combination of negative air ions (NAI) and ozone is more effective than ozone alone in inhibiting microbial growth (Forney et al., 2001; Tanimura et al., 1998). This combination is also more effective in reducing decay and maintaining the quality of fresh fruits than is ozone alone (Li et al., 1989; Tanimura et al., 1998). Li et al. (1989) demonstrated that a weekly 10-minute treatment with a combination of 1 to 5 μL/L ozone and 10^4 to 10^5/cm^3 NAI is more effective in preserving the quality of apples, mandarin oranges, and tomatoes than either treatment alone. Tanaka et al. (1998) reported that a combination of 0.05 µL/L ozone and 5×10^4 NAI/cm^3 at 0°C and 95 percent RH maintains fruit quality and inhibits decay of cherries, peaches, Japanese pears, and grapes for one, two, and three months, respectively. Grapes held for eight months with 0.050 µL/L ozone and 5×10^4 NAI/cm^3 develop little growth of mold (Tanimura et al., 1998). Development of devices to produce NAI in commercial storage facilities could enhance the effectiveness of ozone and warrants further investigation.

Ethylene destruction. The removal of ethylene from the storage environment may have beneficial effects in extending storage life of fresh produce by delaying ripening and senesence (Knee and Hatfield, 1981; Shorter and Scott, 1987). Dickson et al. (1992) described a gas flow reactor that uses ozone to destroy ethylene in storage rooms. However, when this reactor was tested under commercial conditions, it did not effectively reduce ethylene concentrations (Dickson, 1993). The rate of the ethylene-ozone reaction is too slow to effectively destroy ethylene at the ozone concentration of 12 µL/L and exposure time of less than two seconds present in a commercial reactor. When concentrations are increased 100-fold and reaction times increased to 30 seconds, ethylene can be destroyed.

Various attempts have been made to measure ethylene destruction by ozone in storage environments. Crisosto et al. (2000) found that rates of ethylene destruction in an empty 2,135 ft^3 export container containing 3.8 µL/L ethylene average 0.035, 0.038, 0.074, 0.110, and 0.110 µL/L per hour after an ozone generator was run for 24 hours resulting in ozone concentrations of 0, 0.3, 1.0, 3.9, and > 10 µL/L, respectively. When ozone concentrations are maintained constant at 0.10 µL/L in a storage room containing 6 µL/L ethylene, we found the ethylene destruction rate to be 0.28 µL/L per hour (unpublished data). Ethylene destruction by ozone has been attributed to inhibiting banana ripening (Gane, 1934, 1937) and broccoli yellowing (Skog and Chu, 2001). However, we have observed ozone-induced ethylene production concurrent with delayed yellowing in broccoli, suggesting that ozone alters the plant's physiological response to ethylene.

Quality. In most studies, no effects on quality have been observed if phytotoxic concentrations of ozone are avoided. Most of these assessments have been made based on physical appearance. In studies where compositional changes are measured, ozone has some additional effects.

Ozone treatments can affect the flavor of fruits. Schomer (1948) found that long-term exposure of apples to 3.25 µL/L ozone results in a loss of flavor. In strawberries, ozone treatments of 0.35 µL/L for three days at 2°C decrease the production of esters, which are responsible for flavor, by 35 percent compared to controls (Pérez et al., 1999). Ethanol concentrations also are reduced in strawberries, especially following two and four additional days at 20°C, and there is no effect on ethyl acetate or acetaldehyde concentrations suggesting that ozone prevents the development of off-flavors. In cranberries, ozone induces a faint but pleasant flowerlike aroma (Norton et al., 1968). Sucrose, glucose, and fructose are reduced by about 20 percent in strawberry fruit exposed to 0.35 µL/L ozone at 2°C for three days, but after an additional two or four days in air at 20°C, these differences are less apparent and the ozone-treated fruit actually have higher concentrations of sucrose (Pérez et al., 1999). Treatments of 1.0 µL/L ozone for 24 hours have no effect on the soluble solid content of peaches (Ridley and Sims, 1967). Various ozone treatments also do not affect the acid content of strawberry (Pérez et al., 1999), peach (Ridley and Sims, 1967), or apple (Schomer, 1948) fruit.

Anthocyanin metabolism is also affected by ozone exposure. Anthocyanin content of blackberries held in 0.30 µL/L ozone at 2°C has a transient increase one day after treatment, and then returns to control levels (Barth et al., 1995). Anthocyanins decrease by about 20 percent in 'Camarosa' strawberry fruit held for three days in 0.35 µL/L ozone at 2°C, but increase to normal levels after removal into air at 20°C for four days (Pérez et al., 1999).

Firmness of peach fruit is not affected by a 24-hour treatment of 1.0 µL/L ozone (Ridley and Sims, 1967). Similarly, an 80-minute treatment of 4,000 µL/L ozone had no effect on the firmness of grapes (Sarig et al., 1996).

Phytotoxicity

The development of visible injury on plant tissues due to ozone treatments is affected by many factors. Large differences in sensitivity are seen between different crop species and cultivars, and the development of injury is dependent on ozone concentration, duration of exposure, and temperature. Few studies have been conducted that describe the development of injury under different environmental conditions or ozone treatment regimes, making it difficult to identify injury thresholds. A summary of ozone treatments that have been reported to cause injury to stored fresh fruits and vegetables is listed in Table 2.1.

Ozone-induced injury is a surface disorder that is often characterized by discoloration, browning, or bleaching of the epidermis. Injury often occurs around natural openings such as lenticels or stomata where ozone is able to penetrate into the tissue. Injury in apples (Hansen and Berger, 1964; Schomer, 1948) and bananas (Gane, 1934) occurs as discolored spots surrounding the lenticels, and in peaches injury occurs as sunken brown spots around stomata, giving a pebbly appearance (Spalding, 1966).

Fruit tend to be more tolerant of ozone than most leafy vegetables, which may be associated with the thicker cuticle normally found on fruit. In strawberries exposed to 0.50 µL/L of ozone or more, the leafy caps become dried

TABLE 2.1. Damage Caused by Ozone Treatments During Storage of Fresh Horticultural Crops

Commodity	Ozone Conc. (µL/L)	Temp. (°C)	Time (days)	Symptoms	References
Apples	3.25	−0.6	28	Discolored spot at the lenticels becoming dark and sunken	Schomer (1948)
Bananas	1.5	15	10	Dark green areas around the lenticels, becoming brown	Gane (1934)
	5.0	15	8		
	25.0	15	8		
	40.0	15	6		
Beans, green	0.5	7	7	Blotchy brown areas	Spalding (1968)
Carrots	0.53	10	21	Surface browning	Hildebrand et al. (2001)
	1.0	2	28	Pitting with dry white blotches, bleaching of orange color	Liew and Prange (1994)
	1.0	8	28	Surface browning	Hildebrand et al. (2001)
	3.0	16	28		
Celery	0.115	10	21	Surface browning	Hildebrand et al. (2001)
Cranberries	0.6	15	35	Increased decay and weight loss	Norton et al. (1968)
Grapes	4,000	NR[1]	> 0.03	Microscopic veinlike cracks on the epidermis	Sarig et al. (1996)
Lettuce	0.10	2	3	Yellowing of exposed leaves, browning of intravascular areas, some brown flecking, surface flecking, chlorosis, cell clearing, cell collapse, browning	Spalding (1966) Reinert et al. (1972)
	0.40	2	5		
	0.7	~20	0.1		
Mushrooms	0.04	4	14	Surface browning	Skog and Chu (2001)
Peaches	0.9	16	5	Brown freckling	Spalding (1966)
	10	16	5		
Strawberries	0.5	13 - 16	7	Dry and shriveled calyx	Spalding (1966, 1968)
	0.5	1.7	7		
	0.5	5	4	Fruit discoloration	Ikeda et al. (1998)

[1]Not reported.

and shriveled while no injury occurs to the fruit tissue (Spalding, 1966). Lettuce is very sensitive to ozone, developing surface browning and flecking from exposures to only 0.04 µL/L (Reinert et al., 1972; Skog and Chu, 2001; Spalding, 1966). However, many fruit can tolerate exposures to concentrations of 1.0 µL/L or greater for extended periods of time with no visible damage (Schomer, 1948; Norton et al., 1968; Gane, 1934).

Susceptibility to injury varies between cultivars within crops. In apples, 'Golden Delicious' fruit are the most sensitive and 'Winesap' least, with 'Delicious', 'Rome Beauty', and 'Arkansas' being intermediate (Schomer, 1948). Cultivars of lettuce also vary in their sensitivity to ozone. After exposure to 0.70 µL/L for 1.5 h at 27°C, 'Dark Green Boston' and 'Grand Rapids Forcing' develop the most severe injury, while 'Black Seeded Simpson' and 'Great Lakes' develop the least (Reinert et al., 1972). Cultivar variation may be the result of either morphological or physiological differences, or both. In addition to cultivar effects, environmental growing conditions may alter the response of fresh produce to ozone, but these potential effects have not been documented. In leaves, tolerance to ozone is often correlated with stomatal conductance, indicating that the pathway for ozone to enter the leaf is in part controlling tolerance.

CONCLUSION AND TOPICS FOR FURTHER RESEARCH

Ozone can play a significant role in a number of facets of the postharvest handling of fresh horticultural commodities. However, many of these potential roles need to be better defined in order to optimize benefits and avoid detrimental effects. Ozone could be used for a given commodity in any or all of the following areas: sanitation of treatment water, removal of surface microbial contamination, reduction of ethylene effects, reduction of decay, and maintaining or enhancing produce quality.

Ozone is well established as an effective sanitizing agent for water. It is effective over a broad range of microorganisms; its effective concentration is 100-fold lower than that of chlorine; and it leaves no residue. The addition of ozone to water is useful to prevent cross-contamination of pathogens in water used for washing, cooling, and handling produce in flumes and dump tanks.

Sanitation of surfaces, including that of equipment and fresh produce, can be achieved with ozone both in water and in air. However, the effectiveness of these treatments is more variable due to the nature of the contamination. Due to the high reactivity of ozone, some microbial contamination that is not fully exposed may not contact the ozone. This is true for treatments of both water and air. Combining ozone with other treatments may help to improve the effectiveness of these decontamination treatments. Some success has been reported in the use of agitation with ozone-containing water to reduce microbial contamination. Heat treatment could also be combined with ozone-containing water to improve effectiveness. For treatment with ozone in the air, the role of humidity has not been clearly defined. There is some evidence that high relative humidity may enhance the effectiveness of ozone to kill pathogens and inhibit decay of produce. This needs to be confirmed and methods to optimize relative humidity during ozone treatments developed. The inclusion of negative air ions with ozone may enhance its effectiveness to inhibit microbes. Effective methods to apply and distribute the short-lived air ions need to be developed to make this feasible for commercial application. Ozone also must be uniformly distributed in storage rooms or treatment chambers to maintain adequate and uniform contact with produce, which is impeded if produce is packaged or in cartons.

The addition of ozone to storage-room air is effective in reducing ethylene concentrations in the room through ozone oxidation of ethylene. However, the effectiveness of ethylene reduction on preventing its damaging effects is subject to debate. Ozone may induce the production of ethylene, and physiologically active concentrations of ethylene may remain at the cellular level. In some commodities, ozone may reduce the commodity's responsiveness to ethylene and thus reduce its detrimental effects. Additional research is required to determine the mechanism by which ozone affects the role of ethylene in produce deterioration.

Ozone can reduce the postharvest decay of fresh commodities by directly inhibiting the pathogen or by inducing decay resistance. The physiological effects of ozone on fresh produce are not well understood and may vary widely among different commodities. Morphological differences may affect rates of ozone penetration into plant tissues resulting in different tolerance levels. The defense mechanisms present in the apoplast of commodities differ both in the quantities and types of antioxidants present, as well as mechanisms for their maintenance. Different commodities may respond in many different ways to the oxidative stress imposed by ozone, and many of these responses may be concentration dependent. Ozone can induce antioxidant production, ethylene synthesis, and phytoalexin formation. It can also cause damage to membranes resulting in cell death and visible injury. Research is needed to identify and define these physiological responses of different commodities to ozone. Ozone treatments can then be developed to capitalize on the beneficial responses while avoiding those that are detrimental.

Finally, some physiological responses of commodities may improve quality by extending storage life, preventing decay, or enhancing nutritional quality. Since many plant responses to ozone are transient, novel approaches need to be developed to optimize their potential benefits. One approach is to use intermittent or pulsed ozone treatments. This could result in a more sustained response by the plant. With increasing concerns about chemical residues and microbial safety, and growing interest in the nutritional and antioxidant content of food, the responses of fresh produce to ozone may hold opportunities for new postharvest technologies to enhance produce quality and value.

REFERENCES

Anglada, J.M., R. Crehuet, and J.M. Bofill (1999). The ozonolysis of ethylene: A theoretical study for the gas-phase reaction mechanism. *Chemistry—A European Journal* 5:1809-1822.

Aono, M., A. Kubo, H. Saji, T. Natori, K. Tanaka, and N. Kondo (1991). Resistance to active oxygen toxicity of transgenic *Nicotiana tabacum* that expresses the gene for glutathione reductase from *Escherichia coli*. *Plant Cell Physiology* 32:691-697.

Atkinson, R. (1990). Gas-phase tropospheric chemistry of organic compounds: A review. *Atmospheric Environment* 24A:1-41.

Azevedo, R.A., R.M. Alas, R.J. Smith, and P.J. Lea (1998). Response of antioxidant enzymes to transfer from elevated carbon dioxide to air and ozone fumigation, in the leaves and roots of wild-type and a catalase-deficient mutant of barley. *Physiologia Plantarum* 104:280-292.

Bablon, G., W.D. Bellamy, M.M. Bourbigot, F.B. Daniel, M. Doré, F. Erb, G. Gordon, B. Langlais, A. Laplanche, B. Legube, G. Martin, W.J. Masschelein, G. Pacey, D.A. Reckhow, and C. Ventresque (1991). Fundamental aspects. In *Ozone in Water Treatment: Application and Engineering*, B. Langlais, D.A. Reckhow, and D.R. Brink (Eds.) Chelsea, MI: Lewis Publishers, Inc., pp. 11-132.

Badiani, M., G. Schenone, A.R. Paolacci, and I. Fumagalli (1993). Daily fluctuations of antioxidants in bean (*Phaseolus vulgaris* L.) leaves as affected by the presence of ambient air pollutants. *Plant Cell Physiology* 34:271-279.

Baker, C.E. (1933). The effect of ozone upon apples in cold storage. *Ice and Refrigeration* 84:402-404.

Barger, W.R., J.S. Wiant, W.T. Pentzer, A.L. Ryall, and D.H. Dewey (1948). A comparison of fungicidal treatments for the control of decay in California cantaloupes. *Phytopathology* 38:1019-1024.

Barth, M.M., C. Zhou, J. Mercier, and F.A. Payne (1995). Ozone storage effects on anthocyanin content and fungal growth in blackberries. *Journal of Food Science* 60:1286-1288.

Bennett, J.P., R.J. Oshima, and L.F. Lippert (1979). Effects of ozone on injury and dry matter partitioning in pepper plants. *Environmental and Experimental Botany* 19:33-39.

Booker, F.L., E.L. Fiscus, and J.E. Miller (1991). Ozone-induced changes in soybean cell wall physiology. *Current Topics in Plant Physiology* 6:229-232.

Bors, W., C. Langebartels, C. Michel, and H. Sandermann (1989). Polyamines as radical scavengers and protectants against ozone damage. *Phytochemistry* 28:1589-1595.

Boyette, M.D., D.F. Ritchie, S.J. Carballo, S.M. Blankenship, and D.C. Sanders (1993). Chlorination and postharvest disease control. *North Carolina Cooperative Extension Service Publication AG-414-6.*

Broadbent, P., G.P. Creissen, B. Kular, A.R. Wellburn, and P.M. Mullineaux (1995). Oxidative stress responses in transgenic tobacco containing altered levels of glutathione reductase activity. *Plant Journal* 8:247-255.

Broadwater, W.T., R.C. Hoehn, and P.H. King (1973). Sensitivity of three selected bacterial species to ozone. *Applied Microbiology* 26:391-393.

Castillo, F.J. and H. Greppin (1986). Balance between anionic and cationic extracellular peroxidase activities in *Sedum album* L. leaves after ozone exposure. Analysis by high-performance liquid chromatography. *Physiologia Plantarum* 68:201-208.

Castillo, F.J. and H. Greppin (1988). Extracellular ascorbic acid and enzyme activities related to ascorbic acid metabolism in *Sedum album* L. leaves after ozone exposure. *Environmental and Experimental Botany* 28:231-238.

Castillo, F.J., P.R. Miller, and H. Greppin (1987). Extracellular biochemical markers of photochemical oxidant air pollutant damage to Norway spruce. *Experientia* 43:111-115.

Chappelka, A.H. and L.J. Samuelson (1998). Ambient ozone effects on forest trees of the eastern United States: A review. *New Phytologist* 139:91-108.

Chevone, B.I., J.R. Seiler, J. Melkonian, and R.G. Amundson (1990). Ozone-water stress interactions. *Plant Biology* 12:311-328.

Conklin, P.L. and R.L. Last (1995). Differential accumulation of antioxidant mRNAs in *Arabidopsis thaliana* exposed to ozone. *Plant Physiology* 109:203-212.

Conklin, P.L., E.H. Williams, and R.L. Last (1996). Environmental stress sensitivity of an ascorbate deficient *Arabidopsis* mutant. *Proceedings of the National Academy of Sciences of the United States of America* 93:9970-9974.

Craker, L.E. (1971). Ethylene production from ozone injured plants. *Environmental Pollution* 1:299-304.

Crisosto, C.H., D. Garner, J. Smilanick, and J.P. Zoffoli (2000). Ability of the OxtomCav ozone generator to reduce ethylene levels. *Perishables Handling Quarterly* 103:13-14.

Crisosto, C.H., W.A. Retzlaff, L.E. Williams, T.M. DeJong, and J.P. Zoffoli (1993). Postharvest performance evaluation of plum (*Prunus salicina* Lindel., 'Casselman') fruit grown under three ozone concentrations. *Journal of the American Society for Horticultural Science* 118:497-502.

da Silva, M.V., P.A. Fibbs, and R.M. Kirby (1998). Sensorial and microbial effects of gaseous ozone on fresh scad *(Trachurus trachurus). Journal of Applied Microbiology* 84:802-810.

Dickson, R.G. (1993). *Abatement of ethylene by ozone treatment in controlled atmosphere storage of fruits and vegetables.* Master's thesis. University of Georgia.

Dickson, R.G., S.E. Law, S.J. Kays, and M.A. Eiteman (1992). Abatement of ethylene by ozone treatment in controlled atmosphere storage of fruits and vegetables. Presented at the 1992 ASAE International Winter Meeting, Paper No. 92-6571. American Society of Agricultural Engineers.

Dohmen, G.P., A. Koppers, and C. Langebartels (1990). Biochemical response of Norway spruce (*Picea abies* (L.) Karst.), towards 14-month exposure to ozone and acid mist: Effects on amino acid, glutathione and polyamine titers. *Environmental Pollution* 64:375-383.

Eckey-Kaltenbach, H., W. Heller, J. Sonnenbichler, I. Zetl, W. Schäfer, D. Ernst, and H. Sandermann Jr. (1993). Oxidative stress and plant secondary metabolism: 6"-*O*-malonylapiin in parsley. *Phytochemistry* 34:687-691.

Eckey-Kaltenbach, H., E. Kiefer, E. Grosskopf, D. Ernst, and H. Sandermann Jr. (1997). Differential transcript induction of parsley pathogenesis-related proteins and of a small heat shock protein by ozone and heat shock. *Plant Molecular Biology* 33:343-350.

Elford, W.J. and J. van den Ende (1942). An investigation of the merits of ozone as an aerial disinfectant. *Journal of Hygiene* 42:240-265.

Elstner, E.F., W. Osswald, and R.J. Youngman (1985). Basic mechanisms of pigment bleaching and loss of structural resistance in spruce *(Picea abies)* needles: Advances in phytomedical diagnostics. *Experientia* 41:591-597.

Ernst, D., M. Schraudner, C. Langebartels, and H. Sandermann Jr. (1992). Ozone-induced changes of mRNA levels of β-1,3-glucanase, chitinase and 'pathogenesis-related' protein 1b in tobacco plants. *Plant Molecular Biology* 20:673-682.

Evans, P.T. and R.L. Malmberg (1989). Do polyamines have roles in plant development? *Annual Review of Plant Physiology and Molecular Biology* 40:235-269.

Ewell, A.W. (1942). Production, concentration, and decomposition of ozone by ultraviolet lamps. *Journal of Applied Physics* 13:759-767.

Fan, L., J. Song, P.D. Hildebrand, and C.F. Forney (2002). Interaction of ozone and negative air ions to control microorganisms. *Journal of Applied Microbiology* 93: 144-148.

Farooq, S. and S. Akhlaque (1983). Comparative response of mixed cultures of bacteria and virus to ozonation. *Water Research* 17:809-812.

Foegeding, P.M. (1985). Ozone inactivation of *Bacillus* and *Clostridium* spore populations and the importance of the spore coat to resistance. *Food Microbiology* 2:123-134.

Forney, C.F., L. Fan, P.D. Hildebrand, and J. Song (2001). Do negative air ions reduce decay of fresh fruits and vegetables? *Acta Horticulturae* 2(553):421-424.

Foy, C.D., E.H. Lee, R. Rowland, T.E. Devine, and R.I. Buzzell (1995). Ozone tolerance related to flavonol glycoside genes in soybean. *Journal of Plant Nutrition* 18:637-647.

Gane, R. (1934). The effects of ozone on bananas. In *Report of the Food Investigation Board.* Norwich, United Kingdom: His Majesty's Stationary Office, pp. 128-130.

Gane, R. (1937). The respiration of bananas in presence of ethylene. *New Phytologist* 36:170-178.

Graham, D.M. (1997). Use of ozone for food processing. *Food Technology* 51:72-75.

Grimes, H.D., K.K. Perkins, and W.F. Boss (1983). Ozone degrades into hydroxyl radical under physiological conditions. A spin trapping study. *Plant Physiology* 72:1016-1020.

Grimsrud, E.P., H.H. Westberg, and R.A. Rasmussen (1975). Atmospheric reactivity of monoterpene hydrocarbons, nitrogen oxides photooxidation and ozonolysis. *International Journal of Chemical Kinetics* 7:183-195.

Hansen, H. and A. Berger (1964). Ozonation damage to cold stored apples. *Die Gartenbauwissenschaft* 29:517-521.

Harding, P.R. (1968). Effect of ozone on *Penicillium* mold decay and sporulation. *Plant Disease Reporter* 52:245-247.

Heagle, A.S. (1973). Interactions between air pollutants and plant parasites. *Annual Review of Phytopathology* 11:365-388.

Heath, R.L. (1988). Biochemical mechanisms of pollutant stress. In *Assessment of Crop Loss from Air Pollutants*, W.W. Heck, O.C. Taylor, and D.T. Tingey (Eds.). London, United Kingdom: Elsevier Applied Science, pp. 259-286.

Heath, R.L. and P.E. Frederick (1979). Ozone alteration of membrane permeability in *Chlorella* I. Permeability of potassium ion as measured by [86]rubidium tracer. *Plant Physiology* 64:455-459.

Hérouart, D., C. Bowler, H. Willekens, W. Van Camp, L. Slooten, M. Van Montagu, and D. Inzé (1993). Genetic engineering of oxidative stress resistance in higher plants. *Philosophical Transactions of the Royal Society of London* 342: 235-240.

Hibben, C.R. and G. Stotzky (1969). Effects of ozone on the germination of fungus spores. *Canadian Journal of Microbiology* 15:1187-1196.

Hildebrand, P.D., J. Song, C.F. Forney, W.E. Renderos, and D.A.J. Ryan (2001). Effects of corona discharge on decay of fruits and vegetables. *Acta Horticulturae* 2(553):425-426.

Hippeli, S. and E.F. Elstner (1996). Mechanisms of oxygen activation during plant stress: Biochemical effects of air pollutants. *Journal of Plant Physiology* 148: 249-257.

Horie, O. and G.K. Moortgat (1991). Decomposition pathways of the excited Criegee intermediates in the ozonolysis of simple alkenes. *Atmospheric Environment* 25A:1881-1896.

Ikeda, A., Y. Kawai, K. Esaki, and S. Nakayama (1998). Sterilization of vegetables preserved at low temperature with low ozone concentration. *Journal of Society of High Technology in Agriculture* 10:237-242.

Ishizaki, K., N. Shinriki, and H. Matsuyama (1986). Inactivation of *Bacillus* spores by gaseous ozone. *Journal of Applied Bacteriology* 60:67-72.

Kangasjärvi, J., J. Talvinen, M. Utriainen, and R. Karjalainen (1994). Plant defense systems induced by ozone. *Plant, Cell and Environment* 17:783-794.

Keen, N.T. and O.C. Taylor (1975). Ozone injury in soybeans. Isoflavonoid accumulation is related to necrosis. *Plant Physiology* 55:731-733.

Kerstiens, G. and K.J. Lendzian (1989). Interactions between ozone and plant cuticles I. Ozone deposition and permeability. *New Phytologist* 112:13-19.

Khan, M.R., M.W. Khan, and A.A. Khan (1996). Evaluation of the sensitivity of some vegetable crops to ozone. *Test of Agrochemicals and Cutivars* 17:94-95.

Kim, J.G., A.E. Yousef, and G.W. Chism (1999a). Use of ozone to inactivate microorganisms on lettuce. *Journal of Food Safety* 19:17-34.

Kim, J.G., A.E. Yousef, and S. Dave (1999b). Application of ozone for enhancing the microbiological safety and quality of foods: A review. *Journal of Food Protection* 62:1071-1087.

Knee, M. and S.G.S. Hatfield (1981). Benefits of ethylene removal during apple storage. *Annals of Applied Biology* 98:157-165.

Koch, J.R., A.J. Scherzer, S.M. Eshita, and K. R. Davis (1998). Ozone sensitivity in hybrid poplar is correlated with a lack of defense-gene activation. *Plant Physiology* 118:1243-1252.

Kondo, F., K. Utoh, and M. Rostamibashman (1989). Bactericidal effects of ozone water and ozone ice. *Bulletin of Faculty of Agriculture, Miyazaki University* 36:93-98.

Langebartels, C., K. Kerner, S. Leonardi, M. Schraudner, M. Trost, W. Heller, and H. Sandermann Jr. (1991). Biochemical plant responses to ozone. I. Differential induction of polyamine and ethylene biosynthesis in tobacco. *Plant Physiology* 95:882-889.

Larson, R.A. (1988). The antioxidants of higher plants. *Phytochemistry* 27:969-978.

Larson, R.A. (1995). Plant defenses against oxidative stress. *Archives of Insect Biochemistry and Physiology* 29:175-186.

Lehnherr, B., F. Machler, A. Grandjean, and J. Fuhrer (1988). The regulation of photosynthesis in leaves of field-grown spring wheat (*Triticum aestivum* L. cv Albis) at different levels of ozone in ambient air. *Plant Physiology* 88:1115-1119.

Li, J., X. Wang, H. Yao, Z. Yao, J. Wang, and Y. Luo (1989). Influence of discharge products on post-harvest physiology of fruit. *International Symposium on High Voltage Engineering* 6:1-4.

Liew, C.L. (1992). Ozone and postharvest disease control of carrots *(Daucus carota L.)*. Master's thesis, Acadia University.

Liew, C.L. and R.K. Prange (1994). Effect of ozone and storage temperature on postharvest diseases and physiology of carrots (*Daucus carota* L.). *Journal of the American Society for Horticultural Science* 119:563-567.

Luwe, M. and U. Heber (1995). Ozone detoxification in the apoplasm and symplasm of spinach, broad bean and beech leaves at ambient and elevated concentrations of ozone in air. *Planta* 197:448-455.

Luwe, M.W.F., U. Takahama, and U. Heber (1993). Role of ascorbate in detoxifying ozone in the apoplast of spinach (*Spinacia oleracea* L.) leaves. *Plant Physiology* 101:969-976.

Magie, R.O. (1963). Botrytis disease control on gladiolus, carnations, and chrysanthemums. *Proceedings of the Florida State Horticulture Society* 76:458-461.

Mallison, E.D. and D.H. Spalding (1966). Use of ozone in tomato ripening rooms. United States Department of Agriculture ARS52-17, 10 pp.

Manning, W.J., W.A. Feder, and I. Perkins (1970). Ozone and infection of geranium flowers by *Botrytis cinerea*. *Phytopathology* 60:1302.

Manning, W.J., W.A. Feder, and I. Perkins (1972). Effects of *Botrytis* and ozone on bracts and flowers of poinsettia cultivars. *Plant Disease Reporter* 56:814-816.

Mehlhorn, H., G. Seufert, A. Schmidt, and K.J. Kunert (1986). Effect of SO_2 and O_3 on production of antioxidants in conifers. *Plant Physiology* 82:336-338.

Mehlhorn, H. and A.R. Wellburn (1987). Stress ethylene formation determines plant sensitivity to ozone. *Nature* 327:417-418.

Miller, J.D., R.N. Arteca, and E.J. Pell (1999). Senescence-associated gene expression during ozone-induced leaf senescence in *Arabidopsis*. *Plant Physiology* 120:1015-1023.

Miller, J.E. (1987). Effects of ozone and sulfur dioxide stress on growth and carbon allocation in plants. *Recent Advances in Phytochemistry* 21:55-100.

Mudd, J.B., R. Leavitt, A. Ongun, and T.T. McManus (1969). Reaction of ozone with amino acids and proteins. *Atmospheric Environment* 3:669-682.

Mudd, J.B., F. Leh, and T.T. McManus (1974). Reaction of ozone with nicotinamide and its derivatives. *Archives of Biochemistry and Biophysics* 161:408-419.

Nagy, R. (1959). Application of ozone from sterilamp in control of mold, bacteria, and odors. In *Ozone Chemistry and Technology*. Washington, DC: American Chemical Society, pp. 57-65.

Naitoh, S. (1992a). Studies on the application of ozone in food preservation: Effect of metallozeolites and ascorbic acid on the inactivation of *Bacillus subtilis* spores with gaseous ozone. *Journal of Antibacterial and Antifungal Agents* 20:629-632.

Naitoh, S. (1992b). Studies on the application of ozone in food preservation: Synergistic sporicidal effects of gaseous ozone and ascorbic acid, isoascorbic acid to *Bacillus subtilis* spores. *Journal of Antibacterial and Antifungal Agents* 20:565-570.

Nebel, C. (1981). Ozone. In *Encyclopedia of Chemical Technology,* Volume 16. Third Edition. New York: John Wiley and Sons, pp. 683-713.

Norton, J.S., A.J. Charig, and I.E. Demoranville (1968). The effect of ozone on storage of cranberries. *Proceedings of the American Society for Horticultural Science* 93:792-796.

Ogawa, J.M., A.J. Feliciano, and B.T. Manji (1990). Evaluation of ozone as a disinfectant in postharvest dump tank treatments for tomato. *Phytopathology* 80:1020.

Ormrod, D.P. and D.W. Beckerson (1986). Polyamines as antiozonants for tomato. *Horticultural Science* 21:1070-1071.

Örvar, B.L. and B.E. Ellis (1997). Transgenic tobacco plants expressing antisense RNA for cytosolic ascorbate peroxidase show increased susceptibility to ozone injury. *Plant Journal* 11:1297-1305.

Palou, L., J.L. Smilanick, C.H. Crisosto, and M. Mansour (2001). Effect of gaseous ozone exposure on the development of green and blue molds on cold stored citrus fruit. *Plant Disease* 85:632-638.

Pérez, A.G., C. Sanz, J.J.Ríos, R. Olías, and J.M. Olías (1999). Effects of ozone treatment on postharvest strawberry quality. *Journal of Agricultural and Food Chemistry* 47:1652-1656.

Pitcher, L.H., E. Brennan, A. Hurley, P. Dunsmuir, J.M. Tepperman, and B.A. Zilinskas (1991). Overproduction of petunia chloroplastic copper/zinc superoxide dismutase does not confer ozone tolerance in transgenic tobacco. *Current Topics in Plant Physiology* 6:271-273.

Price, A., P.W. Lucas, and P.J. Lea (1990). Age dependent damage and glutathione metabolism in ozone fumigated barley: A leaf section approach. *Journal of Experimental Botany* 41:1309-1317.

Punja, Z.K. and Y.Y. Zhang (1993). Plant chitinases and their roles in resistance to fungal diseases. *Journal of Nematology* 25:526-540.

Rabotti, G. and A. Ballarin-Denti (1998). Biochemical responses to abiotic stress in beech (*Fagus sylvatica* L.) leaves. *Chemosphere* 36:871-875.

Ranieri, A., G. D'Urso, C. Nali, G. Lorenzini, and G.F. Soldatini (1996). Ozone stimulates apoplastic antioxidant systems in pumpkin leaves. *Physiologia Plantarum* 97:381-387.

Rautenkranz, A.A.F., L. Li, F. Mächler, E. Märtinoia, and J.J. Oertli (1994). Transport of ascorbic and dehydroascorbic acids across protoplast and vacuole membranes isolated from barley (*Hordeum vulgare* L. cv Gerbel) leaves. *Plant Physiology* 106:187-193.

Reddy, G.N., Y.R. Dai, F.B. Negm, H.E. Flores, R.N. Arteca, and E.J. Pell (1991). The effect of ozone stress on the levels of ethylene, polyamines, and rubisco gene expression in potato leaves. *Current Topics in Plant Physiology* 6:262-267.

Reich, P.B. (1983). Effects of low concentrations of O_3 on net photosynthesis, dark respiration, and chlorophyll content in aging hybrid poplar leaves. *Plant Physiology* 73:291-296.

Reinert, R.A., D.T. Tingey, and H.B. Carter (1972). Ozone induced foliar injury in lettuce and radish cultivars. *Journal of the American Society for Horticultural Science* 97:711-714.

Rice, R.G. (1999). Ozone in the United States of America—State-of-the-art. *Ozone Science and Engineering* 21:99-118.

Rice, R.G., J.W. Farquhar, and L.J. Bollyky (1982). Review of the applications of ozone for increasing storage times of perishable foods. *Ozone: Science and Engineering* 4:147-163.

Rich, S. and H. Tomlinson (1968). Effects of ozone on conidiophores and conidia of *Alternaria solani*. *Phytopathology* 58:444-446.

Ridley, J.D. and E.T. Sims Jr. (1967). The response of peaches to ozone during storage. *South Carolina Agricultural Experiment Station Technical Bulletin* 1027, 24 pp.

Rowland-Bamford, A.J., A.M. Borland, P.J. Lea, and T.A. Mansfield (1989). The role of arginine decarboxylase in modulating the sensitivity of barley to ozone. *Environmental Pollution* 61:95-106.

Sakaki, T., N. Kondo, and K. Sugahara (1983). Breakdown of photosynthetic pigments and lipids in spinach leaves with ozone fumigation: Role of active oxygens. *Physiologia Plantarum* 59:28-34.

Sakaki, T., N. Kondo, and M. Yamada (1990). Pathway for the synthesis of triacylglycerols from monogalactosyldiacylglycerols in ozone-fumigated spinach leaves. *Plant Physiology* 94:773-780.

Sandermann Jr., H. (1996). Ozone and plant health. *Annual Review of Phytopathology* 34:347-366.

Sandermann Jr., H. (1998). Ozone: An air pollutant acting as a plant-signaling molecule. *Naturwissenschaften* 85:369-375.

Sandermann Jr., H., D. Ernst, W. Heller, and C. Langebartels (1998). Ozone: An abiotic elicitor of plant defense reactions. *Trends in Plant Science* 3:47-50.

Sarig, P., T. Zahavi, Y. Zutkhi, S. Yannai, N. Lisker, and R. Ben-Arie (1996). Ozone for control of post-harvest decay of table grapes caused by *Rhizopus stolonifer. Physiological and Molecular Plant Pathology* 48:403-415.

Schomer, H.A. (1948). Ozone in relation to storage of apples. *United States Department of Agriculture Circular* 765, 24 pp.

Schraudner, M., D. Ernst, C. Langebartels, and H. Sandermann Jr. (1992). Biochemical plant responses to ozone. III. Activation of the defense-related proteins β-1,3-glucanase and chitinase in tobacco leaves. *Plant Physiology* 99:1321-1328.

Schraudner, M., C. Langebartels, and H. Sandermann (1997). Changes in the biochemical status of plant cells induced by the environmental pollutant ozone. *Physiologia Plantarum* 100:274-280.

Sen Gupta, A., R.G. Alscher, and D. McCune (1991). Response of photosynthesis and cellular antioxidants to ozone in *Populus* leaves. *Plant Physiology* 96:650-655.

Sgarbi, E., P. Medeghini Bonatti, R. Baroni Fornasiero, and A. Lins (1999). Differential sensitivity to ozone in two selected cell lines from grape leaf. *Journal of Plant Physiology* 154:119-126.

Shorter, A.J. and K.J. Scott (1987). Controlled atmosphere storage of kiwifruit using an ultra violet scrubber to remove ethylene. *CSIRO Food Research Quarterly* 47:85-93.

Simini, M., J.M. Skelley, D.D. Davis, and J.E. Savage (1992). Sensitivity of four hardwood species to ambient ozone in north central Pennsylvania. *Canadian Journal of Forest Research* 22:1789-1799.

Skog, L.J. and C.L. Chu (2001). Ozone technology for shelf life extension of fruits and vegetables. *Acta Horticuturae* 2(553):431-432.

Smilanick, J.L., C. Crisosto, and F. Mlikota (1999). Postharvest use of ozone on fresh fruit. *Perishables Handling Quarterly* 99:10-14.

Smirnoff, N. (1996). The function and metabolism of ascorbic acid in plants. *Annals Botany* 78:661-669.

Song, J., L. Fan, C.F. Forney, M.A. Jordan, P.D. Hildebrand, W. Kalt, and D.A.J. Ryan (2002). Effect of ozone treatment and controlled atmosphere storage on quality and phytochemicals in highbush blueberries. *Acta Horticuturae* (in press).

Song, J., L. Fan, P.D. Hildebrand, and C.F. Forney (2000). Biological effects of corona discharge on onions in a commercial storage facility. *HortTechnology* 10:608-612.

Spalding, D.H. (1966). Appearance and decay of strawberries, peaches, and lettuce treated with ozone. *United States Department of Agriculture, Marketing Research Report* No. 756. 11 pp.

Spalding, D.H. (1968). Effects of ozone atmospheres on spoilage of fruits and vegetables after harvest. *United States Department of Agriculture, Marketing Research Report* No. 801. 9 pp.

Spotts, R.A. and L.A. Cervantes (1992). Effects of ozonated water on postharvest pathogens of pear in laboratory and packinghouse tests. *Plant Disease* 76:256-259.

Steger-Hartmann, T., U. Koch, T. Dunz, and E. Wagner (1994). Induced accumulation and potential antioxidative function of rutin in two cultivars of *Nicotiana tabacum* L. *Zeitschrift für Naturforschung* 49: Section C, 57-62.

Suslow, T. (1998). Basics of ozone applications for postharvest treatment of fruits and vegetables. *Perishables Handling Quarterly* 94:9-11.

Tadolini, B. (1988). Polyamine inhibition of lipoperoxidation. The influence of polyamines on iron oxidation in the presence of compounds mimicking phospholipid polar heads. *Biochemistry Journal* 249:33-36.

Tanaka, K., T. Askura, Y. Tanimura, N. Muramatsu, Y. Ishikawa-Takano, D. Hirayama, and J. Hirotsuji (1998). Long term keeping of fruit freshness under low temperature and high humidity condition with negative air ions and ozone. *Supplement Journal of the Japanese Society of Horticultural Science* 67:330.

Tanaka, K., H. Saji, and N. Kondo (1988). Immunological properties of spinach glutathione reductase and inductive biosynthesis of the enzyme with ozone. *Plant and Cell Physiology* 29:637-642.

Tanimura, Y., J. Hirotsuji, and K. Tanaka (1998). Food preservation technique using a mixed gas containing negative ions and ozone. *Shokuhin Kgy* (Food Industry) 41(10):71-77.

Tingey, D.T. (1980). Stress ethylene production—A measure of plant response to stress. *HortScience* 25:630-633.

Tingey, D.T. and W.E. Hogsett (1985). Water stress reduces ozone injury via a stomatal mechanism. *Plant Physiology* 77:944-957.

Tingey, D.T., C. Standley, and R.W. Field (1976). Stress ethylene evolution: A measure of ozone effects on plants. *Atmospheric Environment* 10:969-974.

Tingey, D.T. and G.E. Taylor Jr. (1982). Variation in plant response to ozone: A conceptual model of physiological events. In *Effects of Gaseous Air Pollution in Agriculture and Horticulture,* M.H. Unsworth and D.P. Ormrod (Eds.). London, United Kingdom: Butterworth Scientific, pp. 113-138.

Treshow, M., F.M. Harner, H.E. Price, and J.R. Kormelink (1969). Effects of ozone on growth, lipid metabolism, and sporulation of fungi. *Phytopathology* 59:1223-1225.

United States Food and Drug Administration (USFDA) (1997). Substances generally recognized as safe, proposed rule. *Federal Register* 62(74):18937-18964 (April 19).

Van Camp, W., H. Willekens, C. Bowler, M. Van Montagu, D. Inzé, P. Reupold-Popp, H. Sandermann Jr., and C. Langebartels (1994). Elevated levels of superoxide dismutase protect transgenic plants against ozone damage. *Biotechnology* 12:165-168.

Wellburn, A.R. and F.A.M. Wellburn (1996). Gaseous pollutants and plant defense mechanisms. *Biochemistry Society Transactions* 24:461-464.

White, G.C. (1999). Ozone. In G.C. White, *Handbook of Chlorination and Alternative Disinfectants,* Fourth Edition (pp. 1203-1261). New York: John Wiley and Sons, Inc.

Willekens, H., S. Chamnongpol, M. Davey, M. Schraudner, C. Langebartels, M. Van Mantagu, D. Inzé, and W. Van Camp (1997). Catalase is a sink for H_2O_2 and is indispensable for stress defense in C_3 plants. *EMBO Journal* 16:4806-4816.

Xu, L. (1999). Use of ozone to improve the safety of fresh fruits and vegetables. *Food Technology* 53:58-62.

Chapter 3

Low Temperature As a Causative Agent of Oxidative Stress in Postharvest Crops

Wendy V. Wismer

INTRODUCTION

The effect of temperature during growth has been well studied in plants, including the limits of temperature for individual plants and the identification of temperature stress-tolerant and -susceptible plant species. In postharvest situations, high and low temperatures have been used to preserve horticultural products and enhance their shelf life and quality. Heat treatments are used on postharvest crops for disinfestation and disinfection of fungi and insects (Lurie, 1998). Low temperature has been used as a method of preserving the storage life of postharvest crops since antiquity (Paull, 1999), and refrigeration practices for harvested crops were known since the beginning of the Roman Empire (Kays, 1991).

Low temperatures suppress changes due to metabolism and maintain product quality similar to that at which the plant part was harvested (Kays, 1991; Shewfelt and del Rosario, 2000). However, temperature extremes can be stressful for plants and can result in oxidative stress (Inzé and Van Montagu, 1995). Although high temperature has been identified as a causative agent of oxidative stress, moderate heat treatments, such as those used for disinfestation and disinfection, result in the production of heat shock proteins (HSPs) and other modifications that are beneficial to the horticultural product. These heat treatments affect fruit ripening and thermotolerance, and appear to diminish, rather than enhance, the effects of oxidative stress (Lurie, 1998).

A substantial portion of postharvest temperature-related work has focused on the effects of low temperature on postharvest crops because of its extensive use as a preservation method. Each crop and variety has its own limits of low temperature tolerance, which are sometimes exceeded in storage. The stress of low temperature outside a plant's temperature boundaries has generally been characterized as oxidative stress (Inzé and Van Montagu, 1995; Bartosz, 1997; Shewfelt and del Rosario, 2000). This stress can be managed for a brief time by the detached plant part, but eventually promotes accelerated senescence and death when the defensive system is exceeded by prooxidant formation, or there is a decrease in the primary defense mechanisms, or both (Purvis and Shewfelt, 1993). Bartosz (1997) describes a general mechanism for cold-induced oxidative stress. He suggests that cold temperatures decrease the demand for ATP in plants, resulting in an excess of electrons in some steps of electron transport, and increased formation of active oxygen species (AOS). Reduced antioxidant enzyme activity, also an effect of the colder temperatures, diminishes the plant's ability to cope with increased levels of AOS.

Many crop researchers exploring the effect of low temperature on crops have monitored physicochemical changes in membranes because membrane lipid composition, and thus the physical properties of membranes, is affected by low temperature and is associated with low-temperature-induced quality defects. This initial focus on membrane properties was a result of the widely held belief in Lyons' bulk lipid-phase membrane transition hypothesis (Lyons, 1973). Until the early to mid-1990s, low-temperature postharvest crop work focused on cold-induced membrane modification and was not explicitly associated with oxidative stress. Now, however, researchers have expanded their work to include oxidative and antioxidative systems in addition to membrane composition and integrity. Peroxidative changes in membranes are seen as part of the universal mechanism of membrane degradation in chilling, freezing, desiccation, and senescence by some researchers, and by others as a secondary but controllable event in membrane degradation (Shewfelt and Erickson, 1991). A general model of lipid peroxidation of specific susceptible membranes was proposed for chilling injury and other plant tissue membrane disorders (Figure 3.1) (Shewfelt and del Rosario, 2000).

In the postharvest crop, cold temperature can manifest itself in a variety of defects. Two of the most prominent are chilling injury (CI) and low temperature sweetening (LTS). The proposed mechanisms of chilling injury and its effects on produce quality are reviewed elsewhere (Jackman et al., 1988; Parkin et al., 1989), as are the mechanisms of low temperature sweetening (Wismer et al., 1995). This chapter focuses on low temperature as a causative agent of oxidative stress, and quality defects in postharvest crops that are uniquely identified with CI or LTS, and includes low temperature oxidative stress aspects of other crops.

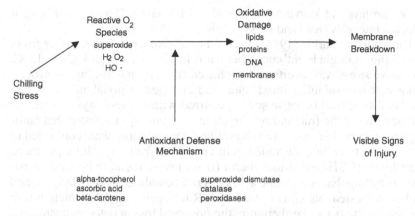

FIGURE 3.1. A conceptual model of the role of chilling injury and other plant tissue membrane disorders (from Shewfelt and del Rosario, 2000).

CHILLING INJURY

Many theories have been advanced to explain the development of CI in postharvest commodities. Parkin and Kuo (1989) proposed that lipid peroxidation may be associated with CI based on their observations of chilling-induced lipid degradation in cucumber fruit, and observations from the literature that lipid peroxidation is a common plant response to stress. After three days' storage at 4°C and eight days' rewarming, ethane evolution was observed from cucumber fruit (*Cucumis sativa* L. cv Hybrid C) while visible external signs of CI were not, indicating that lipid peroxidation occurred before CI became permanent. As decreases in the unsaturation of peel-tissue glycolipids were also observed, Kuo and Parkin (1989) suggested that the plastids were the site of early phases of chilling-induced peroxidation. After seven or more days of chilling followed by rewarming, visible symptoms of CI were observed and ethane evolution (from the peroxidation of linoleic acid) was greater still, indicating increased lipid peroxidation with increased CI. In addition, phospholipase D activity was estimated by the loss of phosphatidylcholine and an increase in phosphatidic acid. Phospholipase activity followed the rate of lipid peroxidation, and thus appeared to be a secondary response to chilling. Kuo and Parkin concluded that peroxidation may have a role in chilling injury and an association with senescence. These roles were further substantiated by monitoring electrolye leakage and ethylene and ethane evolution in cucumber fruits (*Cucumis sativa*

L. cv Carolina and Marketmore) stored at 14°C and 8°C with rewarming at selected intervals (Kuo and Parkin, 1989).

Hariyadi and Parkin (1991) observed oxidative stress in cucumber fruits by placing 'Straight Eight' cucumber fruits in the dark at both 4°C and 13°C. Oxidative stress was ascertained in chilled fruits by monitoring electrolyte leakage, water-soluble antioxidants, and changes in thylakoid lipids and their constituents. Oxidative stress occurred within a few days of chilling, and rewarming the fruit and continuation of the assays indicated that chilling was irreversible after seven days. Plastid membranes were confirmed to be sensitive to chilling degradation in cucumber fruit. In addition, reduced glutathione (GSH) and α-tocopherol (α-toc) were found to be more sensitive to chilling than to ascorbic acid (AA) and catalase (CAT). Hariyadi and Parkin suggested that GSH is used by GR to regenerate AA, which in turn can regenerate α-toc. Furthermore, the observed loss in α-toc indicates that it was oxidized faster than it could be regenerated, and this limited the ability of the cucumber fruit to cope with oxidative stress.

Purvis and coworkers (1995) determined that mitochondria are the source of superoxide in the chilling of green pepper fruit. Mitochondria isolated from the pericarp of chilling-sensitive green bell peppers (*Capsicum annum* L.) demonstrated reduced sensitivity to cyanide and reduced respiratory control ratios, and produced more superoxide with NADH and less with succinate. This increased superoxide production was suggested to be the result of the inability of mitochondria to transfer electrons from NADH to the alternative pathway. Purvis et al. concluded that low temperatures may induce the cyanide-insensitive alternative pathway in some plant tissues and reduce AOS production when succinate is being oxidized by mitochondria.

Isolated microsomal membranes from green bell pepper fruit exhibited greater response to the source of free-radical initiation than temperature when incubated at 6°C, 21°C, and 37°C under five peroxidative challenge systems (Fe^{2+}-ascorate, xanthine oxidase, lipoxygenase with and without phospholipase A_2, and phospholipase A_2) (Cowart et al., 1995). Cowart and colleagues suggested that comparative studies of peroxidative attack in plants could be difficult to interpret due to this varying response to the initiating agent, and that challenge studies based on the use of a single prooxidant at a single temperature cannot adequately measure in vivo membrane susceptibility to oxidative stress.

Citrus fruits are susceptible to chilling injury. Schirra (1992) stored 'Star Ruby' grapefruits both at chilling (4°C) and nonchilling storage temperatures (8°C and 12°C), for one, two, and three months. In addition to estimating chilling injury by examining pitting and fungal growth, he profiled the fatty acids in the juice of the fruit, as chill injured 'Star Ruby' grapefruits develop off-flavors. Both increased fatty acid unsaturation and off-flavor were noted in the fruits stored at 4°C versus 8°C and 12°C. However, Schirra suggested that fatty acid oxidation was a function of not only the lipid profile but the antioxidant profile as well, and did not make any firm conclusions about the relationship between the diminished organoleptic quality and cold-induced lipid oxidation.

Mandarin fruits exhibit chilling injury through peel pitting that turns from brown to black. Sala (1998) performed enzyme assays on the flavedo of mandarins of differing susceptibility to chilling (two each) to observe the role of oxidative stress and resistance in chilling resistance. He found that superoxide dismutase (SOD) activity increased in both chilling-sensitive and chilling-tolerant cultivars, while CAT, ascorbate peroxidase (ASPX), and glutathione reductase (GR) increased in the chilling-tolerant cultivars. It was hypothesized that chilling alters the equilibrium between free-radical production and defense mechanisms in favor of free-radical production. Chilling tolerance is possible in some cultivars because they are able to "mobilize" several antioxidant systems and use them cooperatively to break down H_2O_2.

The general conclusion in the CI literature is that chilling-tolerant species contain initially or generate more antioxidant compounds during stress and/or fewer AOS than their chilling-susceptible counterparts. Antioxidant compounds limit the damage due to lipid peroxidation and the oxidative degradation of other cellular components, and prevent the visible manifestation of CI symptoms.

LOW-TEMPERATURE SWEETENING

Low-temperature sweetening (LTS) is a phenomenon that occurs in detached plant parts, particularly tubers, exposed to lower than optimal temperatures (i.e., 8-10°C). LTS is generally characterized by the accumulation of polysaccharide breakdown products, including sucrose, glucose, and fructose (ap Rees et al., 1981). Much of the research in the LTS area has been performed on processing potatoes, such as those used to manufacture fries and chips. At the lower storage temperatures designed to lower respiration rates of the tubers, slow aging, inhibit sprouting, and discourage microbial pathogen growth, sugars accumulate and combine with free amino acids during the frying process, creating an undesirable brown pigment in the final product (Spychalla and Desborough, 1990a).

Spychalla and Desborough (1990a) stored four varieties of potato tubers at 3°C and 9°C for 40 weeks and assayed rates of tuber membrane electrolyte leakage, total fatty acid composition, free fatty acid composition, and sugar content during storage. All tuber varieties stored at the lower temperature accumulated greater levels of sugars and had increased total fatty acid unsaturation and membrane permeability compared to those at the higher storage temperature. However, two varieties of LTS-tolerant tubers had higher initial or induced levels of free fatty acids. The researchers suggested that higher initial free fatty acid levels could enhance membrane permeability during storage and positively influence storage quality.

In the same study, they observed time- and temperature-related changes in SOD, CAT, and α-toc (Spychalla and Desborough, 1990b). SOD activity increased for each of the four cultivars at 3°C versus 9°C storage. CAT activity decreased for three of the four cultivars, while α-toc levels were variable. It was concluded that relative LTS tolerance of two of the cultivars was the result of high SOD and CAT levels.

Wismer et al. (1998) compared LTS-susceptible 'Norchip' and LTS-tolerant 'ND 860-2' during 55 days of storage at both 4°C and 10°C. Although antioxidant enzymes and compounds were not assayed, there was strong evidence of increased free-radical formation (free fatty acid accumulation and diene conjugation), particularly at low temperatures and for the stress-susceptible cultivar. Wismer and colleagues suggested that the development of a chipping cultivar with low lipid acyl hydrolase (LAH) and lipoxygenase (LOX) levels and/or high antioxidant levels could reduce the affects of LTS during cold storage.

Lojkowska and Holubowska (1989) assayed LAH and LOX activity in potato cultivars resistant and susceptible to postwounding autolysis after cold storage. They found that LAH activities increased over time for four of six potato cultivars, while LOX activities increased throughout 30 weeks of storage, but inconsistently with time and cultivar.

In their review, Duplessis and co-authors (1996) suggested that cold storage of potato tubers initially decreases respiration and increases ATP levels in potato tissue, causing the alternative oxidase pathway (cyanide-resistant respiration) to be turned on. This pathway has been shown to function during periods of high cellular energy or carbohydrate imbalance. This, in turn, decreases ATP levels, and sucrose is catabolized to fructose and glucose by vacuolar acid invertase. If increased ATP production is associated with increased sucrose content (Duplessis et al., 1996) and increased formation of AOS (Bartosz, 1997), then it is plausible that LTS in potato tubers may be an oxidative stress response.

LOW-TEMPERATURE STRESS-RELATED QUALITY DEFECTS

Low-temperature-induced oxidative stress that is not normally a characteristic of CI or LTS has been exhibited in a variety of postharvest horticultural crops, including apples and flower bulbs.

Scald and scaldlike disorders in apples, such as superficial scald, appear to be associated with oxidative stress, and during months of storage at low temperatures, symptoms associated with both CI and senescence occur (Du and Bramlage, 1995). Peroxidation of sesquiterpene α-farnesene in the fruit peel generates conjugated trienes that can disrupt membrane lipids, resulting in disfigured, discolored, and dead surface cells (Du and Bramlage, 1995).

Burmeister and Dilley (1995) observed that scaldlike symptoms did not appear on Empire apples (*Malus domestica* Borkh.) stored in controlled atmosphere (CA) at 1°C and 3°C when treated prestorage with diphenylamine (DPA) scald inhibitor, and proposed that the scaldlike disorder observed on Empire fruit was a free-radical catalyzed oxidation. No oxidative or antioxidative compounds in the apples were measured pre- or poststorage. The researchers based their hypothesis on the antioxidant properties of DPA and the work of Lurie and co-workers (1989), who found that DPA application to Granny Smith apples reduced polyphenol oxidase, peroxidase, and lipoxygenase activity and superficial scald. Prestorage application of DPA is usually used to inhibit superficial scald in apples (Du and Bramlage, 1995).

Work by Du and Bramlage (1995) found that only some of the conjugated trienes produced in cold storage were associated with scald development, and that ethylene production, which can be mediated by free radicals, developed conditions that enhanced scald development. To explore the relationship between lipid peroxidation products and scald development, the researchers undertook a three-year study using a variety of apple selections stored under conditions related to scald development, and assayed them for general peroxidation products and antioxidative enzymes. They did not find any difference in thiobarbituric acid-reactive substances (TBARS), polyphenoloxidase, SOD, catalase or peroxidase activity, or peroxide concentration among apple tissues with different levels of symptom expression, and they suggested that their results did not provide insight into events during scald development. Although they could not establish a clear link between peroxidative activity and scald development, they suggested a subtle relationship in which conjugated trienes disrupted membranes, resulting in cellular disruption and scald symptom expression. Du and Bramlage did observe differences in peroxidative products and enzyme activities in senescent and healthy peel overtissues in Empire apples stored at 0°C for 24 weeks, which suggested that scald development was distinct from general cell breakdown.

In a later study, Ju and Bramlage (1999) found that scald resistance potential in the cuticle and cuticular wax of four apple varieties was similar to total antioxidant activities, defined as the sum of lipophilic (DPA, tocopherol, and other lipid-soluble antioxidants) and hydrophilic (quercitin, gallic acids, and free and bound cuticular phenolics) antioxidant activities. It was suggested that antioxidants in the cuticle likely inhibited the oxidation of α-farnesene, while antioxidants in the epidermal and hypodermal cells provided protection from membrane oxidation, both of which contributed to scald resistance.

Paliyath and co-workers (1997) compared the volatiles generated by scald-susceptible ('Red Delicious' and 'McIntosh'), and scald-resistant ('Empire' and 'Gala') apples stored three months in controlled atmosphere at 0°C (1.5 percent O_2, 1.5 percent CO_2). Solid-phase microextraction of the volatiles coupled with gas chromatography–mass spectroscopy revealed that α-farnesene was the major volatile component and that levels were highest in the scald-susceptible varieties. There were no differences in α-farnesene levels in scalded and unscalded 'Red Delicious', but differences based on geographic growing region were observed. Proton magnetic resonance imaging (MRI) was used to observe free water released upon tissue degradation in scalded and unscalded apples. Free water was observed in the hypodermal and xylem elements, and higher levels developed as the apples developed scald. The MRI results indicated that although scald affects the surface cell layers of apples, cortical areas of the fruit might also be affected.

Treatment with 1-methylcyclopropene (1-MCP) was found to inhibit ethylene production in 'McIntosh' and 'Delicious' apples for 6 to 10 days at 20°C after apples had been stored at 0°C for 60 or 120 days in air or controlled atmosphere (Rupasinghe et al., 2000). Total volatiles and α-farnesene production were limited, and triene alcohol, the α-farnesene catabolite that is believed to be the primary agent in the development of superficial scald, was reduced by 60 to 90 percent relative to the control. The treatment with 1-MCP delayed ethylene-dependent ripening and effectively suppressed superficial scald.

Bonnier and co-workers (1997) investigated oxidative stress as a cause of the loss of regenerative capacity of lily bulbs during long-term cold storage (–2°C in moist peat for five years). The researchers found that after one year of storage, 92 percent of the bulbs regenerated, while after three years just over 50 percent of the bulbs regenerated. At the three-year mark, there was no change in ion leakage, reduced or oxidized glutathione, or free fatty acids, although the variance in the measurements had increased. At the end of the five-year period, the regenerative ability of the bulbs was completely lost. Based on their test results, the researchers determined that they did not have conclusive evidence of oxidative damage until the bulbs had been stored for 3.9 years, although by this time the regenerative ability of the bulbs had continued to decrease yearly. They could not, however, rule out oxidative damage in some of the bulb scales.

Apples and lily bulbs represent two crops in which low-temperature induced stress has been recognized as a cause of postharvest quality loss. It is likely that, in time, more postharvest quality defects will be linked to oxidative stress.

TEMPERATURE PRECONDITIONING TO INHIBIT
LOW-TEMPERATURE OXIDATIVE STRESS

As indicated by Lurie (1998) and the references cited therein, heat treatments can be used to precondition plants to low temperature. Preconditioning has been found to be contingent upon the presence of HSPs, whose formation is induced by exposure of the plant part to temperatures in the range of 35-40°C (Ferguson et al., 1994). In addition, this high temperature exposure results in increased membrane fluidity (increased fatty acid unsaturation) and reduced indiscriminate electrolyte leakage in tomato fruit discs (Saltveit, 1991) and apples (Lurie et al., 1995; Whitaker et al., 1997).

Wang (1995a,b, 1996) preconditioned zucchini squash (*Cucurbita pepo* L. cv. Elite) at 15°C for two days before storage at 5°C. The preconditioning treatment was found to reduce peroxidase activity, reduce the loss of CAT activity, and maintain SOD activity relative to the nonpreconditioned controls (Wang, 1995a). Enzyme activity in the ascorbate metabolic system remained high relative to the activities in the control squash, which increased initially and then decreased after four to eight days of storage (Wang, 1996). In addition, preconditioning zucchini resulted in reduced levels of chilling injury as measured by surface pitting, and a higher ratio of reduced to oxidized glutathione (GSH/GSSG), total nonprotein thiol, and glutathione reductase activity immediately after treatment and during storage (Figure 3.2) (Wang, 1995b). The ratio of GSH/GSSG, rather than the absolute levels of GSH, was believed to be associated with resistance to chilling injury. It is suggested that SH-containing enzymes protect against oxidation and detoxify reactive oxygen species, contributing to the effectiveness of temperature preconditioning to reduce the oxidative effects of chilling injury (Wang, 1995b). From Wang's work it is apparent that temperature preconditioning is effective in reducing cold-induced oxidative stress by enhancing the levels and function of antioxidant systems in cucumber fruit.

Alwan and Watkins (1999) proposed that intermittent warming reduced scald with varying degrees of success in 'Cortland', 'Delicious', and 'Law Rome' apples as warming provided an opportunity for the reduction of chilling-induced accumulated toxic or inhibiting substances.

Thus, work in the literature indicates that temperature preconditioning may alleviate low-temperature oxidative stress through the generation of HSPs, reduced loss of antioxidant enzyme systems, and the removal of toxic oxidation products.

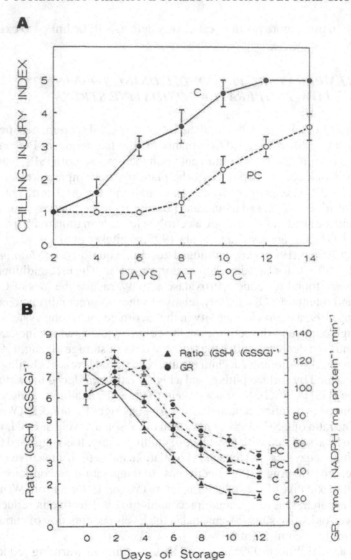

FIGURE 3.2. The effect of temperature preconditioning (15°C for two days) prior to 14 days of storage at 5°C on: (A) visible chilling injury in zucchini squash (chilling injury index; where 1 = no abnormality, 2 = trace, 3 = slight, 4 = moderate, and 5 = severe); and (B) ratio of GSH/GSSG and GR activity (from Wang, 1995b). PC = preconditioned; C = control.

CONCLUSION

A variety of different low-temperature-induced postharvest disorders have been investigated by plant researchers, and the common theme appears to be oxidative stress. Oxidative stress is a complex phenomenon, mediated by the balance between a number of prooxidative and antioxidative compounds, observed in a number of cellular cites and components, and thus observable or measurable by a number of different analytical techniques. The complexity of this phenomenon makes it difficult to elucidate a universal mechanism. However, as research techniques advance, the understanding of low-temperature-induced oxidative stress, and methods for reducing its impact on postharvest losses, also progress. In time, a generalized model may become apparent that, as suggested by Shewfelt and del Rosario (2000), identifies a point or points of commonality among a range of disorders.

ISSUES AND TOPICS FOR FURTHER RESEARCH

Low temperature and a variety of other stresses result in oxidative stress. "Stress response in plant tissue has been associated with lipid peroxidation via a free-radical mechanism" (Cowart et al., 1995, p. 639). As Lichtenthaler (1996) reminds us, plants have the capacity to respond and adapt to changing environmental conditions. However, when the threshold of the plant has been surpassed either by long-term low-level stress or short-term high-level stress, the compensatory abilities of the plant have been exceeded and damage results. Although Lichtenthaler (1996) describes stress effects and responses in whole plants, this holds true for detached plant parts as well (Shewfelt and Erickson, 1991). A small amount of temperature stress may not harm the postharvest material, but sustained or severe stress will result in a damaged product with a shortened shelf life.

As indicated previously in this chapter, most fruit and vegetable postharvest researchers focused solely on physicochemical changes in membranes in determining the cause of quality loss in cold-stressed postharvest crops. In general, it has only been since the early to mid-1990s that oxidative stress has been explored as the primary or secondary event in the mechanism of quality loss. Thus, this area of study is relatively new. A wide variety of approaches to determining the mechanism of cold-induced oxidative stress exists in the literature, resulting in studies that are purely observational to those that are truly investigative and add to the body of scientific knowledge on the topic.

The trend of considering the relationships among the antioxidant systems and evaluation of their synergistic effects will rapidly advance this dynamic area. In the results analysis stage, plant scientists need to carefully analyze project results and determine the phase(s) of the oxidative stress that they have observed (alarm phase, stage of resistance, stage of exhaustion, regeneration phase, or cell death) (Lichtenthaler, 1996). Stress results in a gradual loss of control of the plant's systems. Researchers will glean the most from their results if they view them from a "big picture" perspective to reveal cycling and the loss of recuperative ability that have been shown to occur (Geigenberger and Stitt, 1991; Marangoni et al., 1996, 1997).

Monitoring antioxidative systems and oxidative damage can be resource intensive. Although resources dictate the amount of work that can be performed in a project, researchers must attempt to survey not only what is happening but to determine the mechanisms which underlie the observations.

REFERENCES

Alwan, T.F., and C.B. Watkins (1999). Intermittent warming effects on superficial scald development of 'Cortland', 'Delicious' and 'Law Rome' apple fruit. *Postharvest Biology and Technology* 16:203-212.

ap Rees, W.L. Dixon, C.J. Pollock, and F. Franks (1981). Low-temperature sweetening in higher plants. In *Recent Advances in the Biochemistry of Fruits and Vegetables,* J. Friend and M. C. J. Rhodes (Eds.). New York: Academic Press, pp. 41-61.

Bartosz, G. (1997). Oxidative stress in plants. *Acta Physiologiae Plantarum* 19(1): 47-64.

Bonnier, F.J.M., F.A. Hoekstra, C.H.R. De Vos, and J.M. Van Tuyl (1997). Viability loss and oxidative stress in Lily bulbs during long-term coldstorage. *Plant Science* 122:133-140.

Burmeister, D.M., and D.R. Dilley (1995). A "scald-like" controlled atmosphere storage disorder of Empire apples—A chilling injury induced by CO_2. *Postharvest Biology and Technology* 6:1-7.

Cowart, D.M., M.C. Erickson, and R.L. Shewfelt (1995). Susceptibility of microsomal membranes isolated from bell pepper fruit to peroxidative challenge at different temperatures. *Journal of Plant Physiology* 146:639-644.

Du, Z., and W.J. Bramlage (1995). Peroxidative activity of apple peel in relation to development of poststorage disorders. *HortScience* 30:205-209.

Duplessis, P.M., A.G. Marangoni, and R.Y. Yada (1996). A mechanism for low temperature induced sugar accumulation in stored potato tubers: The potential role of the alternative pathway and invertase. *American Potato Journal* 73(10):483-494.

Ferguson, I.B., S. Lurie, and J.H. Bowen (1994). Protein synthesis and breakdown during heat chock of cultured pear (*Pyrus communis* L.) cells. *Plant Physiology* 104:1429-1437.

Geigenberger, P., and M. Stitt (1991). A "futile" cycle of sucrose synthesis and degradation is involved in regulation partitioning between sucrose, starch and respiration in cotyledons of germinating *Ricinus cummunis* L. seedlings when phloem transport is inhibited. *Planta* 185:81-90.

Hariyadi, P., and K.L. Parkin (1991). Chilling-induced oxidative stress in cucumber fruits. *Postharvest Biology and Technology* 1:33-45.

Inzé, D., and M.V. Montagu (1995). Oxidative stress in plants. *Current Opinion in Biotechnology* 6(2):153-158.

Jackman, R.L., R.Y. Yada, A. Marangoni, K.L. Parkin, and D.W. Stanley (1988). Chilling injury, a review of quality aspects. *Journal of Food Quality* 11:253-278.

Ju, Z., and W.J. Bramlage (1999). Phenolics and lipid-soluble antioxidants in fruit cuticle of apples and their antioxidant activities in model systems. *Postharvest Biology and Technology* 16:107-118.

Kays, S.J. (1991). *Postharvest Physiology of Perishable Products*. New York: Van Nostrand Reinhold, pp. 10, 335-408.

Kuo, S.-J., and K.L. Parkin (1989). Chilling injury in cucumbers (*Cucumis sativa* L.) associated with lipid peroxidation as measured by ethane evolution. *Journal of Food Science* 54(6):488-1491.

Lichtenthaler, H.K. (1996). Vegetation stress: An introduction to the stress concept in plants. *Journal of Plant Physiology* 148:4-14.

Lojkowska, E., and M. Holubowska (1989). Changes of the lipid catabolism in potato tubers from cultivars differing in susceptibility to autolysis during the storage. *Potato Research* 32:463-470.

Lurie, S. (1998). Postharvest heat treatments. *Postharvest Biology and Technology* 14: 257-269.

Lurie, S., J. Klein, and R. Ben-Arie (1989). Physiological changes in diphenylamine-treated 'Granny Smith' apples. *Israel Journal of Botany* 38:199-207.

Lurie, S., S. Othman, and A. Borochov (1995). Effects of heat treatment on plasma membrane of apple fruit. *Postharvest Biology Technology* 8:29-38.

Lyons, J.M. (1973). Chilling injury in plants. *Annual Reviews in Plant Physiology* 24:445-466.

Marangoni, A.G., P.M. Duplessis, RW. Lencki, and R.Y. Yada (1996). Low-temperature stress induces transient oscillations in sucrose metabolism in *Solanum tuberosum*. *Biophysical Chemistry* 61:177-184.

Marangoni, A.G., P.M. Duplessis, and R.Y. Yada (1997). Kinetic model for carbon partitioning in *Solanum tuberosum* tubers stored at 2°C and the mechanism for low temperature stress-induced accumulation of reducing sugars. *Biophysical Chemistry* 65:211-220.

Paliyath, G., M.D. Whiting, M.A. Stasiak, D.P. Murr, and B.S. Clegg (1997). Volatile production and fruit quality during development of superficial scald in Red Delicious apples. *Food Research International* 30:95-103.

Parkin, K. L. and S.-J. Kuo (1989). Chilling-induced lipid degradation in cucumber (*Cucumis sativa* L. cv Hybrid C) fruit. *Plant Physiology* 90:1049-1056.

Parkin, K.L., A. Marangoni, R.L. Jackman, R.Y. Yada, and D.W. Stanley (1989). Chilling injury, a review of possible mechanisms. *Journal of Food Biochemistry* 13:127-153.

Paull, R. E. (1999). Effect of temperature and relative humidity on fresh commodity quality. *Postharvest Biology Technology* 15:263-277.

Purvis, A.C., and R.L. Shewfelt (1993). Does the alternative pathway ameliorate chilling injury in sensitive plant tissues? *Physiologia Plantarum* 88:712-718.

Purvis, A.C., R.L. Shewfelt, and J.W. Gegogeine (1995). Superoxide production by mitochondria isolated from green bell pepper fruit. *Physiologia Plantarum* 94:743-749.

Rupasinghe, H.P.V., D.P. Murr, G. Paliyath, and L. Skogg (2000). Inhibitory effect of 1-MCP on ripening and superficial scald development in 'McIntosh' and 'Delicious' apples. *Journal of Horticultural Science and Biotechnology* 75:271-276.

Sala, J.M. (1998). Involvement of oxidative stress in chilling injury in cold-stored mandarin fruits. *Postharvest Biology and Technology* 13:55-261.

Saltveit, M.E. Jr. (1991). Prior temperature exposure affects subsequent chilling sensitivity. *Physiologia Plantarum* 82:529-536.

Schirra, M. (1992). Behavior of 'Star Ruby' grapefruits under chilling and non-chilling storage temperature. *Postharvest Biology and Technology* 2:315-327.

Shewfelt, R.L., and B.A. del Rosario (2000). The role of lipid peroxidation in storage disorders of fresh fruits and vegetables. *HortScience* 35:575-579.

Shewfelt, R.L., and M.C. Erickson (1991). Role of lipid peroxidation in the mechanism of membrane-associated disorders in edible plant tissue. *Trends in Food Science and Technology* 2:152-154.

Spychalla, J.P., and S.L. Desborough (1990a). Fatty acids, membrane permeability, and sugars of stored potato tubers. *Plant Physiology* 94:1207-1213.

Spychalla, J.P., and S.L. Desborough (1990b). Superoxide dismutase, catalase, and α-tocopherol content of stored potato tubers. *Plant Physiology* 94:1214-1218.

Wang, C.Y. (1995a). Effect of temperature preconditioning on catalase, peroxidase, and superoxide dismutase in chilled zucchini squash. *Postharvest Biology and Technology* 5:67-76.

Wang, C.Y. (1995b). Temperature preconditioning affects glutathione content and glutathione reductase activity in chilled zucchini squash. *Journal of Plant Physiology* 145:148-152.

Wang, C.Y. (1996). Temperature preconditioning affects ascorbate antioxidant system in chilled zucchini squash. *Postharvest Biology and Technology* 8:29-36.

Whitaker, B.D., J.D. Klein, W.S. Conway, and C.E. Sams (1997). Influence of prestorage heat and calcium treatments on lipid metabolism in 'Golden Delicious' apples. *Phytochemistry* 45:465-472.

Wismer, W.V., A.G. Marangoni, and R.Y. Yada (1995). Low-temperature sweetening in roots and tubers. *Horticultural Reviews* 17:230-231.

Wismer, W.V., W.M. Worthing, A.G. Marangoni, and R.Y. Yada (1998). Membrane lipid dynamics and lipid peroxidation in the early stages of low-temperature sweetening of tubers of *Solanum tuberosum*. *Physiologia Plantarum* 102:396-410.

Chapter 4

Effects of Storage Conditions and Postharvest Procedures on Oxidative Stress in Fruits and Vegetables

Peter M. A. Toivonen

INTRODUCTION

Postharvest handling and storage protocols for fruits and vegetables often produce varying degrees of oxidative stress. The effects of oxidative stress can be manifested as accelerated senescence, changes in cellular antioxidant constituents (increases when stress is mild and subsequent declines when stress levels increase), and development of certain postharvest physiological disorders (Leshem, 1988; Du and Bramlage, 1995; Shewfelt and Purvis, 1995; Leshem and Kuiper, 1996). Subacute levels of oxidative stress are manifested as changes in membrane permeability, changes in antioxidant constituent levels in tissue, or transient increases in enzymatic and/or nonenzymatic antioxidant protectant systems. Subsequently, if the postharvest oxidative stress exposure exceeds the harvested product's antioxidant protectant system's capacity, declines in antioxidant system components can occur, resulting in active oxygen species-induced injury and, consequently, postharvest disorders will develop (Shewfelt and Purvis, 1995). Two factors are important to the development of oxidative stress-related disorders: (1) the intrinsic capability of the product to withstand the stress, and (2) the level and duration of the stress which is being imposed on the product (Leshem and Kuiper, 1996).

Several postharvest storage and handling factors may influence the level of oxidative stress in harvested fruits and vegetables. This chapter's discussion will focus on the influence of harvesting maturity, storage duration, storage temperature, postharvest water loss, storage atmosphere, ethylene, and ionizing radiation on levels of oxidative stress in fruits and vegetables.

Before these issues can be discussed, a definition of what constitutes an indicator of oxidative stress or injury must be made. In some cases, the literature very clearly demonstrates that oxidative injury has occurred in the tissues. Direct evidence of oxidative injury is demonstrated with the accumulation of lipid peroxidation products or the loss of membrane integrity (Shewfelt and Purvis, 1995; Hodges et al., 1999). Other indicators of oxidative injury are measures showing the accumulation of oxidation end products such as brown pigments (i.e., oxidized phenols) or the development of other symptoms such as yellowing, surface pitting, or superficial scald (Meir and Bramlage, 1988; Meir et al., 1992; Wang et al., 1992; Shewfelt and Purvis, 1995; Veltman et al., 1999). Early signs of oxidative stress can be demonstrated as changes in antioxidant enzymes such as superoxide dismutase, peroxidases, catalase, or antioxidant compounds such as glutathione, carotenes, and ascorbic acid (Kunert and Ederer, 1985; Toivonen, 1992; Kumar and Knowles, 1993; Dipierro and De Leonardis, 1997; Veltman et al., 1999; Hodges and Forney, 2000). Ascorbic acid levels are extremely labile, and declines in ascorbic acid in tissues in response to postharvest handling treatments are frequently reported (Watada, 1987; Agar et al., 1997), even when few other indicators of oxidative stress are apparent. This may be reflected by ascorbic acid's double role in the direct reduction of oxygen radicals as well as in the regeneration of other cellular antioxidant systems, such as flavonols (Yamasaki et al., 1997) and tocopherols (Kunert and Ederer, 1985; Larsen, 1988) involved with peroxide quenching in plant tissues. Still, under mild stress conditions, the plant tissues can also be induced to accumulate protectant components such as antioxidant enzymes and water-soluble and lipid-soluble antioxidants (Leshem and Kuiper, 1996). Because much of the postharvest literature does not specifically discuss deteriorative processes in relation to oxidative stress, inference of oxidative stress will be made, in some cases, using these indicators as well as actual visual symptoms of active oxygen species-induced injury (e.g., surface pitting, browning).

MATURITY AT HARVEST

The development of a disorder as a consequence of oxidative injury is highly dependent upon the harvest maturity of the fruit or vegetable (Barden and Bramlage, 1994a; Ju et al., 1996; Lentheric et al., 1999). The influence of maturity on susceptibility to oxidative stress is based on the ability of the tissues to scavenge free radicals. Higher levels of scald susceptibility in apples has been associated with lower levels of lipid-soluble antioxidants (Meir and Bramlage, 1988; Barden and Bramlage, 1994a), anthocyanins, and simple phenols (Ju et al., 1996) in peel tissue of less mature fruit. Barden and

Bramlage (1994b) found that lipid-soluble antioxidants increased during storage, but that the relative levels in storage were dependent on the original levels at harvest. Higher levels of lipid-soluble antioxidants at harvest were not themselves adequate to prevent the oxidative reactions leading to scald. However, higher levels of antioxidant in apples from later harvests were required to ensure that sufficiently high levels subsequently accumulated during storage, thus reducing scald susceptibility. The opposite is true for browning of flesh tissue in apples and pears. Braeburn browning disorder susceptibility increases in 'Braeburn' apples with advancing maturity, and this was associated with declines in superoxide dismutase and catalase activity in the flesh tissue (Gong et al., 2000). A similar relationship between core browning and enzymatic and nonenzymatic antioxidant protective systems was found in 'Conference' pears (Lentheric et al., 1999; Veltman et al., 1999). These data suggest that peel tissue and flesh tissue may behave quite differently with respect to maturity and consequent changes in respective tissue antioxidant systems. Further work is certainly justified in the comparative examination of the effects of maturity on peel and flesh tissue antioxidant systems and how they might relate to the development of disorders in pome fruits.

Some limited work has also been done with nonpome fruits. Fully colored bell peppers are chill resistant, whereas green peppers are chill sensitive (Lin et al., 1993). However, fully colored peppers show greater loss in membrane integrity and rates of deterioration than do green fruit (Lurie and Ben-Yehoshua, 1986). Tomatoes that are picked at the mature-green stage developed greater levels of antioxidants (lycopene, β-carotene, and ascorbic acid) than fruit picked at the full-ripe stage (Giovanelli et al., 1999). Papaya picked at the mature-green stage were found to be more resistant to irradiation injury than those picked at one-quarter-yellow stage (Miller and Mc-Donald, 1999). Rogiers et al. (1998) found that saskatoon berries showed greater oxidative stress as they ripened. They found increases in glutathione reductase activity and suggested that these increases were in response to the accumulation of hydroperoxides in the fruit tissue as a consequence of the decline in the superoxide dismutase/catalase protection system at the latter stages of ripening.

Maturity of fruit appears to have a dramatic influence on susceptibility to oxidative injury. The literature provides some evidence to suggest that these differences can be associated with various antioxidant systems, including superoxide dismutase, catalase, peroxidases, glutathione reductase, carotenoids, ascorbic acid, and tocopherols.

STORAGE DURATION

The longer fruits and vegetables are held in storage, the more likely oxidative stress will develop. Meir et al. (1991) showed that lipid peroxidation, as indicated by a sharp rise in lipofuscin-like compounds within one to two days, was an early event in the ripening of avocado during storage in air at 22°C. Similar increases in lipofuscin-like compounds occurred during one to two weeks storage at 5°C; these increases, however, were not associated with ripening at that temperature. Lipid peroxidation, with concomitant loss in membrane integrity, increased during postharvest ripening in nonnetted muskmelon at 25°C and 85 percent relative humidity (RH) (Lacan and Baccou, 1996). A peak in superoxide dismutase and catalase activities occurred at the onset of climacteric ethylene production in postharvest ripening at 1-3°C air storage of 'Fuji' apples, but their levels were relatively stable in a 'Golden Delicious' clone which showed no relationship of these enzymes to climacteric ethylene (Masia, 1998). Du and Bramlage (1994) also found that 'Cortland', 'Delicious', and 'Empire' all showed different patterns of changes in superoxide dismutase activity during air storage at 0°C. They found that 'Cortland' showed a steady decline of superoxide dimutase activity during storage, 'Delicious' an increase, and 'Empire' showed a rise and decline similar to that reported for 'Fuji' apples by Masia (1998). In none of these studies were oxygen free radicals or lipid peroxide levels measured; therefore, the association of climacteric ethylene production to active oxygen species (AOS) is neither proved or disproved with the data.

In nonclimacteric products, other patterns of oxidative stress and antioxidant protectant system responses have been shown during storage. Increases in abscisic acid and membrane leakage in bell peppers were correlated with the length of time in storage at 17°C (Lurie and Ben-Yehoshua, 1986). Increases in abscisic acid levels are considered to be associated with an increase in stress tolerance of plant tissues (Leshem and Kuiper, 1996). Increases in superoxide dismutase and catalase activities as well as in α-tocopherol levels are found in potatoes with increased duration of storage at 3°C and 9°C, although increases were greatest at 3°C (Spychalla and Desborough, 1990). Lojkowska and Holubowska (1989) found that lipid peroxidation in potato tubers increased with time in storage under similar conditions. In other work, potato tubers were shown to accumulate free radicals and this accumulation paralleled increases in peroxidative injury to the membranes with increased time in air storage at 4°C (Kumar and Knowles, 1993). Kumar and Knowles (1993) found these changes also to be concomitant with increases in superoxide dismutase and catalase activities in potato tubers.

Although oxidative stress indicators increase with storage duration, the responses of antioxidant protectant systems do not appear to be consistent among different commodities or among cultivars of a particular commodity. This inconsistency underlies the complexity of the AOS stress response and protectant systems (Shewfelt and Purvis, 1995; Leshem and Kuiper, 1996).

STORAGE TEMPERATURE

Optimal storage temperatures are expected to minimize development of oxidative injury in stored plant products. This statement is best supported by a review of the literature which describes the effects of sub- and superoptimal storage temperatures on deterioration or development of disorders in fruits and vegetables. Suboptimal temperatures lead to the development of chilling injuries (Wismer, 2000), and superoptimal storage temperatures can result in oxidatively induced acceleration of senescence (Zhuang et al.,1997).

Chilling injury has been well studied and is discussed in detail in this compilation (Chapter 3). However, a brief overview of the disorder will be given in this chapter. Cucumbers, apples, tomatoes, zucchini squash, bell peppers, potatoes, pears, and mandarin oranges have all been shown to develop oxidative stress-induced disorders in response to chilling temperatures of 10°C or lower (Kuo and Parkin, 1989; Saltveit, 1989; Hariyadi and Parkin, 1991; Lurie and Klein, 1992; Wang et al., 1992; Cowart, 1993; Ju, Yuan, Liou, et al., 1994; Ju, Yuan, Liu, et al., 1994; Burmeister and Dilley, 1995; Sala, 1998; Wismer et al., 1998; Hakim et al., 1999).

The disorders induced by chilling temperatures are of varied nature. In the case of potatoes, chilling-induced membrane peroxidation has been associated with a phenomenon known as low-temperature sweetening (Dipierro and De Leonardis, 1997; Wismer et al., 1998). Pitting, which has been shown to be related to lipid peroxidation, developed in chill-injured zucchini squash, bell peppers, cucumbers, and mandarin oranges (Kuo and Parkin, 1989; Hariyadi and Parkin, 1991; Wang et al., 1992; Cowart, 1993; Sala, 1998; Hakim et al., 1999). Chill-injured tomatoes ripened abnormally or they lacked the ability to ripen fully (Lurie and Klein, 1992), which is associated with injury to membranes (Saltveit, 1989). Chilling-induced scald in some apple cultivars (Ju, Yuan, Liu, et al., 1994; Burmeister and Dilley, 1995) and flesh browning in some pear cultivars (Ju, Yuan, Liou, et al., 1994) has been correlated with peroxide and free-radical accumulation in fruit tissues. Low-temperature browning of banana skin was related to rapid declines in ascorbic acid and glutathione levels in tissues (Jiang et al., 1991), which is suggestive of oxidative injury in the tissue (Hodges and Forney, 2000).

Superoptimal storage conditions can also lead to increased levels of oxidative stress and resultant injuries in fruits and vegetables. Chlorophyll loss has been associated with lipid peroxidation (Zhuang et al., 1995), and thus reduction of yellowing in various species is mediated by antioxidant protectant systems (Philosoph-Hadas et al., 1994). In broccoli, very little yellowing occurs at optimal storage temperature, even after 17 days (Toivonen, 1997; Zhuang et al., 1997). However, yellowing (chlorophyll loss) proceeds within a few days of holding at 5°C, 13°C, or 23°C (Toivonen, 1997; Zhuang et al., 1997; Toivonen and Sweeney, 1998). These results demonstrate the importance of temperature on the levels of free radicals in tissues and how they may, in some instances, influence the rate of senescence as measured by chlorophyll loss.

Superoptimal storage temperatures can also affect tissue contents of antioxidants and, as such, these changes can indicate oxidative stress in the tissues, even if no symptoms of injury occur. Changes in ascorbic acid (vitamin C) levels have commonly been reported in many postharvest studies and are indicative of oxidative stress in fresh fruits and vegetables (Watada, 1987). Carotenes are lipid-soluble antioxidant constituents, and losses in carotenes can be indicative of lipid peroxidation reactions (Anguelova and Warthesen, 2000). Kalt et al. (1999) found that increasing the storage temperatures from 0°C to 10°C, 20°C, or 30°C resulted in declines in ascorbate levels in raspberries and low-bush blueberries. However, the storage temperature had no significant effect on ascorbate levels in strawberries and high-bush blueberries (Kalt et al., 1999). The optimal temperature for bell pepper storage is between 7°C and 10°C (Lin et al., 1993). Partially colored bell peppers showed an accumulation of ascorbic acid during storage, and the levels were greater in 15°C than in 8°C storage (Kosson et al., 1998). In contrast, fully colored bell pepper fruits showed a decline in ascorbic acid levels in storage, and these declines were similar at both 8°C and 15°C storage. Similar results were found for senescing apple fruit held at 0°C or 20°C where no differences in superoxide dismutase activity in the tissues was found for the two storage temperatures (Du and Bramlage, 1994). Increasing storage temperature from 0°C to 10°C or 21°C resulted in greater losses in ascorbic acid levels in kale, cabbage, and snap beans (Ezell and Wilcox, 1959), and carotene in kale, collards, and turnip greens (Ezell and Wilcox, 1962). In potatoes stored at 15°C and 20°C, carotenoids decreased with time, whereas those stored at 5°C showed an increase in carotenoids with time (Bhushan and Thomas, 1990). These limited results indicate that superoptimal storage temperatures can lead to significant declines in antioxidant constituents of plant tissues, suggesting that free-radical accumulation is enhanced at higher storage temperatures.

POSTHARVEST WATER LOSS

Increasing humidity around a product can reduce chilling injury since a part of the process involving expression of chilling injury involves water loss (Forney and Lipton, 1990). In nonchilling situations, wilting can be an indicator of losses in antioxidant constituents. Wilting, induced by lowering of storage humidity, has been shown to enhance the loss of both ascorbic acid in kale, cabbage, and snap beans (Ezell and Wilcox, 1959), and carotene in kale, collards, and turnip greens (Ezell and Wilcox, 1962). Misting at shelf temperatures to reduce water loss promoted the retention of ascorbic acid and total carotenes in broccoli (Barth et al., 1990; Barth and Zhuang, 1996). Pesis et al. (2000) found that packaging provided a high humidity atmosphere, which prevented the development of chilling injury in mangoes. Wrapping lemons and bell peppers to reduce water losses results in reduction of membrane injury (Ben-Yehoshua et al., 1983). These examples suggest that oxidative injury is one consequence of tissue dessication in postharvest handling systems. Some direct information in the environmental plant literature states that injuries induced with dessication stress are a consequence of oxidative stress (Senaratna et al., 1985; Price and Hendry, 1991; Pastori and Trippi, 1992; Simontacchi and Puntarulo, 1992).

CONTROLLED AND MODIFIED ATMOSPHERES

Modified atmosphere storage has been shown to affect tissue antioxidant contents. Both 2 percent O_2 and 12 percent CO_2 were required to optimally preserve carotenoids in persimmon slices, while an atmosphere with 2 percent O_2 with no CO_2 best retained carotenoid levels in peach slices (Palmer Wright and Kader, 1997a). However, neither 2 percent O_2 nor 12 percent CO_2, alone or in combination, had any effect on ascorbic acid retention in sliced persimmons and strawberries (Palmer Wright and Kader, 1997b). A modified atmosphere of 5 percent O_2 and 4 percent CO_2 preserved carotenoid and ascorbic acid levels in jalapeno peppers better than air controls (Howard and Hernandez-Brenes, 1998). Modified atmosphere packaging (MAP) storage better maintained the vitamin C content in fresh-cut spinach as compared with air storage (Gil et al., 1999). An atmosphere of 2 percent O_2 and 10 percent CO_2 at a temperature of 5°C better retained ascorbic acid than air atmosphere for cubed honeydew melon (Qi et al., 1999). Serrano et al. (1997) found that several types of modified atmosphere packages inhibited increases in putrescine and abscisic acid levels in peppers in response to chilling when compared with controls. This inhibition of increases in putrescine and abscisic acid levels suggest a lowering of oxidative stress (Leshem and Kuiper, 1996) in response to chilling with the use of modified atmosphere packaging.

Effects of modified or controlled atmospheres on oxidative stress in broccoli has been studied to a greater degree than in other fruits or vegetables. Modified atmosphere packages (7.5 percent CO_2 + 11.2 percent O_2) have been shown to improve the retention of ascorbic acid, carotenoids, chlorophyll, and polyunsaturated fatty acids levels compared with air storage (Zhuang et al., 1994; Barth and Zhuang, 1996). Tocopherol levels did not change in MAP but did increase in air atmospheres (Barth and Zhuang, 1996). The loss of ascorbic acid and increase in tocopherol in the air treatment was attributed to the role of ascorbic acid in regenerating tocopherol (Kunert and Ederer, 1985). A modified atmosphere of 0.1 percent O_2 and 21 percent CO_2 did not preserve glutathione levels but did preserve chlorophyll and ascorbic acid levels better than air controls (Hirata et al., 1995). Further work showed that 2 percent O_2 with 4 to 10 percent CO_2 would improve preservation of chlorophyll, carotenoids, glutathione, and ascorbic acid (Ishikawa et al., 1998). Abe et al. (1995) reported that 10 percent CO_2 in controlled atmospheres reduced yellowing in broccoli. They also found that the yellowing of broccoli was associated with the accumulation of peroxides in the tissue, although the beneficial high carbon dioxide treatment did not mediate the response via the reduction of peroxide levels. Makhlouf et al. (1990) found that an atmosphere of 2.5 percent O_2 and 8.5 percent CO_2 could control membrane deterioration in stored broccoli, and this was associated with a delay in senescence. It is clear that optimal storage atmospheres will reduce the losses of antioxidant components and reduce membrane degradation in broccoli. However, the effect of modified or controlled atmospheres on reducing yellowing in broccoli has not been adequately explained, which may be due to the lack of a comprehensive study that should include evaluation of the complete oxidant protectant system and onset of oxidative reactions leading to yellowing.

Lowering O_2 levels to 1.2 to 3.0 percent significantly increased retention of ascorbic acid, reduced onset of membrane leakage, and prevented skin blackening of 'Niitaka' pears compared with air controls (Yang, 1997). CO_2 had no significant effect on ascorbic acid, membrane leakage, or skin blackening. Chen et al. (1993) found that both 0.5 percent O_2 and 2.7 percent CO_2 were required to control accumulation and oxidation of α-farnesene in 'd'Anjou' pears, and this resulted in reduction of superficial scald. Farnesene production and oxidation in 'Cortland' apples is also inhibited with low O_2 and high CO_2 (Blanpied and Creasy, 1993).

Low oxygen injury is often expressed only after the product is removed to air atmosphere and this injury has been related to lipid peroxidation (Blokhina et al., 1999). Du and Bramlage (1995) suggested that scald was related to post–low oxygen storage peroxidation of α-farnesene. Superficial scald can be induced by anaerobic treatment in 'Granny Smith' apples (Bauchot et al., 1999). Monk, Fagerstedt, et al. (1987) found that superoxide dismutase activity was increased under anaerobic stress and hypothesized that such increases were important to recovery once a normal atmosphere was restored. Catalase activity was apparently responsible for a postanoxic upsurge in acetaldehyde through its peroxidatic oxidation of ethanol (Zuckermann et al., 1997), and catalase activity is also enhanced in response to the anaerobic stress (Monk, Braendle, et al., 1987). Wang and Kramer (1989) found that low oxygen storage increased polyamine levels in Chinese cabbage, zucchini, and 'McIntosh' apples and delayed senescence in these products. Polyamines are known to increase in response to oxidative stress and are an important component of the plants' resistance mechanisms (Leshem and Kuiper, 1996). The existing information on low oxygen effects in relation to antioxidant protectant systems and oxidative injury is poor and the lack of understanding post-low oxygen storage disorders indicate that much work needs to be done.

High CO_2 (10 percent) results in increased tissue leakage in cauliflower, suggesting membrane injury (Romo-Parada et al., 1989). Elgar et al. (1998) suggested that a CO_2-related internal browning problem in 'Braeburn' apples was associated with antioxidant protectant systems. They also found that delayed controlled atmosphere storage, which reduces the disorder, enhanced lipid-soluble antioxidant levels in the flesh of 'Braeburn' apples. High CO_2 levels have been shown to decrease vitamin C levels in berry fruits (Agar et al., 1997). They found that the lowering of O_2 in the presence of high CO_2 had no effect on vitamin C content. These limited data suggest that high levels of CO_2 may lead to oxidative injury, and losses in antioxidant vitamins or damage to cell membranes may result.

ETHYLENE

Ethylene continues to be a concern for postharvest handling in fruits and vegetables due to the continued use of propane-powered forklifts to handle produce during shipping and distribution as well as from the risks associated with handling ethylene-producing products alongside ethylene-sensitive products in the same facilities and shipping containers (Saltveit, 1999). An association between ethylene production and active oxygen species in plant tissues has been demonstrated (McRae et al., 1982). Postharvest exposure of green tissues to ethylene results in accumulation of lipid peroxidation products, and this has been related to chlorophyll losses (Meir et al., 1992; Hodges and Forney, 2000). It has been suggested that the process of yellowing is regulated by hydrogen peroxide metabolism in the tissues and that ethylene enhances the yellowing process (Yamauchi and Watada, 1991; Hodges and Forney, 2000). Application of exogenous ethylene, however, does not appear to modify the pathway for chlorophyll degradation, but it accelerates the rate of normal senescence (Yamauchi and Watada, 1991). Ethylene has also been associated with α-farnesene synthesis in apple peel (Watkins et al., 1993), and hence the development of superficial scald, which is currently considered an oxidation-induced injury (Whitaker et al., 1997). Ethylene has also been implicated in the development of russet spotting lesions in lettuce (Peiser et al., 1998), and such lesions can be attributed to oxidative membrane injury—as similarly demonstrated in chilling injury (Cowart, 1993).

Direct association of ethylene with oxidative injury has not been made. In the case of superficial scald in apples, the accumulation of α-farnesene has been correlated to the disorder and this accumulation process has been shown to be enhanced by postharvest ethylene exposures (Meir et al., 1992; Ju and Curry, 2000; Watkins et al., 2000). However, it is not known whether ethylene also plays a role in directly modifying oxidative processes leading to the scald symptoms (Watkins et al., 2000). Russet spotting in lettuce is another tissue browning disorder that is associated with ethylene exposure. It has been reported that ethylene is associated with lesion development, and this was separated from a secondary browning process leading to russet spotting in lettuce (Peiser et al., 1998). Ethylene is known to increase phenylalanine ammonia-lyase activity in lettuce, resulting in accumulation of phenolics in lettuce tissue (Hyodo et al., 1978). The role of ethylene in causing russet spotting was previously attributed to the accumulation of phenolics, which can be oxidized to produce brown pigments (Hyodo et al., 1978). However, the evidence presented by Peiser et al. (1998) suggests that the initial ethylene-dependent process leading to lesion development (i.e., oxidative membrane injury) is key to the phenomenon of russet spotting. Therefore, the browning reactions are a secondary process for that disorder.

Similarly, ethylene-induced chlorophyll loss in green leafy vegetables may be a consequence of secondary events. Even though a general relationship exists between ethylene, yellowing, and lipid peroxidation, losses in chlorophyll are not always directly linked with the ethylene-induced onset of increased lipid peroxidation, or the rate of ethylene-induced chlorophyll loss may not closely reflect the rate of accumulation of lipid peroxidation products (Zhuang et al., 1995; Hodges and Forney, 2000). Recently, it has been reported that ethylene-induced chlorophyll catabolism is directly related to a hydrogen peroxide-peroxidase reaction (Adachi et al., 1999) and that chlorophyll degradation can be associated with ethylene-induced accumulations of hydrogen peroxide in tissues (Hodges and Forney, 2000).

Therefore, it is possible that ethylene induces a hydrogen peroxide accumulation, which in turn accelerates a peroxidase-associated chlorophyll bleaching, which is distinct in nature from an ethylene-enhanced lipid peroxidation process. This view is supported by several reports which demonstrate that phenolic catalyzed peroxidase-hydrogen peroxide reactions are an important factor in chlorophyll degradation (Huff, 1982; Kato and Shimizu, 1985; Yamauchi and Minamide, 1985; Thompson et al., 1987). Another complication in the chlorophyll degradation phenomenon is that there is also an alternate, direct oxidative process associated with the early processes that leads to lipoxygenase-catalyzed lipid peroxidation (Martinoia et al., 1982; Thompson et al., 1987). This process is not a consequence of lipid peroxidation (Larson, 1988) per se but rather an oxidative event in early processes that may also lead to lipid peroxidation (Thompson et al., 1987; Baardseth and von Elbe, 1989). Therefore, it appears that ethylene-induced chlorophyll loss can be associated with various oxidative processes.

IRRADIATION

Ionizing radiation has been developed as an approach to disinfect fruits and vegetables in order to increase shelf life and to meet phytosanitary regulations when exporting fresh produce (Kader, 1986). Ionizing irradiation induces the development of oxygen radicals that can lead to oxidative injuries in fruit and vegetable tissues. The fact that gamma and electron beam irradiation produce ozone in air and any oxygen-containing atmospheres led to the conclusion that the mode of action of irradiation-induced injuries is mediated by ozone generation in the air-filled cavities of the tissue (Maxie and Abdel-Kader, 1966). Urbain (1986) reviewed other possible short-lived radicals such as hydrogen peroxide, which is a product of water radiolysis. Irradiation also produces stable radicals in plant tissues, such as cell walls, and these can be detected using electron spin resonance (ESR) (Deighton et al., 1993; Tabner and Tabner, 1996; de Jesus et al., 1999). However, irradiation exposure of tissues has been shown to produce many other oxygen radicals, and direct radiation damage in tissues and DNA fragmentation, perhaps not surprisingly, has been shown to occur as a result of irradiation (Jones and Bulford, 1990). At 0.6 kilogray (kGy) and above, impairment of ethylene production and enhanced softening occurred in 'Stayman' apples stored for five months (Kushad and Myron, 1989). A dose of 1.15 kGy reduced scald development in 'Stayman' apples whereas 2.30 and 3.45 kGy doses not only increased the incidence and severity of scald, they also caused tissue breakdown (Kushad and Myron, 1989). Hanotel et al. (1995) found that a 3 kGy dose of gamma irradiation resulted in increases in the glutathione protectant system immediately after treatment, followed by significant reductions in that system, and this latter decline was associated with tissue browning for witloof chicory. They inferred that the changes in browning were largely associated with membrane injury, which allowed polyphenol oxidase (PPO) to mix with normally compartmentalized phenols in the cells. Doses of gamma irradiation greater than 0.2 kGy have been shown to cause severe membrane injuries in the peel of green bananas, including deterioration of granal structure in the chloroplasts (Strydom et al., 1991). They noted that a 0.2 kGy dose would delay yellowing, and 0.4 and 0.6 kGy doses would cause slightly less delay in yellowing; a 1 kGy dose would have yellowing similar to the control. Doses between 0.6 to 1.0 kGy were shown to have lower sensitivity to ethylene, which is likely indicative of severe membrane injury (Abeles, 1973). This was similar to results of Couture et al. (1990), who found that maximal effects of irradiation on ethylene production was 1 kGy in strawberries. They suggested that the recommended dose for strawberries should be 0.3 kGy, since minimal injury occurred in response to treatment at that dose.

Larrigaudiere et al. (1990) found that 1 kGy would significantly inhibit the ethylene-forming enzyme in early climacteric cherry tomatoes. This enzyme is membrane bound, which suggests that the irradiation dose is causing significant membrane injury as indicated by the work of Strydom and Whitehead (1990). Increased durations of exposure to 0.25 kGy of gamma irradiation were correlated to increases in electrolyte leakage of 'Ruby Red' grapefruit (Lester and Wolfenbarger, 1990). Relatively high doses of gamma irradiation (2 and 4 kGy) have been shown to cause significant changes in microsomal membranes that were directly associated with electrolyte leakage measurements of the tissues (Voisine et al., 1991). They showed that irradiation treatment resulted in chemical de-esterification of phospholipids via the generated free radicals. This process was also associated with protein losses from the membrane, suggesting a loss in functionality that led to tissue senescence. Gholap et al. (1990) also showed significant changes in fatty acid composition on mangoes irradiated a 0.25 kGy dose of gamma irradiation.

Antioxidant constituents in tissues may or may not be affected by irradiation. Mitchell et al. (1990) found that using doses of 0.75 and 3.0 kGy on red capsicums and 0.75, 3.0, and 6.0 kGy on mangoes had no significant effects on carotene levels. Irradiation with 0.1 kGy resulted in greater losses in carotenoids for potatoes which were subsequently stored at 15°C and 20°C than for controls stored at those temperatures (Bhushan and Thomas, 1990). Raghava and Nisha-Raghava (1990) also reported the 0.5 kGy reduced the carotenoid content in husk tomato, but that lower doses did not. Lu et al. (1989) found that increasing dose rates to 1 kGy actually increased ascorbic acid and carotenoid levels in 'Jewel' sweet potatoes. Ascorbic acid levels were shown to decline with gamma irradiation doses from 0.25 to 1.5 kGy for grapefruit stored at 12°C to 15°C (Garcia Yanez et al., 1990). Reduced glutathione, an endogenous antioxidant in plant tissues, has been shown to decrease exponentially for grapefruit that are exposed to a range of doses from 0.3 to 2.0 kGy of gamma irradiation (Toyo'oka et al., 1989); these declines, however, were not associated with visual injury.

Although there are many indications of oxidative injury induced by ionizing irradiation, the effects on perceptible quality or nutritional value may be considered marginal if the appropriate doses are used. Kader (1986) outlines the recommendations for the safe use of irradiation for fresh fruit and vegetable products.

FUTURE DIRECTIONS

The effects of storage conditions and postharvest handling have been shown to have tremendous impacts on specific indicators of oxidative stress or on the development of oxidative stress-induced disorders in fruits and vegetables. Several factors appear to be important in determining susceptibility to oxidative injury such as maturity, cultivar, storage temperature, storage atmosphere, and water loss. Although much information has been published in this area of study, it is relatively superficial in nature; this probably reflects the fact that oxidative injury research in plants has only recently become an area of significant research activity (Alscher et al., 1997). However, information regarding decline in antioxidant vitamin levels, such as vitamin A precursors and vitamin C, may be sufficient for development of postharvest handling recommendations to preserve nutritional quality in fruit and vegetable products. In contrast, the current understanding of the mechanisms for the development of postharvest oxidatively induced disorders is incomplete. A comprehensive model has been forwarded to explain disorders in plant tissues in relation to lipid peroxidation (Shewfelt and Purvis, 1995). It is clear that a model regarding lipid peroxidation and the development of postharvest disorders cannot be too simplistic because several apparent inconsistencies have not been resolved at this time (Du and Bramlage, 1995; Shewfelt and Purvis, 1995; Alscher et al., 1997).

In relation to oxidative stress in postharvest situations, changes in specific components of endogenous antioxidant systems can be shown, but these changes may not be consistent between commodities or even within a commodity. Moreover, the relationship of oxidative stress-related disorders to cellular injury or to levels of antioxidant protectant systems cannot be consistently demonstrated. To complicate the issue, low levels of antioxidants may reflect either the lack of oxidative stress or, in other cases, a decline in content due to oxidation by active oxygen species in the tissues. It is clear that mild oxidative stress appears to enhance levels of antioxidants in tissues. The lack of a clear relationship among stress, tissue response, and the disorder suggests that the model to explain the observations is far from complete (Shewfelt and Purvis, 1995). It is clear that current models for oxidatively induced injuries should be reevaluated with regard to the roles of different oxidative stress reactions and their integration into the process, which ultimately leads to measurable injury. Although lipid peroxidation may be a common process for many oxidatively induced injuries (Leshem and Kuiper, 1996), perhaps the development of postharvest disorders/senescence involves other secondary oxidative processes in addition to lipid peroxidation. The challenge is whether primary and secondary oxidative events and their time lines can be distinguished. Better understanding of oxidative injury event sequences may lead to the improved ability to predict the development of physiological postharvest disorders. It is obvious from the current literature that we have a long way to go in this regard. Until better models can be constructed, good understanding and reliable control of oxidative stress in the postharvest situation cannot be achieved.

REFERENCES

Abe, K., Y. Gong, and K. Chachin (1995). Effects of O_2 and CO_2 concentrations and ethyl alcohol and acetaldehyde on the yellowing of broccoli (*Brassica oleracea* L. var. italica Plenck) sections. *Journal of the Japanese Society for Cold Preservation of Food* 21: 205-209 [English Translation].

Abeles, F. B. (1973). *Ethylene in Plant Biology*. New York: Academic Press, Inc.

Adachi, M., K. Nakabayashi, R. Azuma, H. Kurata, Y. Takahashi, and K. Shimokawa (1999). The ethylene-induced chlorophyll catabolism of radish (*Rhaphanus sativus* L.) cotyledons: Production of colorless fluorescent chlorophyll catabolite (FCC) in vitro. *Journal of the Japanese Society for Horticultural Science* 68: 1139-1145.

Agar, I.T., J. Streif, and F. Bangerth (1997). Effect of high CO_2 and controlled atmosphere (CA) on the ascorbic acid and dehydroascorbic acid content of some berry fruits. *Postharvest Biology and Technology* 11: 47-55.

Alscher, R.G., J.L. Donahue, and C.L. Cramer (1997). Reactive oxygen species and antioxidants: Relationships in green cells. *Physiologia Plantarum* 100: 224-233.

Anguelova, T. and J. Warthesen (2000). Degradation of lycopene, α-carotene, and β-carotene during lipid peroxidation. *Journal of Food Science* 65: 71-75.

Baardseth, P. and J.H. von Elbe (1989). Effect of ethylene, free fatty acid, and some enzyme systems on chlorophyll degradation. *Journal of Food Science* 54: 1361-1363.

Barden, C.L. and W.J. Bramlage (1994a). Accumulation of antioxidants in apple peel as related to preharvest factors and superficial scald susceptibility of the fruit. *Journal of the American Society for Horticultural Science* 119: 264-269.

Barden, C.L. and W.J. Bramlage (1994b). Relationships of antioxidants in apple peel to changes in α-farnesene and conjugated trienes during storage, and to superficial scald development after storage. *Postharvest Biology and Technology* 4: 23-33.

Barth, M.M., A.K. Perry, S.J. Schmidt, and B.P. Klein (1990). Misting effects on ascorbic acid retention in broccoli during cabinet display. *Journal of Food Science* 55: 1187-1188, 1191.

Barth, M.M. and H. Zhuang (1996). Packaging design affects antioxidant vitamin retention and quality of broccoli florets during postharvest storage. *Postharvest Biology and Technology* 9: 141-150.

Bauchot, A.D., S.J. Reid, G.S. Ross, and D.M. Burmeister (1999). Induction of apple scald by anaerobiosis to naturally occurring superficial scald in 'Granny Smith' apple fruit. *Postharvest Biology and Technology* 16: 9-14.

Ben-Yehoshua, S., B. Shapiro, Z.E. Chen, and S. Lurie (1983). Mode of action of plastic film in extending life of lemon and bell pepper fruits by alleviation of water stress. *Plant Physiology* 73: 87-93.

Bhushan, B. and P. Thomas (1990). Effects of gamma irradiation and storage temperature on lipoxygenase activity and carotenoid disappearance in potato tubers (*Solanum tuberosum* L.). *Journal of Agricultural and Food Chemistry* 38: 1586-1590.

Blanpied, G.D. and L.L. Creasy (1993). Concentrations of farnesene and conjugated trienes in the skin of Cortland apples stored in 2 percent and 4 percent oxygen with 1 percent, 3 percent and 4 percent carbon dioxide at 2.2°C. *Proceedings of the 6th International Controlled Atmosphere Research Conference* 2: 481-486. Ithaca, NY: Northeast Regional Agricultural Engineering Service, Cornell University.

Blokhina, O.B., K.V. Fagerstedt, and T.V. Chirkova (1999). Relationships between lipid peroxidation and anoxia tolerance in a range of species during post-anoxic reaeration. *Physiologia Plantarum* 105: 625-632.

Burmeister, D.M., and D.R. Dilley (1995). A "scald-like" controlled atmosphere storage disorder of Empire apples—A chilling injury induced by high CO_2. *Postharvest Biology and Technology* 6: 1-7.

Chen, P.M., R.J. Varga, and Y.Q. Xiao (1993). Inhibition of α-farnesene biosynthesis and its oxidation in the peel tissue of "d'Anjou" pears by low-O_2/elevated CO_2 atmospheres. *Postharvest Biology and Technology* 3: 215-223.

Couture, R., J. Makhlouf, F. Cheqour, and C. Willemot (1990). Production of CO_2 and C_2H_4 after gamma irradiation of strawberry fruit. *Journal of Food Quality* 13:385-393.

Cowart, D.M. (1993). *Mechanistic studies on the role of membrane lipid peroxidation during the development of chilling injury in bell pepper fruit.* Master's thesis, University of Georgia.

de Jesus, E.F.O., A.M. Rossi, and R.T. Lopes (1999). An ESR study on identification of gamma-irradiated kiwi, papaya and tomato using fruit pulp. *International Journal of Food Science and Technology* 34: 173-178.

Deighton, N., S.M. Glidwell, B.A. Goodman, and I.M. Morrison (1993). Electron paramagnetic resonance of gamma-irradiated cellulose and lignocellulose material. *International Journal of Food Science and Technology* 28: 45-55.

Dipierro, S. and S. De Leonardis (1997). The ascorbate system and lipid peroxidation in stored potato (*Solanum tuberosum* L.) tubers. *Journal of Experimental Botany* 48: 779-783.

Du, Z. and W.J. Bramlage (1994). Superoxide dismutase activities in senescing apple fruit (*Malus domestica* Borkh.). *Journal of Food Science* 59: 581-584.

Du, Z. and W.J. Bramlage (1995). Peroxidative activity of apple peel in relation to development of poststorage disorders. *HortScience* 30: 205-208.

Elgar, H.J., D.M. Burmeister, and C.B. Watkins (1998). Storage and handling effects on a CO_2-related internal browning disorder of 'Braeburn' apples. *HortScience* 33: 719-722.

Ezell, B.D. and M.S. Wilcox (1959). Loss of vitamin C in fresh vegetables as related to wilting and temperature. *Journal of Agricultural and Food Chemistry* 7: 507-509.

Ezell, B.D. and M.S. Wilcox (1962). Loss of carotene in fresh vegetables as related to wilting and temperature. *Journal of Agricultural and Food Chemistry* 10: 124-126.

Forney, C.F. and W.J. Lipton (1990). Influence of controlled atmospheres and packaging on chilling sensitivity. In *Chilling Injury of Horticultural Crops,* ed. C.Y. Wang (Ed.). Boca Raton, FL: CRC Press, pp. 257-267.

Garcia-Yanez, M., A. Garcia-Arteaga, J. Fernandez-Miranda, A. Paradoa, E. Sampere, E. Castillo, and G. Serrano (1990). Stability of vitamin C content in grapefruit treated with doses of gamma irradiation. *Revista de Agroquimica y Tecnologia de Alimentos* 30: 409-415.

Gholap, A.S., C. Bandypadhyay, and P.M. Nair (1990). Lipid composition and flavor changes in irradiated mango (var. Alphonso). *Journal of Food Science* 55: 1579-1580.

Gil, M.I., F. Ferreres, and F.A. Tomás-Barberán (1999). Effect of postharvest storage and processing on the antioxidant constituents (flavonoids and vitamin C) of fresh-cut spinach. *Journal of Agricultural and Food Chemistry* 47: 2213-2217.

Giovanelli, G., V. Lavelli, C. Peri, and S. Nobili (1999). Variation in antioxidant components of tomato during vine and post-harvest ripening. *Journal of the Science of Food and Agriculture* 79: 1583-1588.

Gong, Y., P.M.A. Toivonen, P.A. Wiersma, and O.L. Lau (2000). *Development of a biochemical prediction for susceptibility of B.C.-grown 'Braeburn' apples to "Braeburn Browning Disorder."* Pacific Agri-Food Research Center Final Project Report. 69 pp.

Hakim, A., A.C. Purvis, and B.G. Mullinix (1999). Differences in chilling sensitivity of cucumber varieties depends on storage temperature and the physiological dysfunction evaluated. *Postharvest Biology and Technology* 17: 97-104.

Hanotel, L., A. Fleuriet, and P. Boisseau (1995). Biochemical changes involved in browning of gamma-irradiated cut witloof chicory. *Postharvest Biology and Technology* 5: 199-210.

Hariyadi, P. and K.L. Parkin (1991). Chilling-induced oxidative stress in cucumber fruits. *Postharvest Biology and Technology* 1: 33-45.

Hirata, T., A. Nakatani, Y. Ishikawa, C. Yamada, and S. Katsuura (1995). Changes in chlorophylls, carotenoids, ascorbic acid and glutathiones of broccoli during storage in modified atmosphere packaging (trans.). *Nippon Shokuhin Kagaku Kogaku Kaishi* 42: 996-1002.

Hodges, D.M., J.M. DeLong, C.F. Forney, and R.K. Prange (1999). Improving the thiobarbituric acid-reactive-substances assay for estimating lipid peroxidation in plant tissues containing anthocyanin and other interfering compounds. *Planta* 207: 604-611.

Hodges, D.M. and C.F. Forney (2000). The effects of ethylene, depressed oxygen and elevated carbon dioxide on antioxidant profiles of senescing spinach leaves. *Journal of Experimental Botany* 51: 645-655.

Howard, L.R. and C. Hernandez-Brenes (1998). Antioxidant content and market quality of jalapeno pepper rings as affected by minimal processing and modified atmosphere packaging. *Journal of Food Quality* 21: 317-327.

Huff, A. (1982). Peroxidase-catalyzed oxidation of chlorophyll by hydrogen peroxide. *Phytochemistry* 21: 261-265.

Hyodo, H., H. Kuroda, and S.F. Yang (1978). Induction of phenylalanine ammonia-lyase and increase in phenolics in lettuce leaves in relation to the development of russet spotting caused by ethylene. *Plant Physiology* 2: 31-35.

Ishikawa, Y., C. Wessling, T. Hirata, and Y. Hasegawa (1998). Optimum broccoli packaging conditions to preserve glutathione, ascorbic acid, and pigments. *Journal of the Japanese Society for Horticultural Science* 67: 367-371.

Jiang, Y.M., M.. Chen, and Z.F. Lin (1991). Enzymatic browning of banana during low temperature storage (trans.). *Acta Phytophysiologica Sinica* 17: 157-163.

Jones, J.L. and B.B. Bulford (1990). *A feasibility study of the use of DNA fragmentation as a method for detecting irradiation of food.* Technical memorandum no. 584 of the Campden Food and Drink Research Association, 33 pp.

Ju, Z. and E.A. Curry (2000). Lovastatin inhibits α-farnesene synthesis without affecting ethylene production during fruit ripening in 'Golden Supreme' apples. *Journal of the American Society for Horticultural Science* 125: 105-110.

Ju, Z., Y. Yuan, C. Liou, S. Zhan, S. Xin (1994). Effects of low temperature on H_2O_2 and heart browning of Chili and Yali (*Pyrus bret schneideri* R.). *Scientia Agricultura Sinica* 27(5): 77-81.

Ju, Z., Y. Yuan, C. Liu, H. Dai, and R.J. Liu (1994). Effect of low temperature on H_2O_2 level during storage of apples (trans.). *Journal of Fruit Science* 11: 10-13.

Ju, Z., Y. Yuan, C. Liu, S. Zhan, and M. Wang (1996). Relationships among simple phenol, flavonoid and anthocyanin in apple fruit peel at harvest and scald susceptibility. *Postharvest Biology and Technology* 8: 83-93.

Kader, A.A. (1986). Potential applications of ionizing radiation in postharvest handling of fresh fruits and vegetables. *Food Technology* 40(6): 117-121.

Kalt, W., C.F. Forney, A. Martin, and R.L. Prior (1999). Antioxidant capacity, vitamin C, phenolics, and anthocyanins after fresh storage of small fruits. *Journal of Agricultural and Food Chemistry* 47: 4638-4644.

Kato, M. and S. Shimizu (1985). Chlorophyll metabolism in higher plants. VI. Involvement of peroxidse in chlorophyll degradation. *Plant and Cell Physiology* 26: 1291-1301.

Kosson, R., F. Adamicki, M. Horbowicz, and J. Dobrazañska (1998). The effect of storage conditions on weight losses, quality and marketable value of breaker and red pepper fruits (*Capsicum annuum* L.). *Vegetable Crops Research Bulletin, Research Institute of Vegetable Crops—Skerniewice* 49: 121-129.

Kumar, G.N.M. and N.R. Knowles (1993). Changes in lipid peroxidation and lipolytic and free-radical scavenging enzyme activities during aging and sprouting of potato (*Solanum tuberosum*) seed tubers. *Plant Physiology* 102: 115-124.

Kunert, K.J. and M. Ederer (1985). Leaf aging and lipid peroxidation: The role of the antioxidants vitamin C and E. *Physiologia Plantarum* 65: 85-88.

Kuo, S.-J. and K.L. Parkin (1989). Chilling injury in cucumbers (*Cucumis sativa* L.) associated with lipid peroxidation as measured by ethane evolution. *Journal of Food Science* 54: 1488-1491.

Kushad, M.M. and J. Myron (1989). Effect of ionized radiation on quality of 'Stayman' apple. *Proceedings of the 5th International Controlled Atmosphere Research Conference.* Wenatchee, WA: Washington State University. pp. 263-272.

Lacan, D. and J.C. Baccou (1996). Changes in lipids and electrolyte leakage during nonnetted muskmelon ripening. *Journal of the American Society for Horticultural Science* 121: 554-558.

Larrigaudiere, C., A. Latche, J.C. Pech, and C. Triantaphylides (1990). Short-term effects of gamma-irradiation on 1-aminocyclopropane-1-carboxylic acid metabolism in early climacteric cherry tomatoes. *Plant Physiology* 92: 577-581.

Larson, R.A. (1988). The antioxidants of higher plants. *Phytochemistry* 27: 969-978.

Lentheric, I., E. Pinto, M. Vendrell, and C. Larrigaudiere (1999). Harvest date affects the antioxidant systems in pear fruits. *Journal of Horticultural Science and Biotechnology* 74: 791-795.

Leshem, Y.Y. (1988). Plant senescence processes and free radical. *Free Radical Biology and Medicine* 5: 39-49.

Leshem, Y.Y. and P.J.C. Kuiper (1996). Is there a GAS (general adaptation syndrome) response to various types of environmental stress? *Biologia Plantarum* 38: 1-18.

Lester, G.E. and D.A. Wolfenbarger (1990). Comparisons of cobalt-60 gamma irradiation dose rates on grapefruit flavedo tissue and on Mexican fruit fly mortality. *Journal of Food Protection* 53: 329-331.

Lin, W.C., J.W. Hall, and M.E. Saltveit Jr. (1993). Ripening stage affects the chilling sensitivity of greenhouse-grown peppers. *Journal of the American Society for Horticultural Science* 118: 791-795.

Lojkowska, E. and M. Holubowska (1989). Changes in the lipid catabolism in potato tubers from cultivars differing in susceptibility to autolysis during the storage. *Potato Research* 32: 463-470.

Lu, J.Y., S. White, P. Yakubu, and P.A. Loretan (1989). Effects of gamma irradiation on nutritive and sensory quality of sweet potato storage roots. *Proceedings of the 1987 Singapore Institute of Food Science and Technology Conference.* Singapore: Singapore Institute of Food Science and Technology. pp. 224-228.

Lurie, S. and S. Ben-Yehoshua (1986). Changes in membrane properties and abscisic acid during senescence of harvested bell pepper fruit. *Journal of the American Society for Horticultural Science* 111: 886-889.

Lurie, S. and J.D. Klein (1992). Ripening characteristics of tomatoes stored at 12°C and 2°C following a prestorage heat treatment. *Scientia Horticulturae* 51: 55-64.

Makhlouf, J., C. Willemot, R. Couture, J. Arul, and F. Castaigne (1990). Effect of low temperature and controlled atmosphere storage on the membrane lipid composition of broccoli flower buds. *Scientia Horticulturae* 42: 9-19.

Martinoia, E., M.J. Dalling, and P. Matile (1982). Catabolism of chlorophyll: Demonstration of chloroplast-localized peroxidative and oxidative activities. *Zeitschrift für Pflanzenphysiologie* 107: 269-279.

Masia, A. (1998). Superoxide dismutase and catalase activities in apple fruit during ripening and post-harvest and with special reference to ethylene. *Physiologia Plantarum* 104: 668-672.

Maxie, E.C. and A. Abdel-Kader (1966). Food irradiation—Physiology of fruits as related to feasibility of the technology. *Advances in Food Research* 15: 105-145.

McRae, D.G., J.E. Baker, and J.E. Thompson (1982). Evidence for involvement of the superoxide radical in the conversion of 1-aminocyclopropane-1-carboxylic acid to ethylene by pea microsomal membranes. *Plant and Cell Physiology* 23: 375-383.

Meir, S. and W.J. Bramlage (1988). Antioxidant activity in 'Cortland' apple peel and susceptibility to superficial scald after storage. *Journal of the American Society for Horticultural Science* 113: 412-418.

Meir, S., S. Philisoph-Hadas, and N. Aharoni (1992). Ethylene-increased accumulation of fluorescent lipid-peroxidation products detected during senescence of parsley by a newly developed method. *Journal of the American Society for Horticultural Science* 117: 128-132.

Meir, S., S. Philosoph-Hadas, G. Zauberman, Y. Fuchs, M. Akerman, and N. Aharoni (1991). Increased formation of fluorescent lipid-peroxidation products in avocado peels precedes other signs of ripening. *Journal of the American Society for Horticultural Science* 116: 823-826.

Miller, W.R. and R.E. McDonald (1999). Irradiation, stage of maturity at harvest, and storage temperature during ripening affect papaya fruit quality. *HortScience* 34: 1112-1115.

Mitchell, G.E., R.L. McLauchlan, T.R. Beattie, C. Banos, and A.A. Gillen (1990). Effect of gamma irradiation on the carotene content of mangos and red capsicums. *Journal of Food Science* 55: 1185-1186.

Monk, L.S., R. Braendle, and R.M.M. Crawford (1987). Catalase activity and post-anoxic injury in monocotyledonous species. *Journal of Experimental Botany* 38: 233-246.

Monk, L.S., K. Fagerstedt, and R.M.M. Crawford (1987). Superoxide dismutase as anaerobic polypeptide. A key factor in recovery from oxygen deprivation in *Iris pseudacorus*. *Plant Physiology* 85: 1016-1020.

Palmer Wright, K. and A.A. Kader (1997a). Effect of controlled atmosphere storage on the quality and carotenoid content of sliced persimmons and peaches. *Postharvest Biology and Technology* 10: 89-97.

Palmer Wright, K. and A.A. Kader (1997b). Effect of slicing and controlled atmosphere storage on the ascorbate content and quality of strawberries and persimmons. *Postharvest Biology and Technology* 10: 39-40.

Pastori, G.M. and V.S. Trippi (1992). Oxidative stress induced high rate of glutathione reductase synthase in drought resistant maize strain. *Plant and Cell Physiology* 33: 957-961.

Peiser, G., G. López-Gálvez, M. Cantwell, and M.E. Saltveit (1998). Phenylalanine ammonia-lyase inhibitors do not prevent russet spotting lesion development in lettuce midribs. *Journal of the American Society for Horticultural Science* 123: 687-691.

Pesis, E., D. Aharoni, Z. Aharon, R. Ben-Arie, N. Aharoni, and Y. Fuchs (2000). Modified atmosphere and modified humidity packaging alleviates chilling injury symptoms in mango fruit. *Postharvest Biology and Technology* 19: 93-101.

Philosoph-Hadas, S., S. Meir, B. Akiri, and J. Kanner (1994). Oxidative defense systems in leaves of three edible herb species in relation to their senescence rate. *Journal of Agricultural and Food Chemistry* 42: 2376-2381.

Price, A.H. and G.A.F. Hendry (1991). Iron catalyzed oxygen free radical formation and its possible contribution to drought damage in nine grasses and three cereals. *Plant, Cell and Environment* 14: 477-484.

Qi., L., T. Wu, and A.E. Watada (1999). Quality changes of fresh-cut honeydew melons during controlled atmosphere storage. *Journal of Food Quality* 22: 513-521.

Raghava, R.P. and Nisha-Raghava (1990). Carotenoid content of husk tomato under the influence of growth regulators and gamma rays. *Indian Journal of Plant Physiology* 33: 87-89.

Rogiers, S.Y., G.N.M. Kumar, and N.R. Knowles (1998). Maturation and ripening of *Amelanchier alnifolia* Nutt. are accompanied by increasing oxidative stress. *Annals of Botany* 81: 203-211.

Romo-Parada, L., C. Willemot, F. Castaigne, C. Gosselin, and J. Arul (1989). Effect of controlled atmospheres (low oxygen, high carbon dioxide) on storage of cauliflower (*Brassica oleracea* L., Botrytis Group). *Journal of Food Science* 54: 122-124.

Sala, J.M. (1998). Involvement of oxidative stress in chilling injury in cold-stored mandarin fruits. *Postharvest Biology and Technology* 13: 255-261.

Saltveit Jr., M.E. (1989). A kinetic examination of ion leakage from chilled tomato pericarp disks. *Acta Horticulturae* 258: 617-622.

Saltveit, M.E. (1999). Effect of ethylene on quality of fresh fruits and vegetables. *Postharvest Biology and Technology* 15: 279-292.

Senaratna, T., B.D. McKersie, and R.H. Stinson (1985). Antioxidant levels in germinating soybean axes in relation to free radical and dehydration tolerance. *Plant Physiology* 78: 168-171.

Serrano, M., M.C. Martinez-Madrid, M.T. Petrel, F. Riquelme, and F. Romojaro (1997). Modified atmosphere packaging minimizes increases in putrescine and abscisic levels caused by chilling injury in pepper fruit. *Journal of Agricultural and Food Chemistry* 45: 1668-1672.

Shewfelt, R.L. and A.C. Purvis (1995). Toward a comprehensive model for lipid peroxidation in plant tissue disorders. *HortScience* 30: 213-218.

Simontacchi, M. and S. Puntarulo (1992). Oxy-radical generation by isolated microsomes from soybean seedligs. *Plant Physiology* 100: 1263-1268.

Spychalla, J.P. and S.L. Desborough (1990). Superoxide dismutase, catalase, and α-tocopherol content of stored potato tubers. *Plant Physiology* 94: 1214-1218.

Strydom, G.J., J. van Staden, and M.T. Smith (1991). The effect of gamma radiation on the ultrastructure of the peel of banana fruits. *Environmental and Experimental Botany* 31: 43-49.

Strydom, G.J. and C.S. Whitehead (1990). The effect of ionizing radiation on ethylene sensitivity and postharvest ripening of banana fruit. *Scientia Horticulturae* 41: 293-304.

Tabner, B.J. and V.A. Tabner (1996). Stable radicals observed in the flesh of irradiated citrus fruits by electron spin resonance spectroscopy for the first time. *Radiation Physics and Chemistry* 47: 601-605.

Thompson, J.E., R.L. Legge, and R.F. Barber (1987). The role of free radicals in senescence and wounding. *New Phytologist* 105: 317-344.

Toivonen, P.M.A. (1992). Chlorophyll fluorescence as a nondestructive indicator of freshness in harvested broccoli. *HortScience* 27: 1014-1015.

Toivonen, P.M.A. (1997). The effects of storage temperature, storage duration, hydro-cooling, and micro-perforated wrap on shelf life of broccoli (*Brassica oleracea* L., Italica group). *Postharvest Biology and Technology* 10: 59-65.

Toivonen, P.M.A. and M. Sweeney (1998). Differences in chlorophyhll loss at 13°C for two broccoli (*Brassica oleracea* L.) cultivars associated with antioxidant enzyme activities. *Journal of Agricultural and Food Chemistry* 46: 20-24.

Toyo'oka, T., S. Uchiyama, and Y. Saito (1989). Effect of gamma-irradiation on thiol compounds in grapefruit. *Journal of Agricultural and Food Chemistry* 37: 769-775.

Urbain, W.M. (1986). Fruits, vegetables and nuts. In *Food Irradiation*, B.S. Schweigert (Ed.). Orlando, FL: Academic Press, Inc., pp. 170-216.

Veltman, R.H., M.G. Sanders, S.T. Persijn, H.W. Peppelenbos, and J. Osterhaven (1999). Decreased ascorbic acid levels and brown core development in pears (*Pyrus communis* L. cv. Conference). *Physiologia Plantarum* 107: 39-45.

Voisine, R., L.P. Vezina, and C. Willemot (1991). Induction of senescence-like deterioration of microsomal membranes from cauliflower by free radicals generated during gamma irradiation. *Plant Physiology* 97: 545-550.

Wang, C.Y. and G.F. Kramer (1989). Effect of low-oxygen storage on polyamine levels and senescence in Chinese cabbage, zucchini squash and McIntosh apples. *Proceedings of the 5th International Controlled Atmosphere Research Conference.* Wenatchee, WA: Washington State University. pp. 19-27.

Wang, C.Y., G.F. Kramer. B.D. Whitaker, and W.R. Lusby (1992). Temperature preconditioning increases tolerance to chilling injury and alters lipid composition in zucchini squash. *Journal of Plant Physiology* 140: 229-235.

Watada, A.E. (1987). Vitamins. In *Postharvest Physiology of Vegetables*, J. Weichmann (Ed.). New York: Marcel Decker Inc., pp. 455-468.

Watkins, C.B., C.L. Barden, and W.J. Bramlage (1993). Relationships among α-farnesene, conjugated trienes, and ethylene production with superficial scald development of apples. *Acta Horticulturae* 343: 155-160.

Watkins, C.B., J.H. Nock, and B.D. Whitaker (2000). Responses of early, mid and late season apple cultivars to postharvest application of 1-methylcyclopropene (1-MCP) under air and controlled atmosphere storage conditions. *Postharvest Biology and Technology* 19: 17-32.

Whitaker, B.D., T. Solomos, and D.J. Harrison (1997). Quantification of α-farnesene and its conjugated trienol oxidation products from apple peel by C18-HPLC with UV detection. *Journal of Agricultural and Food Chemistry* 45: 760-765.

Wismer, W.V., W.M. Worthing, R.Y. Yada, and A.G. Marangoni (1998). Membrane lipid dynamics and lipid peroxidation in the early stages of low-temperature sweetening in tubers of *Solanum tuberosum*. *Physiologia Plantarum* 102: 396-410.

Yamasaki, H., Y. Sakihama, and N. Ikehara (1997). Flavanoid-peroxidase reaction as a detoxification mechanism of plant cells against H_2O_2. *Plant Physiology* 115: 1405-1412.

Yamauchi, N. and T. Minamide (1985). Chlorophyll degradation by peroxidase in parsley leaves. *Journal of the Japanese Society for Horticultural Science* 54: 265-271.

Yamauchi, N. and A.E. Watada (1991). Regulated chlorophyll degradation in spinach leaves during storage. *Journal of the American Society for Horticultural Science* 116: 58-62.

Yang, Y.-J. (1997). Effect of controlled atmospheres on storage life in 'Niitaka' pear fruit. *Journal of the Korean Society for Horticultural Science* 38: 734-738.

Zhuang, H., M.M. Barth, and D.F. Hildebrand (1994). Packaging influenced total chlorophyll, soluble protein, fatty acid composition and lipoxygenase activity in broccoli florets. *Journal of Food Science* 59: 1171-1174.

Zhuang, H., D.F. Hildebrand, and M.M. Barth (1995). Senescence of broccoli buds is related to changes in lipid peroxidation. *Journal of Agricultural and Food Chemistry* 43: 2585-2591.

Zhuang, H., D.F. Hildebrand, and M.M. Barth (1997). Temperature influenced lipid peroxidation and deterioration in broccoli buds during postharvest storage. *Postharvest Biology and Technology* 10: 49-58.

Zuckermann, H., F.J.M. Harren, J. Reuss, and D.H. Parker (1997). Dynamics of acetaldehyde production during anoxia and post-anoxia in red bell pepper studied by photoacoustic techniques. *Plant Physiology* 113: 925-932.

Chapter 5

Superficial Scald—A Postharvest Oxidative Stress Disorder

John M. DeLong
Robert K. Prange

Superficial scald is a postharvest physiological disorder of apples and pears characterized by light bronzing to deep browning of the peel during or following a refrigerated air (RA) [ambient oxygen (O_2) and carbon dioxide (CO_2)] or a controlled atmosphere (CA) (0.7-3.0 kilopascals [kPa] O_2; 2.0-5.0 kPa CO_2) storage period (Ingle and D'Souza, 1989; Wang and Dilley, 1999). The disorder is largely confined to the chloroplast-containing hypodermis located immediately below the epidermis. Browning of cell contents initially occurs in the outer hypodermal region and then extends throughout the remaining five or six cell layers of the hypodermis. As the disorder progresses, the color of the affected tissue changes from bronze or light brown to dark brown. In severely scalded fruit, the affected hypodermal cells collapse radially resulting in sunken areas; the outer cells of the cortex can become distorted and epidermal browning may also occur (Bain, 1956; Bain and Mercer, 1963).

The potential for fruit to develop superficial scald has been linked to various environmental or cultural influences encountered during seasonal growth and development as well as the conditions within the storage environment following harvest (Emonger et al., 1994; Bramlage and Weis, 1997; Weis et al., 1998, 1999). For example, fruit maturity can strongly influence scald incidence, with greater maturity at harvest being correlated with less scald development (Smock, 1961; Meir and Bramlage, 1988). Larger fruit—those with low calcium concentrations and those from orchards that receive high levels of nitrogen fertilizer—tend to show more incidence of scald than smaller fruit with adequate and balanced mineral levels within the peel and cortical regions (Smock, 1961; Romaniuk et al., 1981; Raese and Drake, 1993). Climate markedly influences scald ontogeny as hot, dry weather during the last weeks of the growing season predisposes susceptible fruit to more scald expression than a cooler late season having normal quantities of rainfall and sunshine (Barden and Bramlage, 1994c; Bramlage and Weis,

1997; Thomai et al., 1998). Low-O_2 storage environments can reduce the severity of scald and maintain the quality of fruit so beneficially that the modern apple industry routinely employs O_2 partial pressures of 0.7-3.0 kPa for long-term apple storage (Anderson, 1967; Lau et al., 1998; Prange et al., 1998). Although much is known about the conditions that predispose susceptible cultivars toward scald, predictive scald models have not proven dependable due to the complex interaction of cultivar, geographic, growing season, fruit maturity, and harvest and storage-related factors that influence disorder development (Ingle and D'Souza, 1989; Emonger et al., 1994; Wang and Dilley, 1999). The inability to reliably predict scald development arises from the correlative nature of much of the presently available data and reflects the fact that the definitive biochemical cause of scald is not known.

Superficial scald is an economically important disorder as affected fruit must be made into juice, sauce, or bakery products because they are not acceptable in the more profitable fresh market. Apple cultivars that are particularly susceptible to superficial scald include: Baldwin, Ben Davis, Cortland, Delicious, Granny Smith, Law Rome, Stayman, and Winesap as well as Anjou pears (Meheriuk et al., 1994). Unless controlled by chemical and/or cultural means, scald has the potential to reduce the annual market value of millions of tonnes of apples worldwide. Compounding this potential loss of economic return for susceptible cultivars is the restriction of chemical choices as antiscald treatments. At present, a postharvest, prestorage drench of the antioxidant diphenylamine (DPA) (N-phenylbenzenamine) is the only effective chemical registered for preventing scald expression in apples in North America, while ethoxyquin (6-ethoxy-1,2-dihydro-2,2,4-trimethyl quinoline) can be used for scald control on pears in the states of Washington and Oregon (John Kelly, personal communication, October 2000).* However, with consumer safety concerns over pesticide usage and chemical additives in food (Baker, 1996, 1999), the possible future loss of DPA registration as a preventative scald treatment has fueled a renewed search for scald-controlling processes and DPA alternatives (Scott, Yuen, and Ghahramani, 1995; Scott, Yuen, and Kim, 1995; Lau et al., 1998; Ghahramani et al., 1999; Wang and Dilley, 1999, 2000b).

CHEMISTRY OF SCALD

Although much research has been directed to understanding the ontogeny of scald, elucidation of the basic biochemistry has only moderately pro-

---

*DPA is also registered for use on apples in Chile, France, Greece, Ireland, Israel, Italy, Portugal, South Africa, Spain, and the United Kingdom. Ethoxyquin is also registered for use on apples and pears in Canada.

gressed beyond the findings of nearly four decades ago. The prevailing paradigm is that superficial scald occurs as the sesquiterpene hydrocarbon, α-farnesene, is oxidized to several degradation products, which eventually cause the browning of hypodermal cells; the chemistry that directly leads to the browning reactions is not known. Although scald is viewed as an oxidative storage disorder, its development and expression can be strongly influenced by the growing season prior to the storage period (Emonger et al., 1994) as well as the shelf-life environment (e.g., temperature) following storage removal (Prange and Harrison, unpublished data; Whitaker and Saftner, 2000).

The role of α-farnesene as a causal agent of superficial scald was first forwarded by Murray et al. (1964) following isolation of the volatile sesquiterpene from the cuticles of 'Granny Smith' apples. It was soon discovered that α-farnesene readily oxidizes to conjugated trienes (CTs), ketones, and alcohol radicals, and that scald appears after a portion of the α-farnesene is oxidized (Huelin and Coggiola, 1970a; Anet, 1972a,c). In general, CTs are positively correlated with scald expression (Huelin and Coggiola, 1970a,b; Abdallah et al., 1997; Whitaker, Solomos, et al., 1997), while α-farnesene levels may be positively (Meir and Bramlage, 1988; Watkins et al., 1995) or poorly associated with scald incidence (Watkins et al., 1993; Du and Bramlage, 1994; Rao et al., 1998; Rupasinghe et al., 2000). Hence, the CTs are viewed as either being the scald-inducing agents or as having a more direct biochemical linkage with the browning reactions than does α-farnesene.

Quantification of CTs is routinely accomplished by placing the whole fruit or excised peel segments in hexane for two or three minutes followed by the spectrophotometric measurement of the extract at 258, 269, and 281 nanometers (nm). For nearly 30 years, the absorbance (or optical density) at 281 nm minus that at 290 nm for removal of interference artifacts (i.e., $OD_{281-290}$), has been routinely used for CT determination in hexane extracts (Huelin and Coggiola, 1970a; Anet, 1972b; Du and Bramlage, 1993, 1994; Rao et al., 1998). In some cases, however, the association of $OD_{281-290}$ with scald expression is not strong (Du and Bramlage, 1993, 1994; Rao et al., 1998; Alwan and Watkins, 1999). A CT absorbance peak at 258 nm, and its negative correlation with scald, has led to the modified theory that a number of CT components having absorbance maxima at 258 and 281 nm differ in their metabolic influence on scald development: a $(OD_{258-290}) / (OD_{281-290})$ ratio of ≥ 2.0 is associated with little or no scald, while a ratio of ≤ 1.0 is associated with significant scald expression (Du and Bramlage, 1993, 1994).

A significant recent development in the theory of CTs as scald-inducing metabolites is the determination that the primary oxidation products of α-farnesene are neither hydro nor endoperoxide CT metabolites, but a conjugated trienol isomer. Rowan et al. (1995) identified 2,6,10-trimethyldodeca-2,7(E),9(E),11-tetraen-6-(ol) and its 9Z isomer as constituting   88   to   95 percent and 5 to 11 percent, respectively, of the trienes in apple skin wax. These two compounds accounted for only 12 to 35 percent of the triene concentration measured in apple skin via the conventional UV spectrophotometric method. Whitaker, Solomos, et al. (1997) corroborated these findings in 'Granny Smith' and 'Delicious' fruit with incidence of scald being well correlated with trienol content. In a subsequent study, Whitaker et al. (2000) observed that scald susceptibility in 'White Angel' x 'Rome Beauty' peel tissue is not correlated with conjugated trienol content, which led to the conclusion that if trienol isomers do play a role in scald development, other factors such as the state of the antioxidant system, the free radical challenge, and toxic volatiles generated by α-farnesene oxidation have equally important roles.

A recent shift in thinking regarding the causal agent of scald is that the ketone 6-methyl-5-hepten-2-one (MHO), rather than CTs, may be the putative scald-inducing oxidation product of α-farnesene. Although this compound has been reported to be the main volatile produced during the autooxidation of α-farnesene (Filmer and Meigh, 1971; Anet, 1972a), it was not viewed as the scalding compound. The generation of MHO is thought to arise from the oxidation of CTs (Mir, Perez, and Beaudry, 1999) or conjugated trienols (CTols) (Whitaker and Saftner, 2000) and not from α-farnesene directly. Interestingly, Song and Beaudry (1996) observed that applying MHO to apples causes a scaldlike browning, which provided the observational basis for rethinking the CTs causal paradigm. In addition, Mir, Perez, and Beaudry (1999) found that the concentration of MHO in DPA-treated 'Cortland' peels is 8,000-fold lower than in scalded control fruit and that the increase in vapor phase MHO and peel browning is concomitant. A time course graph of scald development and MHO generation in their study shows striking similarity, suggesting causal linkage of MHO with scald expression. Whitaker and Saftner (2000) show that autooxidation of CTols generates MHO; they emphasize the theory that scald-susceptible tissue becomes increasingly sensitized to MHO toxicity during storage. Contrastingly, Rupasinghe et al. (2000) note that MHO does not show any relation to scald expression in 'Delicious' skin following storage, although an MHO alcohol increased 20 to 60 percent in fruit with varying degrees of scald severity. The high performance liquid chromatography (HPLC)-derived metabolite data in this latter study should be confirmed with mass spectroscopy techniques in order to resolve any questions regarding compound identification.

SCALD AND OXIDATIVE STRESS THEORIES

At least three schools of thought can be discerned regarding the cellular ontogeny of scald: (1) scald develops as the toxic metabolites of α-farnesene oxidation damage hypodermal cells; (2) scald occurs as the antioxidant systems within the hypodermis are overwhelmed by cell-damaging oxidative stress, which oxidizes α-farnesene, causing an increase in toxic metabolite levels; and (3) α-farnesene, and its metabolites are not causally linked to scald, but scald occurs as the antioxidant capacity of the hypodermal cells is overwhelmed and damaged by oxidative stress. Other theories, such as scald being an expression of chilling injury or anoxia-induced O_2 sensitivity, have been put forth; however, the postchilling and postanoxia mechanism of cellular damage has been attributed to uncontrolled cellular oxidation (Watkins et al., 1995; Bauchot et al., 1999), which would fall within the second or third school of thought noted in this paragraph.

The established theory—that the basic chemistry of scald involves oxidation of cellular molecules—rests on several areas of consistent observation: (1) as scald develops, there is a concomitant increase in various CTs, carbonyl, and related compounds, presumably arising from α-farnesene oxidation (Huelin and Coggiola, 1970a; Filmer and Meigh, 1971; Anet, 1972a; Whitaker, Solomos, et al., 1997); (2) scald is controlled by the amine-type antioxidants DPA and ethoxyquin; and (3) scald can be suppressed by storing fruit in environments with low O_2 partial pressures (0.5-3.0 kPa) and is expressed when the fruit are placed in ambient O_2 environments following low O_2 storage. As stated previously in this monograph, the oxidation of α-farnesene and the concomitant rise of its oxidative metabolites have been associated with scald since the late 1960s and early 1970s. In an oxygenic environment, α-farnesene rapidly breaks down to its CT degradation products (Anet, 1972a); however, the correlation of α-farnesene, CTs, and the incidence of scald among resistant and susceptible cultivars is too variable to consider the α-farnesene oxidation theory as the only plausible scald causal model.

What role does oxygen have in the ontogeny of superficial scald? Certainly O_2 is a critical requirement for the generation of the CTs, ketones, and alcohol radicals that have been identified during α-farnesene degradation (Anet, 1972c). In addition, the active forms of oxygen (singlet O_2, $^1O_2$; superoxide anion radical, $O_2 \cdot^-$; hydrogen peroxide, H_2O_2; hydroxyl radical, OH·) may contribute to, or be largely responsible for, the chemistry causing the cellular browning reactions. This active oxygen theory is particularly plausible in light of the hypodermal tissue being metabolically active and containing chloroplasts and mitochondria. In a dark storage environment, it is more likely that the mitochondria are the sites of active oxygen generation, as chilling-induced impairment of mitochondrial function has been demonstrated in maize (*Zea mays* L.) seedlings held for seven days at 4°C (Prasad et al., 1994). Stressed mitochondria generate AOS, which can initiate oxidative damage within the organelle or diffuse across membranes causing disruption in other intracellular locations (Purvis and Shewfelt, 1993; Shewfelt and Purvis, 1995; McLennan and Degli Esposti, 2000).

In a recent comparison between scald-resistant and scald-susceptible apple fruit, the former had lower H_2O_2 concentrations, less membrane and protein damage, along with higher activities of catalase at harvest and following 16 weeks of storage (Rao et al., 1998). Interestingly, α-farnesene and its CT oxidation products were poorly related to scald incidence. Rao et al. concluded that it is the ability of a particular cultivar to efficiently metabolize AOS which confers resistence to scald. Upon finding poor associations between measurements of α-farnesene synthase, α-farnesene, CT alcohols, and MHO content with scald susceptibility in 11 apple cultivars, Rupasinghe et al. (2000) reasoned that the catabolism of α-farnesene is a secondary event in scald development and that scald susceptibility is primarily due to the generation of more oxidative stress in the hypodermal cells than the antioxidant systems can effectively control. Also, Abdallah et al. (1997) observed that a wounded epidermal area on 'Granny Smith' did not develop scald symptoms, likely due, they surmised, to a measurable increase in cinnamic acid derivatives which are known to possess antioxidant activity (Pratt, 1992; Ziegler et al., 1995).

The assumption that superficial scald is inversely related to the level of endogenous antioxidant molecules in the hypodermal/peel tissue has been an integral part of some early α-farnesene oxidation models (Huelin and Coggiola, 1970b; Anet, 1974). Anet (1974) observed variable quantities of 11 antioxidants isolated from apple peels and concluded that scald does not occur if the antioxidant levels are high enough to inhibit the oxidation of α-farnesene. It has been observed that green epidermal areas manifest scald whereas red-colored regions do not—the implication being that anthocyanins or polyphenolic compounds impart greater antioxidant capacity compared with less colored areas (D'Souza et al., 1994; Ju et al., 1996). Barden and Bramlage (1994b) found that higher antioxidant concentrations increase resistance toward scald; however, scald expression is not associated with a loss of antioxidant activity during storage, nor is it linked to any one antioxidant molecule. Gallerani et al. (1990) reported that nonscalded tissue has more α-tocopherol than scalded tissue on the same fruit. In contrast, Barden and Bramlage (1994a) showed that α-tocopherol levels are not strongly related to scald resistance in 'Cortland' and 'Delicious' fruit.

A family of phenolic fatty acid esters having a UV absorbance maximum at 258 nm, initially thought to be novel isoflavone conjugates, was isolated from the epicuticular wax of scald-resistant 'Gala' and scald-susceptible

'Delicious', with levels being much higher in the former compared with the latter cultivar (Whitaker, 1998). The phenolic moieties of these compounds were recently identified as *cis* and *trans* isomers of *p*-coumaryl alcohol rather than isoflavones. Peel levels of these phenolic esters are 20-fold higher in 'Gala' compared with those in 'Delicious' and 'Granny Smith' fruit, but are undetectable in scald-resistant 'Empire' apples (Whitaker, personal communication, October 2000). The hypothesis that these endogenous lipophilic phenolic esters confer scald resistance via antioxidant function is plausible but likely not universal among scald-resistant cultivars.

A period of anoxia during storage can induce scaldlike hypodermal browning upon exposure of the fruit to ambient air and may involve some of the same biochemical mechanisms responsible for superficial scald development (Bauchot et al., 1999). Bauchot et al. (1999) speculated that O_2-induced postanoxic stress, mediated through active oxygen species, is responsible for the cellular browning and not the up-regulation of polyphenol oxidase (PPO). The reexposure of anoxic plant tissue to air has been shown to cause extensive cellular damage mediated through enhanced active oxygen production and ethanol conversion to acetaldehyde (Pfister-Sieber and Brändle, 1994; Crawford and Braendle, 1996). Whitaker and Saftner (2000) recently proposed that a weakened antioxidant system is part of the biochemical mechanism of superficial scald induction, which likely involves free-radical species. Whether strict anoxia or a critically low O_2 threshold in the storage environment disposes the hypodermis toward scald expression through anaerobic metabolism or subsequent oxidative stress is unknown, but, if resolved, it would provide valuable insight into scald chemistry.

ETHYLENE AND SCALD

For at least 50 years the relationship between ethylene and scald development in apples has been investigated. Findings are often conflicting, with data generally demonstrating either a positive (Fidler and North, 1969; Liu, 1977; Knee and Hatfield, 1981; Little et al., 1985) or absence of (Fidler, 1950; Patterson and Workman, 1962; Lau, 1990) association. To date, little biochemical evidence has been found for a direct ethylene role in the oxidation of α-farnesene to CTs or to other putative scald-inducting volatiles, or for a direct role of ethylene in other scald causal models. The observation that ethylene and α-farnesene production increase concomitantly may be simply due to increased metabolic rates or to the stage of fruit ripening/senescence. Whether its influence is direct or indirect, ethylene likely plays an important role in scald ontogeny.

The use of ethephon ([2-chloroethyl] phosphonic acid) to study scald development is based on the premise that the release of ethylene advances fruit maturity and thus reduces scald expression (Couey and Williams, 1973), although the mechanism of action is not known. Such studies often present contrasting results: ethephon-treated fruit exhibit reduced (Lurie at al., 1989; Curry, 1994; Ju and Bramlage, 2000), increased (Greene et al., 1974; Windus and Shutak, 1977), or relatively unchanged (Watkins et al., 1993; Barden and Bramlage, 1994b) scald incidence compared with untreated apples.

Du and Bramlage (1994) forwarded an ethylene-mediated, α-farnesene oxidation/scald model in which ethylene has a rapid or delayed effect—the former favoring the accumulation of scald-inducing CTs (CT281), the latter favoring the accumulation of scald-inhibiting CTs (CT258). The development of scald symptoms is thus contingent upon the balance of the rapid and delayed ethylene-mediated responses as influenced by fruit maturity, growing conditions, or pre- and postharvest treatments such as DPA application.

Reduced α-farnesene production following aminoethoxyvinylglycine (AVG) treatment on 'Jonagold', 'Delicious', and 'Granny Smith' fruit indicate that inhibition of ethylene synthesis results in a reduction of scald-inducing volatile production (Mir, Perez, Schwallier, et. al, 1999; Ju and Curry, 2000a). Fan et al. (1999) demonstrated ethylene influence in scald ontogeny as the ethylene action inhibitor 1-methylcyclopropene (1-MCP) suppresses scald development as well as respiration and MHO in 'Granny Smith' and 'Red Chief Delicious' apples following long-term RA storage. Similarly, Watkins, Nock, et al. (2000) observed that 1-MCP reduces ethylene, α-farnesene, and conjugated trienol production as well as scald incidence in 'Law Rome' fruit during 24 to 32 weeks of storage. Through the use of inducers and inhibitors of ethylene synthesis, α-farnesene biosynthetic precursors, and transcription and translation inhibitors, Ju and Curry (2000a) concluded that initiation of α-farnesene biosynthesis requires ethylene function and involves gene expression and de novo enzyme synthesis. These recent findings suggest a direct role for ethylene in scald development. In addition, the hypothesis that ethylene exerts regulatory control on the activity of a key mevalonic acid pathway enzyme, hydroxymethylglutaryl coenzyme-A reductase, needs to be thoroughly tested, as this enzyme appears to regulate α-farnesene synthesis during fruit ripening (Ju and Curry, 2000b).

POSTHARVEST SCALD TREATMENTS TO CONTROL OXIDATIVE STRESS

In recent years there has been a growing consumer preference for fewer chemical additives in food (Baker, 1996, 1999). Superficial scald has been largely controlled by one of two lipophilic antioxidant compounds—DPA and ethoxyquin—which are applied as prestorage drenches. Often, a fungicide is added to the drench solution to stop the proliferation of rot-inducing fungi while the fruit are held in storage. Although DPA has been recently re-registered in the United States for scald control on apples, ethoxyquin is not used in Europe and can be applied only on pears in two American states, although it is registered for both apples and pears in Canada (John Kelly, personal communication, October 2000). Hence, a large portion of the apple industry has only one or two possible chemical choices for scald control.

The future possibility of losing DPA has spurred a renewed quest for effective, alternative scald control methods. Success in scald reduction results from retarding the degree of hypodermal oxidation through the use of the lowest possible O_2 partial pressures during long-term storage, the application of fruit coatings that presumably create O_2 barriers, treatment with substances having antioxidant activity, and acclimation regimes that confer greater tissue resistance to temperature and O_2-related stresses.

Low O_2 Environments

The use of low O_2 concentrations in apple storage has become common industry practice since benefits to fruits and vegetables in altered gas atmospheres were observed during the 1920s and 1930s (Kidd and West, 1927; Laties et al., 1995; Kays, 1997). Besides reduction of rates of respiration and ethylene generation and reduced metabolic activity in general, low O_2 environments reduce scald incidence. A recent North American study shows that storing 'Starkrimson Delicious' at 0.7 kPa O_2 for six to eight months markedly suppresses scald incidence compared with air-stored fruit in most of the participating research labs (Lau et al., 1998). Dynamic O_2 storage regimes have been shown to more strongly inhibit scald development than static O_2 storage environments. With an initial CA regime of 0.5 kPa O_2 for 9 to 14 days followed by an O_2 partial pressure of 1.5 kPa for the remainder of the storage period, scald is strongly reduced in 'Granny Smith' apples compared with fruit held in a 3.5 kPa static O_2 environment (Little et al., 1982). Wang and Dilley (2000b) observed that an initial low oxygen stress (ILOS) of 0.25 or 0.50 kPa for 14 days at 0°C followed by a static O_2 regime of 1.5 kPa for eight months effectively controls scald in 'Law Rome' and 'Granny Smith' fruit without off-flavor development. They surmised that ethanol production and its subsequent metabolism during the ILOS period may have suppressed the activity of the biochemical mechanism required for scald expression—a view that has been corroborated by other investigators (Ghahramani and Scott, 1998; Ghahramani et al., 1999, 2000). Other research has shown that initial low O_2 treatments and ultra-low static O_2 storage environments do not markedly reduce scald expression (Truter et al., 1994;

Lau et al., 1998) as the stage of fruit maturity at harvest coupled with seasonal environmental influences overrides the antiscald benefits of low O2 storage atmospheres (Emonger et al., 1994; Bramlage and Weis, 1997).

Another low O_2 method for scald control is the establishment of a hypobaric environment by ventilating the storage area with air at less than atmospheric pressure. Wang and Dilley (2000a) found that if 'Law Rome' and 'Granny Smith' fruit were placed in a hypobaric atmosphere of 5 kPa total pressure within one month of harvest, scald does not develop after a seven-day, postharvest shelf life at 20°C. However, if hypobaric storage was delayed for three months following harvest, scald incidence was similar to air-stored fruit. They propose that hypobaric conditions remove scald-inducing volatile compounds, thus preventing their accumulation in the epicuticular wax. Their work also demonstrates that a low O_2 environment must be initiated within the first few weeks following harvest or scald development will not be suppressed. Although the hypobaric storage system is a valuable research tool for investigations where oxygen-related phenomena are being studied, it is largely impractical from an industrial application standpoint.

Antioxidants and Other Coatings

Some success in scald reduction for upward to four months in RA has been achieved by dipping apples in sucrose ester-based coatings formulated with food grade antioxidants; compared with DPA, the degree of scald control varies from similar to less effective (Bauchot et al., 1995; Bauchot and John, 1996). Canola, castor, cottonseed, corn, linseed, olive, palm, peanut, soybean, and purified petroleum oils have been shown to control scald in 'Delicious' and 'Granny Smith' fruit (Scott, Yuen, and Kim, 1995; Ju et al., 2000), likely by forming an O_2-limiting modified atmosphere on the epidermis. Interestingly, Ju et al. (2000) dipped 'Delicious' fruit in 6 or 9 percent emulsions of mono-, di-, or triacylglycerols or phospholipids and observed that scald is suppressed similarly to DPA following six months of RA storage. However, nonstripped oils containing α-tocopherol as well as α-tocopherol alone are not as effective as DPA in controlling scald, and they even enhance scald incidence. The potential alteration of taste and the impartation of a greasy epidermis would likely render oil-based fruit coatings unacceptable to the consumer, although in some cases taste and greasiness problems are not apparent following a lengthy period of storage (Ju et al., 2000).

Exogenous application of numerous antioxidant compounds have been tested over the years as potential in vivo scald inhibitors; most, except the amine type, have not proved feasible due to the lack of scald suppression, the potential residue toxicity, the impartation of off-flavors or injury to the fruit (Smock, 1957, 1961; Anet and Coggiola, 1974; Bauchot et al., 1995; Bauchot and John, 1996). An anti-scald antioxidant compound will not only have to control the disorder as well as DPA with no undesirable side effects, but also be a product that satisfies consumer and regulatory concerns for fewer chemical food additives. To date, no novel antioxidant treatment(s) meeting these requirements has been reported.

Alcohol Vapors

The use of alcoholic vapors or dips to reduce superficial scald has shown promise as an effective treatment (Chellew and Little, 1995; Scott, Yuen, and Ghahramani, 1995; Ghahramani et al., 2000). Ghahramani et al. (1999) exposed 'Granny Smith' apples to six organic alcohols—ethanol (2C), propanol (3C), butanol (4C), pentanol (5C), hexanol (6C), and α-terpineol (10C)—and observed that scald incidence is fully controlled by the two to five carbon species. A decline in α-farnesene and CT production, concomitant with scald reduction, suggests that the alcoholic vapors generally suppress metabolism or act as antioxidant molecules. In a subsequent study, however, pentanol was observed to cause severe skin injury resembling soft or deep scald, although it did control superficial scald (Ghahramani et al., 2000). Organic alcohols possess antioxidant properties (Gutteridge and Quinlan, 1993; Shen et al., 1997; Alvarez et al., 1998); ethanol vapors, via conversion to acetaldehyde (Beaulieu et al., 1997), inhibit fruit ripening (Ritenour et al., 1997) and suppress ethylene production (Saltveit, 1989; Wu et al., 1992; Ghahramani et al., 1999).

Fruit Warming and Heat Treatments

Heating or warming fruit prior to storage or as an intermittent storage treatment reduces scald severity (Hardenburg and Anderson, 1965; Lurie et al., 1990; Watkins et al., 1993; Klein and Lurie, 1994). Alwan and Watkins (1999) found that warming 'Cortland', 'Delicious', and 'Law Rome' apples to 20°C for 24 hours every one, two, and four weeks for 16 weeks, generally reduces scald incidence but also increases the rate of fruit softening. They hypothesized that warming reduces the accumulation of toxic, scald-related metabolites before cellular damage occurs or enhances the degradation of CTs. Watkins et al. (1995) observed that a 20°C warming period for five days following two to eight weeks of 0°C storage, reduces scald severity similarly to DPA in 'Granny Smith'. They speculated that warming ameliorates cellular chilling-type injury in the stored fruit by increasing: (a) endogenous antioxidants; (b) the degree of membrane fatty acid unsaturation; and/or (c) the catabolism of toxic substances that accumulate at lower temperatures. The scald-controlling benefit conferred by a warming period(s) must outweigh the degree of fruit softening, the expression of disorders that advanced senescence can cause and the cost of breaking and re-establishing a CA environment before warming period protocols become practically feasible (Watkins, Bramlage, et al., 2000).

Following a prestorage heat treatment of 38°C for four days, Lurie at al. (1990) observed inhibited scald development in 'Granny Smith' fruit as well as lower α-farnesene, CTs, and PPO levels after three, but not five, months of RA storage. They speculated that the heat treatment reversibly inhibits the enzymes involved in α-farnesene synthesis and oxidation as well as those responsible for cellular browning reactions. Heating 'Golden Delcious' fruit for four days at 38°C prior to storage causes pronounced changes in lipid metabolism, including a greater degree of fatty acid unsaturation, decreased electrolyte leakage, and microviscosity when examined four to five months following 0°C storage (Lurie et al., 1995; Whitaker, Klein, et al., 1997). These modifications in membrane chemistry are similar to those observed during tissue acclimation to low temperatures (Alberdi and Corcuera, 1991; Kratsch and Wise, 2000), suggesting that scald susceptibility or resistance is related to the ability of a cultivar to adapt to the low temperatures (0-3°C) experienced during extended storage periods (Bramlage and Meir, 1990). Hence, heating or warming the fruit prior to, or during storage, induces processes that lead to low temperature acclimation which may include the production of heat shock proteins (HSPs) (Lurie, 1998; Woolf et al., 2000). These proteins are a universal organismal response to elevated temperatures and other environmental stresses (Lee et al., 2000; O'Mahoney and Burke, 2000), although it is presently unknown what influence HSPs have on scald development and expression.

Calcium

Although not a novel control per se, repeated sprays of calcium chloride ($CaCl_2$) during the growing season or dips of $CaCl_2$ prior to storage show some success in controlling scald in apples (O'Loughlin and Jotic, 1978; Hardenburg and Anderson, 1980; Bramlage et al., 1985). In some cases, a $CaCl_2$ dip combined with heat treatments synergistically reduces scald severity (Klein and Lurie, 1994). Klein and Lurie (1994) hypothesize that calicum helps stabilize membranes against free-radical-induced lipid oxidation by binding to the membrane as permeability is altered during heating. Calcium ions aid in stabilizing membranes by binding to the phosphate and carboxylate groups of phospholipids at membrane surfaces and maintain cell wall rigidity by linking pectin molecules (Palta, 1996). Calcium treatments have also been shown to alter lipid metabolism and delay lipid catabolism in 'Golden Delicious' fruit (Picchioni et al., 1995, 1998).

One of the most effective methods for improving fruit tissue calcium levels is pressure infiltration in which the fruit are held in a $CaCl_2$ brine while a positive pressure is applied (Beavers et al., 1994; Conway et al., 1994). Although this method has successfully improved calcium levels in fruit tissue, the cost and inconvenience of using this technology industrially will likely limit its broad adoption. Repeated $CaCl_2$ sprays during the growing season, although less effective at increasing fruit calicum levels than postharvest pressure infiltration, have been shown to reduce superficial scald levels in 'McIntosh' (Bramlage et al., 1985) and 'Delicious' (O'Loughlin and Jotic, 1978) apples. The more traditional calcium spray application method also controls bitter pit, senescence, and internal browning disorders, and due to its low-cost and easy integration into orchard spray schedules, will likely remain the most popular calcium augmentation method for some time.

CONCLUSION AND FUTURE RESEARCH DIRECTION

Historically, the majority of superficial scald causal models indicate that the disorder is characterized by cellular oxidation within the hypodermis, although the definite biochemical pathway has not been elucidated. Conflicting views on the role of α-farnesene and its various oxidation products, juxtaposed with the theories that scald is a more generalized response to overwhelming oxidative stress, confound the presentation of a simple causal explanation. However, with the aid of recently developed or improved biochemical tools (e.g., the specific inhibitors of key enzymes of the mevalonic acid pathway, precise measurements of active oxygen species, and better indicators of the cellular damage they cause) and the ability to accurately control storage atmospheres, the next decade may see the development of a more unified scald causal model.

From a practical standpoint, the potential loss of DPA without a suitable replacement would mean substantial economic loss for the apple industry worldwide. Fortunately, with the reregistration of DPA in the United States, the loss of DPA will not likely occur for some time yet. Nonetheless, the search for a suitable alternative chemical or process that confers DPA-like scald control fuels much of present-day research. With the recent public controversy over genetically modified foods, genetic manipulation of susceptible cultivars to render them more scald resistant may not receive the highest research priority. Traditional plant breeding methods could be utilized to develop new scald-resistant cultivars; however, the long-term commitment for thorough testing of a new cultivar's commercial utility will need to be conducted alongside research oriented toward finding quicker solutions for the important scald-susceptible cultivars presently grown.

The two most pressing needs in scald research are (1) elucidation of the definitive biochemical cause of the disorder, and (2) development of acceptable pre- and postharvest practices and treatments that provide DPA-like scald suppression for the worldwide apple industry. Following identification of the biochemical causal agent(s) of scald, better predictive indices could be developed so that particularly susceptible fruit can be recognized while on the tree or immediately following harvest. Thus, the scald-potentiating influences of genetic susceptibility and growing season could be accurately gauged and the appropriate preventative action taken if a fruit lot were deemed likely to express the disorder. Knowing the immediate biochemical cause of scald would also allow researchers to conduct more relevant explorations for control at the cellular/pathway level. Perhaps inhibitors of key enzymes, which have no toxic or undesirable side effects, could be developed for blocking scald initiation at the pathway level.

The need to develop acceptable postharvest treatments that confer DPA-like control is particularly pressing. It is possible that initial low oxygen treatments, intermittent warming periods at room temperature, heat treatments prior to storage, and ultra-low O_2 regimes during long-term storage can markedly reduce scald incidence. However, the apple industry is not likely to adopt any atmosphere manipulation technique that requires a CA environment to be prematurely broken once the desired O_2 and CO_2 concentrations are attained. In addition, state-of-the-art CA technology will allow for an operator to confidently achieve storage-room O_2 concentrations down to approximately 1 kPa; prolonged control at 0.5-1.0 kPa or holding O_2 levels at 0.25-0.50 kPa even in the short term may not be possible in some CA rooms. Moreover, the potential for fermentation-related off-flavors and internal browning disorders at these ultra-low O_2 levels may make damage/risk factors unacceptable to storage operators. A prestorage dip or drench comprised of innocuous or even healthful compounds (e.g., antioxidants) which consistently, effectively, and affordably control scald is an ideal toward which practical research often gravitates. A scald-control dip foregoes the necessity of conditioning the fruit intermittently during storage and obviates the need to push CA systems beyond their limits and the apples toward potentially damaging ultra-low O_2 regimes. The low-input technology necessary for prestorage drenching is presently available in many fruit storage facilities; therefore, the introduction of a new dipping material makes no demands for new mechanical infrastructure in the warehouse. However, fungal infection of the fruit is a constant threat in any drenching and storage operation; the financial costs and the public perception issues that revolve around the use of fungicides, along with potential fungicidal resistance, must be considered as DPA alternatives are examined.

In short, it is an exciting time for scald research; public pressure for fewer chemical food additives may eventually toll the death knell for DPA usage, a possibility which has resulted in a renewed search for effective alternatives. Concomitant with efforts to find new chemical controls is an increase in basic and applied research resources being committed to the scald problem. Optimistically, this will lead to a clearer understanding of the causal biochemistry as well as refine the storage protocols for susceptible cultivars. The combination of these efforts will hopefully culminate in better control of the disorder thereby mitigating the economic devastation scald can cause.

REFERENCES

Abdallah, A.Y., M.I. Gil, W. Biasi, and E.J. Mitcham (1997). Inhibition of superficial scald in apples by wounding: Changes in lipis and phenolics. *Postharvest Biology and Technology* 12:203-212.

Alberdi, M. and L.J. Corcuera (1991). Cold acclimation in plants. *Phytochemistry* 30:3177-3184.

Alvarez, B., G. Ferrer-Sueta, and R. Radi (1998). Slowing of peroxynitrite decomposition in the presence of mannitol and ethanol. *Free Radical Medicine Biology* 24:1331-1337.

Alwan, T.F. and C.B. Watkins (1999). Intermittent warming effects on superficial scald development in 'Cortland', 'Delicious' and 'Law Rome' apple fruit. *Postharvest Biology and Technology* 16:203-212.

Anderson, R.A. (1967). Experimental storage of eastern-grown 'Delicious' apples in various controlled atmospheres. *Proceedings of the American Society for Horticultural Science* 91:810-820.

Anet, E.F.L.J. (1972a). Superficial scald, a function disorder of stored apples. VIII. Volatile products form the autooxidation of α-farnesene. *Journal of the Science of Food and Agriculture* 23:605-608.

Anet, E.F.L.J. (1972b). Superficial scald, a function disorder of stored apples. IX. Effect of maturity and ventilation. *Journal of the Science of Food and Agriculture* 23:763-769.

Anet, E.F.L.J. (1972c). Superficial scald, a function disorder of stored apples. XI. Apple antioxidants. *Journal of the Science of Food and Agriculture* 25:299-304.

Anet, E.F.L.J. (1974). Superficial scald, a functional disorder of stored apples. XI. Apple antioxidants. *Journal of the Science of Food and Agriculture* 25:299-304.

Anet, E.F.L.J. and I.M. Coggiola (1974). Superficial scald, a functional disorder of stored apples. X. Control of α-farnesene autoxidation. *Journal of the Science of Food and Agriculture* 25:293-298.

Bain, J.M. (1956). A histological study of the development of superficial scald in Granny Smith apples. *Journal of Horticultural Science* 31:234-238.

Bain, J.M. and F.V. Mercer (1963). The submicroscopic cytology of superficial scald, a physiological disease of apples. *Australian Journal of Science* 16:442-449.

Baker, G.A. (1996). Strategic implications of consumer food safety preferences. *International Food and Agribusiness Management Review* 1:451-463.

Baker, G.A. (1999). Consumer preference for food safety attributes in fresh apples: Market segments, consumer characteristics, and marketing opportunities. *Journal of Agricultural and Resource Economics* 24:80-97.

Barden, C.L. and W.J. Bramlage (1994a). Acclimation of antioxidants in apple peel as related to preharvest factors and superficial scald susceptibility of the fruit. *Journal of the American Society for Horticultural Science* 119:264-269.

Barden, C.L. and W.J. Bramlage (1994b). Relationships of antioxidants in apple peel to changes in α-farnesene and conjugated trienes during storage and to superficial scald development after storage. *Postharvest Biology and Technology* 4:23-33.

Barden, C.L. and W.J. Bramlage (1994c). Separating the effects of low temperature, ripening, and light on loss of scald susceptibility in apples before harvest. *Journal of the American Society for Horticultural Science* 119:54-58.

Bauchot, A.D. and P. John (1996). Scald development and the levels of α-farnesene and conjugated triene hydroperoxides in apple peel after treatment with sucrose ester-based coatings in combination with food-approved antioxidants. *Postharvest Biology and Technology* 7:41-49.

Bauchot, A.D., P. John, Y. Soria, and I. Recasens (1995). Carbon dioxide, oxygen and ethylene changes in relation to the development of scald in Granny Smith apples after cold storage. *Journal of Agricultural Food and Chemistry* 43:3007-3011.

Bauchot, A.D., S.J. Reid, G.S. Ross, and D.M. Burmeister (1999). Induction of apple scald by anaerobiosis has similar characteristics to naturally occurring superficial scald in 'Granny Smith' apple fruit. *Postharvest Biology and Technology* 16:9-14.

Beaulieu, J.C., G. Peiser, and M.E. Saltveit (1997). Acetaldehyde is a causal agent responsible for ethanol-induced ripening inhibition in tomato fruit. *Plant Physiology* 113:431-439.

Beavers, W.B., C.E. Sams, W.S. Conway, and G.A. Brown (1994). Calcium source affects calcium content, firmness, and degree of injury of apples during storage. *HortScience* 29:1520-1523.

Bramlage, W.J., M. Drake, and S.A. Weis (1985). Comparisons of calcium chloride, calcium phosphate, and a calcium chelate as foliar sprays of 'McIntosh' apple tress. *Journal of the American Society for Horticultural Science* 110(6):786-789.

Bramlage, W.J. and S. Meir (1990). Chilling injury of crops of temperate origin. In *Chilling Injury of Horticultural Crops,* C.Y. Wang (Ed.). Boca Raton, FL: CRC Press, Inc. pp. 38-49.

Bramlage, W.J. and S.A. Weis (1997). Effects of temperature, light, and rainfall on superficial scald susceptibility in apples. *HortScience* 32:808-811.

Chellew, J.P and C.R. Little (1995). Alternative methods of scald control in 'Granny Smith' apples. *Journal of Horticultural Science* 70:109-115.

Couey, H.M. and M.X. Williams (1973). Preharvest application of ethephon on scald and quality of stored 'Delicious' apples. *HortScience* 8:56-57.

Conway, W.S., C.E. Sams, G.A. Brown, W.B. Beavers, R.B. Tobias, and L.S. Kennedy (1994). Pilot test for the commercial use of postharvest pressure infiltration of calcium into apples to maintain fruit quality in storage. *HortTechnology* 4:239-243.

Crawford, R.M.M. and R. Braendle (1996). Oxygen deprivation stress in a changing environment. *Journal of Experimental Botany* 47:145-159.

Curry, E.A. (1994). Preharvest applications of ethephon reduce superficial scald of 'Fuji' and 'Granny Smith' apples in storage. *Journal of Horticultural Science* 69:1111-1116.

D'Souza, M.C.D., M. Ingle, and S. Singha (1994). Superficial scald of 'Rome Beauty' apples after storage is correlated with chromaticity values at harvest. *HortTechnology* 4:144-146.

Du, Z. and W.J. Bramlage (1993). A modified hypothesis on the role of conjugated trienes in superficial scald development on stored apples. *Journal of the American Society for Horticultural Science* 118:807-813.

Du, Z. and W.J. Bramlage (1994). Roles of ethylene in the development of superficial scald in 'Cortland' apples. *Journal of the American Society for Horticultural Science* 119:516-523.

Emonger, V.E., D.P. Murr, and E.C. Lougheed (1994). Preharvest factors that predispose apples to superficial scald. *Postharvest Biology and Technology* 4:289-300.

Fan, X., J.P. Mattheis, and S. Blankenship (1999). Development of apple superficial scald, soft scald, core flush, and greasiness is reduced by MCP. *Journal of Agricultural Food and Chemistry* 47:3063-3068.

Fidler, J.C. (1950). Studies of the physiologically-active volatile organic compounds produced by fruits. II. The rate of production of carbon dioxide and of volatile organic compounds by King Edward VII apples in gas storage, and the effect of removal of the volatiles from the atmosphere of the store on the incidence of superficial scald. *Journal of Horticultural Science* 25:81-110.

Fidler, J.C. and C.J. North (1969). Production of volatile organic compounds by apples. *Journal of the Science of Food and Agriculture* 20:521-526.

Filmer, A.A.E. and D.F. Meigh (1971). Natural coating of the apple and its influence on scald in storage. IV. Oxidation products of α-farnesene. *Journal of the Science of Food and Agriculture* 22:188-190.

Gallerani, G., G.C. Pratella, and R.A. Budini (1990). The distribution and role of natural antioxidant substances in apple fruit affected by superficial scald. *Advances in Horticultural Science* 4:144-146.

Ghahramani, F. and K.J. Scott (1998). Oxygen stress of 'Granny Smith' apples in relation to superficial scald, ethanol, α-farnesene and conjugated trienes. *Australian Journal of Agricultural Research* 49:207-210.

Ghahramani, F., K.J. Scott, K.A. Buckle, and J.E. Paton (1999). A comparison of the effects of ethanol and higher alcohols for the control of superficial scald in apples. *Journal of Horticultural Science and Biotechnology* 74:87-93.

Ghahramani, F., K.J. Scott and R. Holmes (2000). Effects of alcohol vapors and oxygen stress on superficial scald and red color of stored 'Delicious' apples. *HortScience* 35:1292-1293.

Greene, D.W., W.J. Lord, W.J. Bramlage, and F.W. Southwick (1974). Effects of low ethephon concentrations on quality of 'McIntosh' apples. *Journal of the American Society for Horticultural Science* 99:239-242.

Gutteridge, J.M.C. and G.J. Quinlan (1993). Antioxidant protection against organic and inorganic oxygen radicals by normal human plasma: The important primary role for iron-binding and iron-oxidizing proteins. *Biochimica et Biophysica Acta* 1156:144-150.

Hardenburg, R.E. and R.E. Anderson (1965). Postharvest chemical, hot water and packaging treatments to control apple scald. *Proceedings of the American Society for Horticultural Science* 87:93-99.

Hardenburg, R.E. and R.E. Anderson (1980). Keeping qualities of 'Stayman' and 'Delicious' apples treated with calcium chloride, scald inhibitors and other chemicals. *HortScience* 15:425.

Huelin, F.E. and I.M. Coggiola (1970a). Superficial scald, a functional disorder of stored apples. V. Oxidation of α-farnesene and its inhibition by diphenylamine. *Journal of the Science of Food and Agriculture* 21:44-48.

Huelin, F.E. and I.M. Coggiola (1970b). Superficial scald, a functional disorder of stored apples. VII. Effect of applied α-farnesene, temperature and diphenylamine on scald and the concentration and oxidation of α-farnesene in the fruit. *Journal of the Science of Food and Agriculture* 21:584-589.

Ingle, M. and M.C. D'Souza (1989). Physiology and control of superficial scald of apples: A review. *HortScience* 24:28-31.

Ju, Z. and W.J. Bramlage (2000). Cuticular phenolics and scald development in 'Delicious' apples. *Journal of the American Society for Horticultural Science* 125:498-504.

Ju, Z. and E.A. Curry (2000a). Evidence that α-farnesene biosynthesis during fruit ripening is mediated by ethylene regulated gene expression in apples. *Postharvest Biology and Technology* 19:9-16.

Ju, Z. and E.A. Curry (2000b). Lovastatin inhibits α-farnesene biosynthesis and scald development in 'Delicious' and 'Granny Smith' apples and 'd'Anjou' pears. *Journal of the American Society for Horticultural Science* 125:626-629.

Ju, Z., Y. Duan, and Z. Ju (2000). Mono-, di-, and tri-acylglycerols and phosopholipids from plant oils inhibit scald development in 'Delicious' apples. *Postharvest Biology and Technology* 19:1-7.

Ju, Z., Y. Yuan, C. Liu, S. Zhan, and M. Wang (1996). Relationships among simple phenol, flavonoid and anthocyanin in apple fruit peel at harvest and scald susceptibility. *Postharvest Biology and Technology* 8:83-93.

Kays, S.J. (1997). Science and practise of postharvest plant physiology. In S.J. Kays, *Postharvest Physiology of Perishable Plant Products*. Athens, GA: Exon Press, pp. 1-22.

Kidd, F. and C. West (1927). A relation between the concentration of oxygen and carbon dioxide in the atmosphere, rate of respiration, and length of storage of apples. In *Food Investigation Board Report London for 1925, 1926*. pp 41-42.

Klein, J.D. and S. Lurie (1994). Time, temperature, and calcium interact in scald reduction and firmness retention in heated apples. *HortScience* 29:194-195.

Knee, M. and S.G.S. Hatfield (1981). Benefits of ethylene removal during apple storage. *Annals of Applied Biology* 98:157-165.

Kratsch, H.A. and R.R. Wise (2000). The ultrastructure of chilling stress. *Plant, Cell and Environment* 23:337-350.

Laties, G.F., Kidd, C. West, and F.F. Blackman (1995). The start of modern postharvest physiology. *Postharvest Biology and Technology* 5:1-10.

Lau, O.L. (1990). Efficacy of diphenylamine, ultra-low oxygen, and ethylene scrubbing on scald control in 'Delicious' apples. *Journal of the American Society for Horticultural Science* 115:959-961.

Lau, O.L., C.L. Barden, S.M. Blankenship, P.M. Chen, E.A. Currry, J.R. DeEll, L. Lehman-Salada, E.J. Mitcham, R.K. Prange, and C.B. Watkins (1998). A North American cooperative survey of 'Starkrimson Delicious' apple responses to 0.7 percent O_2 storage on superficial scald and other disorders. *Postharvest Biology and Technology* 13:19-26.

Lee, B., S. Won, H. Lee, M. Miyao, W. Chung, I. Kim, and J. Jo (2000). Expression of the chloroplast-localized small heat shock protein by oxidative stress in rice. *Gene* 245:283-290.

Little, C.R., J.D. Faragher, and H.J. Taylor (1982). Effects of initial oxygen stress treatments in low oxygen modified atmosphere storage of 'Granny Smith' apple. *Journal of the American Society for Horticultural Science* 107:320-323.

Little, C.R., H.J. Taylor, and F. MacFarlene (1985). Postharvest and storage factors affecting superficial scald and core flush of 'Granny Smith' apples. *HortScience* 20:1080-1082.

Liu, F.W. (1977). Varietal and maturity differences of apples in response to ethylene in controlled atmosphere storage. *Journal of the American Society for Horticultural Science* 102:93-95.

Lurie, S. (1998). Postharvest heat treatments. *Postharvest Biology and Technology* 14:257-269.

Lurie, S., J.D. Klein, and R. Ben Arie (1990). Postharvest heat treatment as a possible means of reducing superficial scald of apples. *Journal of Horticultural Science* 65:503-509.

Lurie, S., S. Meir, and R. Ben Arie (1989). Preharvest ethephen sprays reduce superficial scald of 'Granny Smith' apples. *HortScience* 24:104-106.

Lurie, S., S. Othman, and A. Borochov (1995). Effects of heat treatment on plasma membrane of apple fruit. *Postharvest Biology and Technology* 5:29-38.

McLennan, H. and M. Degli Esposti (2000). The contribution of mitochondrial respiratory complexes to the production of reactive oxygen species. *Journal of Bioenergetics and Biomembranes* 32:153-162.

Meheriuk, M., R.K. Prange, P.D. Lidster, and S.W. Porritt (1994). Postharvest disorders of apples and pears. *Agriculture Canada Publication 1737/E.*

Meir, S. and W.J. Bramlage (1988). Antioxidant activity in 'Cortland' apple peel and susceptibility to superficial scald after storage. *Journal of the American Society for Horticultural Science* 113:412-418.

Mir, N.A., R. Perez, and R. Beaudry (1999). A poststorage burst of 6-methyl-5-hepten-2-one (MHO) may be related to superficial scald development in 'Cortland' apples. *Journal of the American Society for Horticultural Science* 124:173-176.

Mir, N.A., R. Perez, P. Schwallier, and R. Beaudry (1999). Relationship between ethylene response manipulation and volatile production in Jonagold variety apples. *Journal of Agricultural Food and Chemistry* 47:2653-2659.

Murray, K.E., F.E. Huelin, and J.B. Davenport (1964). Occurrence of farnesene in the natural coating of apples. *Nature* 204:80.

O'Loughlin, J.B. and P. Jotic (1978). The relative effects of rootstocks and calcium sprays on the appearance of internal breakdown and superficial scald of 'Red Delicious' apples during storage. *Scientia Horticulturae* 9:245-249.

O'Mahoney, P. and J. Burke (2000). A ditelosomic line of 'Chinese Spring' wheat with augmented acquired thermotolerance. *Plant Science* 158:147-154.

Palta, J. (1996). Role of calcium in plant responses to stresses: Linking basic research to the solution of practical problems. *HortScience* 31:51-57.

Patterson, M.E. and M. Workman (1962). The influence of oxygen and carbon dioxide on the development of apple scald. *Proceedings of the American Society for Horticultural Science* 80:130-136.

Pfister-Sieber, M. and R. Brändle (1994). Aspects of plant behavior under anoxia and post-anoxia. *Proceedings of the Royal Society of Edinburgh* 102B:313-324.

Picchioni, G.A., A.E. Watada, W.S. Conway, B.D. Whitaker, and C.E. Sams (1995). Phospholipid, galactolipid and steryl lipid composition of apple fruit cortical tissue following postharvest $CaCl_2$ infiltration. *Phytochemistry* 39:763-769.

Picchioni, G.A., A.E. Watada, W.S. Conway, B.D. Whitaker, and C.E. Sams (1998). Postharvest calcium infiltration delays membrane lipid catabolism in aple fruit. *Journal of Agricultural and Food Chemistry* 46:2452-2457.

Prange, R.K., J.M. DeLong, and P.A. Harrison (1998). 1998 recommended storage conditions for Nova Scotia apples. *Storage Notes* 5(1) (September 16): 3.

Prasad, T.K., M.D. Anderson, and C.R. Stewart (1994). Acclimation, hydrogen peroxide, and abscisic acid protect mitochondria against irreversible chilling injury in maize seedlings. *Plant Physiology* 105:619-627.

Pratt, D.E. (1992). Natural antioxidants from plants. *American Chemical Society Symposium Series* 507:54-71.

Purvis, A.C. and R.L. Shewfelt (1993). Does the alternative pathway ameliorate chilling injury in sensitive plant tissue? *Physiologia Plantarum* 88:712-718.

Raese, J.T. and S.R. Drake (1993). Nitrogen fertilization and elemental composition affects fruit quality of 'Fuji' apples. *Journal of Plant Nutrition* 20:1797-1809.

Rao, M.V., C.B. Watkins, S.K. Brown, and N.F. Weeden (1998). Active oxygen species metabolism in 'White Angel' × 'Rome Beauty' apple selections resistant and susceptible to superficial scald. *Journal of the American Society for Horticultural Science* 123:299-304.

Ritenour, M.A., M.E. Mangrich, J.C. Beaulieu, A. Rab, and M.E. Saltveit (1997). Ethanol effects on the ripening of climacteric fruit. *Postharvest Biology and Technology* 12:35-42.

Romaniuk, J., Z. Soczek, and J. Machnik (1981). Influence of Alar (succinic acid 2,2-dimethyl hydrazide), nitrogen fertilization and the picking date on the incidence of storage diseases and the climacteric of McIntosh apples. *Fruit Science Reports* 8:163-172.

Rowan, D.D., J.M. Allen, S. Fielder, J.A. Spicer, and M.A. Brimble (1995). Identification of conjugated triene oxidation products of α-farnesene in apple skins. *Journal of Agricultural Food and Chemistry* 43:2040-2045.

Rupasinghe, H.P.V., G. Paliyath, and D.P. Murr (2000). Sesquiterpene α-farnesene synthase: Partial purification, characterization, and activity in relation to superficial scald development in apples. *Journal of the American Society for Horticultural Science* 125:111-119.

Saltveit, M.E. (1989). Effects of alcohols and their interaction with ethylene on the ripening of epidermal pericarp discs of tomato fruit. *Plant Physiology* 90:167-174.

Scott, K.J., C.M.C. Yuen, and F. Ghahramani (1995). Ethanol vapor—A new antiscald treatment for apples. *Postharvest Biology and Technology* 6:201-208.

Scott, K.J., C.M.C. Yuen, and G.H. Kim (1995). Reduction of superficial scald of apples with vegetable oils. *Postharvest Biology and Technology* 6:219-223.

Shen, B., R.G. Jensen, and H.J. Bohnert (1997). Mannitol protects against oxidation by hydroxyl radicals. *Plant Physiology* 115:527-532.

Shewfelt, R.L. and A.C. Purvis (1995). Toward a comprehensive model for lipid peroxidation in plant tissue disorders. *HortScience* 30:213-218.

Smock, R.M. (1957). A comparison of treatments for control of the apple scald disease. *Proceedings of the American Society for Horticultural Science* 69:91-100.

Smock, R.M. (1961). Methods of scald control on the apple. *Cornell University Agricultural Experimental Station Bulletin* No. 970.

Song, J. and R.M. Beaudry (1996). Rethinking apple scald: New hypothesis on the causal reason for development of scald in apples. *HortScience* 31:605.

Thomai, T., E. Sfakiotakis, G. Diamantidis, and M. Vasilakakis (1998). Effects of low preharvest temperature on scald susceptibility and biochemical changes in 'Granny Smith' apple peel. *Scientia Horticulturae* 76:1-15.

Truter, A.B., J.C. Combrink, and S.A. Burger (1994). Control of superficial scald in 'Granny Smith' apples by ultra-low and stress levels of oxygen as an alternative to diphenylamine. *Journal of Horticultural Science* 69:581-587.

Wang, Z. and D. Dilley (1999). Control of superficial scald of apples by low-oxygen atmospheres. *HortScience* 34:1145-1151.

Wang, Z. and D. R. Dilley (2000a). Hypobaric storage removes scald-related volatiles during the low temperature induction of superficial scald of apples. *Postharvest Biology and Technology* 18:191-199.

Wang, Z. and D. R. Dilley (2000b). Initial low oxygen stress controls superficial scald of apples. *Postharvest Biology and Technology* 18:201-213.

Watkins, C.B., C.L. Barden, and W.J. Bramlage (1993). Relationships between α-farnesene, ethylene production and superficial scald development of apples. *Acta Horticulturae* 343:155-160.

Watkins, C.B., W.J. Bramlage, P.L. Brookfield, S.J. Reid, S.A. Weis, and T.F. Alwan (2000). Cultivar and growing region influence efficacy of warming treatments for amelioration of superficial scald development on apples after storage. *Postharvest Biology and Technology* 19:33-45.

Watkins, C.B., W.J. Bramlage, and B.A. Cregoe (1995). Superficial scald of 'Granny Smith' apples is expressed as a typical chilling injury. *Journal of the American Society for Horticultural Science* 120:88-94.

Watkins, C.B., J.F. Nock, and B.D. Whitaker (2000). Responses of early, mid and late season apple cultivars to postharvest application of 1-methylcyclopropene (1-MCP) under air and controlled atmosphere storage conditions. *Postharvest Biology and Technology* 19:17-32.

Weis, S.A., W.J. Bramlage, and W.J. Lord (1998). An easy and reliable procedure for predicting scald and DPA requirement for New England Delicious apples. *Fruit Notes* 63:1-8.

Weis, S.A., W.J. Bramlage, and W.J. Lord (1999). Characteristics of scald susceptibility and development on Cortland apples in New England. *Fruit Notes* 64:1-5.

Whitaker, B.D. (1998). Phenolic fatty-acid esters from the peel of 'Gala' apples and their possible role in resistance to superficial scald. *Postharvest Biology and Technology* 13:1-10.

Whitaker, B.D., J.D. Klein, W.S. Conway, and C.E. Sams (1997). Influence of prestorage heat and calcium treatments on lipid metabolism in 'Golden Delicious' apples. *Phytochemistry* 45:465-472.

Whitaker, B.D., J.F. Nock, and C.B. Watkins (2000). Peel tissue α-farnesene and conjugated trienol concentrations during storage of 'White Angel' × 'Rome Beauty' hybrid apple selections susceptible and resistant to superficial scald. *Postharvest Biology and Technology* 20:231-241.

Whitaker, B.D. and R.A. Saftner (2000). Temperature-dependent autooxidation of conjugated trienols from apple peel yields 6-methyl-5-hepten-2-one, a volatile implicated in scald induction. *Journal of Agricultural and Food Chemistry* 48:2040-2043.

Whitaker, B.D., T. Solomos, and D.J. Harrison (1997). Quantification of α-farnesene and its conjugated trienol oxidation products from apple peel by C_{18}-HPLC with UV detection. *Journal of Agricultural and Food Chemistry* 45:760-765.

Windus, N. and V. Shutak (1977). Effect of ethephon, diphenylamine and diaminozide on the incidence of scald development on 'Cortland' apples. *Journal of the American Society for Horticultural Science* 102:715-718.

Woolf, A.B., A. Wexler, D. Prusky, E. Kobiler, and S. Lurie (2000). Direct sunlight influences postharvest temperature responses and ripening of five avocado cultivars. *Journal of the American Society for Horticultural Science* 125:370-376.

Wu, M.J., L. Zacarias, M.E. Saltveit, and M.S. Reid (1992). Alcohols and carnation senescence. *HortScience* 27:136-138.

Ziegler, G.C.H., R. Volpert, W. Osswald, and E.F. Elstner (1995). Effects of cinnamic acid derivatives on indole acetic acid oxidation by peroxidase. *Phytochemisty* 38:19-22.

Chapter 6

Oxidative Stress Affecting Fruit Senescence

Gene Edward Lester

INTRODUCTION

Economic Impact of Fruit Senescence

The demand for fresh fruits and vegetables in the United States and abroad has increased steadily in the past decade, and this trend is projected to continue well into this new century as more and more people become aware of the wellness benefits of fresh produce (Shewfelt and Prussia, 1993). Fresh produce is highly perishable and subject to extensive postharvest losses through microbial decay, physical injury, and senescence during handling and marketing. The monetary value of fresh produce more than doubles between the time it is harvested and the time it is sold at the retail level due to processing, storage, distribution, and marketing costs (Kays, 1991). Postharvest losses due to senescence and other storage, handling, and shipping processes are, therefore, substantially more costly than losses that occur during production. Maintaining and enhancing postharvest quality and reducing postharvest senescence of fresh fruits and vegetables improves their market availability and contributes toward meeting domestic and export demands for high-quality fresh produce, as well as augmenting the health of the world's population.

Defining Fruit Senescence

Fruit senescence involves lipid peroxidation, often the earliest detectable process in the loss of membrane integrity leading to membrane leakage, ultimately resulting in cellular or organ death. Fruit senescence can be differentiated from fruit ripening, which involves developmental changes in color, firmness, sugar, flavor, and proteins, but no loss of membrane integrity (Noodén, 1988). Differentiating the metabolic changes that occur when fruits are senescing versus ripening may be difficult as cells within individual fruit tissues can ripen and/or senesce at different times. For example, netted muskmelon and honey dew fruits begin senescing in the inner-mesocarp tissues surrounding the seed cavity, and then senescence progresses outward toward the epidermis (Lester, 1988). The inner-mesocarp melon tissue becomes senescent seven days prior to adjacent middle-mesocarp tissue and 14 days prior to the outer hypodermal-mesocarp tissue. The stage between the onset of fruit tissue breakdown (i.e., the completion of ripening) and tissue death (i.e., the completion of senescence) can be greatly extended in comparison to the ripening process. Depending upon the fruit, it can take days, as in the case of cucumbers or tomatoes, or weeks to months, as in the case of squash or pumpkins (Ryall and Lipton, 1979). The prolonged period between fruit ripening and the onset of senescence can be referred to as fruit maturity, and occurs when most fruits are considered edible. The presence of this maturation period, which is also part of senescence, raises an important question: what is the initiation reaction of senescence? Defining the cellular initiation reactions of senescence and their possible molecular and genetic controls is a primary objective of all plant senescence research and is treated as such in this chapter.

Cellular Localization of Senescence

Generally, it has been thought that fruit senescence begins in the chloroplasts, resulting in color change. However, because isolated choroplasts can remain green, carrying on photosynthesis for more than three days longer than those in the plant, this may not be the starting point (Noodén, 1988). Thus, the initiation of senescence is probably localized somewhere within the cytoplasm (Dörnenburg and Davies, 1999; Paliyath and Droillard, 1992). Sorting out the enzyme changes and hormonal regulators of fruit senescence versus fruit ripening can be difficult. Many of the conditions used to describe fruit senescence, such as color change that occurs when chloroplasts change to chromoplasts (Gross, 1991), and fruit softening, which can be both cell wall (McCullum et al., 1989) and cell membrane modulations (Lester, 1996), may be attributed more to fruit ripening than senescence. Senescence, therefore, as affected by oxidative stress, is cytoplasmic in origin and involves microsomal membrane—initially those of the mitochondria (Monk et al., 1989).

Role of Cell Membranes in Fruit Senescence

Senescence can often be marked by the development of leakiness (i.e., loss of integrity) in cellular membranes (Dörnenburg and Davies, 1999). The physical and biochemical reactions leading to membrane deterioration can be measured by a decline in total phospholipid content and either a simultaneous slight increase or no change in neutral lipids such as sterols (Lester and Whitaker, 1996). Membrane deterioration leads to an increase in the ratios of sterols:phospholipids and saturated:unsaturated fatty acids and levels of free unsaturated fatty acids (especially linoleic C18:2 and linolenic C18:3) and peroxidized lipids (Paliyath and Droillard, 1992). During senescence, as compared to ripening, the membrane lipid alterations proceed to the point where the system cannot revert to homeostasis, but instead proceeds to complete deterioration of the membrane (Leshem, 1988)—homeostasis being defined as the ability of the organism to actively repair or replace membrane deterioration events. The membrane deterioration process, termed *enzymatic senescence,* is affected directly by enzymes such as phospholipase-D (EC 3.1.4.4) and lipoxygenase (EC 1.13.11.12), and the process termed *nonenzymatic senescence* is affected by reactive oxygen species or free radicals, and both are coincident with a decline in antioxidants (such as carotenoids, α-tocopherol, superoxide dismutase [EC 1.15.1.1] and catalase [EC 1.11.1.6]) (Leshem, 1988).

To date, little direct research has been published on oxidative stress of fresh fruits and its impact on fruit senescence, as compared to oxidative stress of vegetables, leaves, fats, and oils. Two excellent recent reviews discuss in detail the association of oxidative stress and lipid oxidation in postharvest senescing leaves and vegetables (Dörnenburg and Davies, 1999; Buchanan-Wollaston, 1997). This chapter will, thus, focus on nonchlorophilic membranes, which are abundant in postharvest fruits, rather than the chloroplastic membranes prevalent in leafy vegetables, and which can have a different molecular biology (Buchanan-Wollaston, 1997) than fruit microsomal membranes (Leshem, 1992).

Autocatalytic Lipid Degradation

Paliyath and Droillard (1992), in their review on the mechanisms of membrane deterioration, outlined the autocatalytic lipid degradation pathway (or enzymatic senescence pathway) common in all higher plants. This pathway "begins" when membrane phospholipids are converted by phospholipase-D (which is stimulated by elevated levels of cytoplasmic Ca^{2+} and in some fruit by ethylene) to phosphatidic acid (Figure 6.1). Phosphatidic acid is immediately converted by phosphatidic acid phosphatase to produce diacylglycerols. Diacylglycerols are deacylated by lipolytic acyl hydrolases to produce free fatty acids, particularly unsaturated fatty acids with 1,4-pentadiene configurations (i.e., C18:2 and C18:3). These two unsaturated fatty acids are the specific substrates, along with oxygen, for lipooxygenase (Lester, 1990) to produce fatty acid hydroperoxides. Fatty

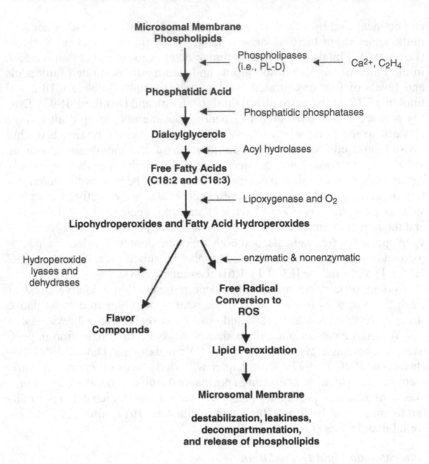

FIGURE 6.1. Schematic of the autocatalytic lipid degradation or enzymatic senescence pathway.

acid hydroperoxides are reactive and undergo various nonenzymatic and enzymatic reactions including the generation of free radicals. Active oxygen species (AOS) involved in oxidative damage are: superoxide ion ($O_2 \cdot^-$), hydroperoxyl radical (perhydroxy) ($HO_2 \cdot$), hydroxyl radical ($OH \cdot$), alkoxy radical ($RO \cdot$), hydroperoxy radical ($ROO \cdot$), organic hydroperoxide (ROOH), singlet oxygen ($^1O_2$), and excited carbonyl ($RO \cdot$) ions. When fatty acid hydroperoxides are acted on by hydroperoxide lyase enzymes, they are converted to volatile aldehydes, such as hexanal and hexenals (Vick and Zimmerman, 1989). These volatile compounds provide many of the fruit aroma and flavor attributes so highly desired by consumers of ripened or mature, edible fruits. Fatty acids also are broken down by α-oxidation during senescence, generating CO_2 and fatty aldehydes or alkanes (Paliyath and Droillard, 1992). As a result, there is a 50 percent decline in membrane phospholipid content in autocatalytic lipid degradation during senescence, producing a vast number of reactive intermediates, some of which cause oxidative stress. An interesting aspect of the autocatalytic lipid degradation pathway is that these same enzymes are also required during normal fruit growth, development, and ripening when membrane biosynthesis is active, and when catabolic products of membrane components are recycled to regenerate the membrane (i.e., homeostasis).

During fruit maturation and senescence, regeneration of the membrane declines and catabolism continues virtually unchecked. The biochemical reactions surrounding membrane deterioration, although well detailed in the autocatalytic lipid degradation pathway, do not elucidate what causes the pathway to change from catabolism and recycling to catabolism only. To date, the initiation of unchecked membrane catabolism that disrupts membrane regeneration is unclear. It is highly likely that initiation involves the decline in Ca^{2+}-ATPase (EC 3.6.1.38) and H^+-ATPase (EC 3.6.1.36) activities (Lester et al., 1998) in association with a decline in ion selective membrane permeability (i.e., loss of integrity). An increase in cytoplasmic Ca^{2+} concentration and acidity (i.e., H^+ concentration) results. Oxidative stress-induced membrane lipid peroxidation has been shown to inhibit Ca^{2+}-ATPase activity, which functions solely to rid the cytoplasm of excessive Ca^{2+} (Evans et al., 1991). A combination of excessively high (within the cytoplasm) or excessively low (outside the plasma membrane) levels of Ca^{2+} and the resultant inhibition of Ca^{2+}-ATPase activity most likely accelerates the unregulated autocatalytic deteriorative process. Compartmentation and regulation of ions, especially Ca^{2+} and H^+, are important to membrane homeostasis (Paliyath and Droillard, 1992). Under homeostatic conditions, Ca^{2+} is maintained at high nM to low μM levels and H^+ is maintained near neutral pH in the cytoplasm by active ATPase pumping of Ca^{2+} and H^+ into the vacuole or across the plasma membrane to the apoplast. A change in the concentration of either ion can be deleterious to the cell by affecting ATPase selective permeability functions, provoking membrane leakage. For example, artificial calcium starvation of cucumber root apoplasts leads to reduced Ca^{2+}-ATPase activity and increased cytoplasmic Ca^{2+} concentration, eventually resulting in the heightened production of phosphatidic acid (Yapa et al., 1986). Acidification of the cytoplasm also occurs inhibiting H^+-ATPase pumping, further lowering cellular pH; and low cytoplasmic pH, along with heightened Ca^{2+} concentration, are stimulatory to phospholipase-D activity—the initiating enzyme in the autocatalytic degradation pathway.

Natural calcium starvation occurs in postharvest melon fruits and is directly associated with a loss in plasma membrane integrity due to peroxidation. In postharvest honey dew and netted muskmelon fruits, calcium migrates from outside of the plasma membrane of the outer mesocarp layer, near the peel, to the seeds (Lester and Grusak, 1999). Replacing the calcium-starved outer mesocarp tissue (the region that regulates melon postharvest shelf life) with exogenous amino acid chelated calcium retards hypodermal-mesocarp plasma membrane senescence by reducing phospholipase-D perturbation of the membrane, which results in maintaining H^+-ATPase pumping activity. Maintaining active H^+-ATPase pumping in melon fruit reduces a buildup of cytoplasmic acidity, which subsequently maintains a low free sterol to phospholipid ratio in the plasma membrane (i.e., maintaining membrane integrity) and keeps the peroxidative enzyme (i.e., lipoxygenase) activities low (Lester and Grusak, 1999). A question arises, however, as to what might initiate the natural decline of both Ca^{2+}- and H^+-ATPase pumping systems in melons or other fruits. Ethylene may be involved and, indeed, it was shown to occur after ethylene treatment of carnation flowers (Sylvestre et al., 1989). But what initiates unchecked catabolism in fruits that do not emit ethylene? Oxidative stress via AOS is the likely agent.

OXIDATIVE STRESS

Nature of Oxidative Stress

Oxidative stress is hypothesized to be a direct agent in the alteration of the autocatalytic lipid degradation pathway. Superoxide and H_2O_2 can disrupt Ca^{2+}-ATPase activity, causing a prolonged release of calcium and elevation of cytoplasmic calcium content (Paliyath and Droillard, 1989). The elevated cytoplasmic calcium content promotes decompartmentation, causing an increase in cytoplasmic hydrogen ion concentration, and this cytoplasmic acidity stimulates phospholipase-D activity, promoting phosphatidic acid formation. Both $O_2{}^{\cdot-}$ and H_2O_2, if not catabolized, will eventually be transformed into active oxygen species (AOS) that can induce lipid peroxidation and accelerate the deteriorative cycle (AOS perturbing the membrane lipid matrix, thus generating more AOS) associated with oxidative stress and fruit senescence (Bowler et al., 1992).

Oxidative stress is a prominent factor in all aerobic organisms, and has been investigated to a limited extent in several fruits, including apple (Du and Bramlage, 1995; Masia, 1998), pear (Brennan and Frenkel, 1977), avocado (Meir et al., 1991), and tomato (Brüggemann et al., 1999; Suurmeijer et al., 2000).

Free Radicals

Oxidative stress will not often occur as a direct result of the generation of $O_2\cdot^-$ and H_2O_2 as these two species are relatively unreactive (Bowler et al., 1992). However, when $O_2\cdot^-$ and H_2O_2 are transformed into $OH\cdot$ via the Haber-Weiss reaction (H_2O_2 and $O_2\cdot^-$ in the presence of Fe^{3+} and Fe^{2+} generate $OH^- + OH\cdot + O_2$), the derived $OH\cdot$ species is extremely reactive. Hydroxyl radicals react rapidly to denature proteins, mutate DNA, and peroxidize lipids, the measurement of which acts as a common and reliable indicator of oxidative stress (Thompson et al., 1987; Hodges et al., 1999). Oxidative stress occurs when the rate of formation of reactive oxygen species is greater than the rate of removal and, as mentioned previously, this increases the membrane autocatalytic lipid degradation pathway such that the system cannot revert to homeostasis.

Antioxidants—Enzymatic and Nonenzymatic

Fruits have evolved both enzymatic and nonenzymatic antioxidative mechanisms to eliminate the less reactive $O_2\cdot^-$ and H_2O_2 molecules and thus inhibit the rate of postharvest senescence. For example, in a nonenzymatic reaction, α-tocopherol serves as an antioxidant to reduce oxy radicals and singlet oxygen. Similarly carotenoids act on singlet oxygen within the lipid matrix, and ascorbic acid acts on peroxides in a nonlipid environment. Enzymatic antioxidant reactions involve such enzymes as superoxide dismutase which dismutates $O_2\cdot^-$, catalase, and peroxidase (EC 1.11.1.7) which catabolize H_2O_2, and glutathione reductase (EC 1.6.4.2) which reduces both $O_2\cdot^-$ and H_2O_2. Both enzymatic and nonenzymatic antioxidative reactions serve to maintain reactive oxidant species concentration at a low steady-state level and allow for homoeostatic autocatalytic lipid degradation (Pastori and del Rio, 1997; Mathieu et al., 1998). The role of antioxidants α-tocopherol, ascorbic acid, and, to a lesser extent, carotenoids in reducing oxidative damage in senescing fruits is largely diminished following fruit harvest by high temperatures, fruit moisture loss, and altered surrounding gas atmospheres (Hultin and Milner, 1978).

Superoxide dismutase (SOD) activity in senescing apple fruits has been studied for its association with physiological breakdown (senescence) and superficial scald (Du and Bramlage, 1995). Although the correlation of apple superficial scald with SOD activity is poor, the apple studies have shown that there is cultivar (genetic) variability for sustained enzymatic antioxidant activity of SOD.

Oxidative Stress Localization

In soybean nodules, senescence occurs due to an increase in reactive oxygen species, rather than a decline in nonenzymatic antioxidant content (Evans et al., 1999). Nonenzymatic antioxidants are ubiquitous in both plants and animals and would, therefore, appear to be important in regulating AOS (Kappus, 1985). In fruits, however, it is doubtful if nonenzymatic antioxidants are universally available throughout the cytoplasm. Nonenzymatic antioxidants are extremely effective scavengers, but in fruits they are primarily linked to the thylakoid membrane (Bowler et al., 1992), which is not the principal site of oxy radical production. In fruit, oxy radical production primarily occurs, and thus oxidative stress initiated, in the mitochondria and, to a lesser degree, in the endoplasmic reticulum (Monk et al., 1989). Fruit oxidative stress is proposed to occur when AOS form and accumulate primarily due to a decline in catabolizing antioxidant enzymes rather than due to changes in nonenzyme antioxidant content (Evans et al., 1999).

FACTORS AFFECTING OXIDATIVE STRESS AND SENESCENCE

Stresses

Kays (1991) has published a comprehensive review on stress in postharvest produce. Stress is defined as a wide range of biotic and abiotic conditions that interrupts, restricts, or accelerates normal metabolic processes, such as senescence, and often in an adverse or negative way. The extent of stress-induced response by the fruit is dependent upon the severity of the stress, the duration of exposure, and the fruit's natural (inherent) resistance mechanisms. Typical environmental or naturally occurring stress factors listed by Kays (1991) are carbon dioxide (stress occurs when toxic levels of acetaldehyde and ethanol develop), ethylene (can cause increased membrane permeability and alter cellular compartmentalization), excessive ions (altered normal ionic balance within the cell can affect cellular selective permeability), ionizing radiation (even at low dosages, 1 kGy or less, radiation can increase membrane permeability and alter cellular compartmentalization), oxygen (stress occurs when anaerobic conditions develop), salt (can cause toxic effect on cells and alter turgor pressure), temperature (high temperature, low temperature, or chilling and freezing are very disruptive to membrane permeability and cellular compartmentalization), ultraviolet radiation, excessive ozone, visible light (resulting in excess heat or dehydration stress), water loss (fruit dehydration is a strain when cell turgor pressure drops below its maximum value), and harvest stress (mechanical, physical, gravitational, or chemicals can disrupt membrane integrity or rupture the fruit affecting tissue water loss). All of these stresses can disrupt the membrane in such a way that permeability is altered and/or cellular pH decreased due to reduced ATPase pumping capacity. This will often result in the activation of the autocatalytic degradation pathway.

Numerous studies, mostly on nonfruit tissues, report changes in activities of specific antioxidant enzymes during stress assaults (Shewfelt and del Rosario, 2000). However, in these studies, the various plant tissue responses to salt, herbicide, drought, heat, chilling, etc., appears to be more related to the fruit, vegetable, or leaf tissues having multiple antioxidant systems rather than relying on just one antioxidant. These researchers suggest that any study of lipid peroxidation/antioxidant defense mechanisms and their relationships to abiotic or biotic stresses must combine an investigation of the organelles and their membranes and all accompanying enzymatic antioxidants and their related isozymes.

Controlled or Modified Atmospheres

According to Kader (1992), modified atmospheres or controlled atmospheres are used to remove or add atmospheric gases surrounding the fruit such that they become different from ambient air. The benefits of altering fruit storage atmospheres that decrease fruit respiration and metabolism are retardation of oxidative stress and senescence, reduced sensitivity to ethylene gas, reduction or inhibition of low temperature injury, and reduced pathogen and/or insect activity. The harmful effects of altered atmospheres include increased physiological disorders attributed to toxic levels of carbon dioxide, off-flavors due to low oxygen levels, membrane damage due to stressful levels of too-low oxygen or too-high carbon dioxide, and inhibition of periderm development resulting in tissue water-loss stress. Thus, controlled or modified atmospheres, if administered correctly, would greatly reduce or suspend oxidative stress by lowering the activation of the autocatalytic degradation pathway.

Implications for Harvested Fruits

Postharvest storage or preservation is the substitution of one set of stresses (e.g., heat, light, water, and insects) for a second set (e.g., chilling injury, modified atmospheres [diminished O_2 and elevated CO_2], ethylene, and pathogens) (Lurie et al., 1994). Although optimized storage conditions, via significant breakthroughs in refrigerated storage technology, have been very effective at prolonging the postharvest life and minimizing disorders of fresh fruits, oxidative stress still occurs (Shewfelt and del Rosario, 2000). Foyer et al. (1997) suggests that AOS are the signaling molecules in response to stresses; cellular membrane peroxidation may, therefore, provide the link between these stresses and storage disorders. Shewfelt and del Rosario (2000) state that lipid peroxidation is the mediating response between stress-to-stress interactions, such as chilling injury and senescence; and future advances in preventing oxidative stress or other postharvest storage disorders will come from a basic understanding of cellular metabolism. An example of stress-to-stress interactions has been found with zucchini, whereby preconditioning fruits with sub stress-inducing doses of high temperatures reduced chilling injury by elevating the levels of catalase, peroxidase, and superoxide dismutase antioxidant enzymes (Wang, 1995). Another example is the suppression of chilling injury in 'Charentais' muskmelon fruit through antisense the ACC oxidase gene, a precursor to ethylene (Ben-Amor et al., 1999). Treating antisense and wild-type melons to chilling temperatures (2°C), followed by rewarming, resulted in improved tolerance to chilling in antisense fruit as assessed by lowered accumulations of ethanol, acetaldehyde, and membrane deterioration. A higher potential capacity to remove AOS through heightened catalase, peroxidase, and superoxide dismutase enzyme activities was also observed. In wild-type melon fruit, severe chilling injury was correlated with lower activities of AOS scavenging enzymes. Theses results demonstrate that postharvest stress-to-stress interactions do act in concert to trigger oxidative-induced metabolic shifts in fruit membranes, thus hastening senescence.

RECENT RESEARCH INTO FRUIT OXIDATIVE STRESS

Genetic Variation

Active oxygen species are highly involved in stress tolerance and, as expected, plants exhibit genetic variations in regulating or controlling oxidative stress. Utilizing cultivars of mandarin fruit selected for differences in susceptibility to cold-induced peel damage (chilling injury), cultivar responses were associated with differences in oxidative activity (Sala, 1998). The main difference separating chilling-injured versus noninjured mandarin fruit was the ability of the noninjured fruit to produce a more effective antioxidant system; i.e., higher ascorbate peroxidase (EC 1.11.1.11), catalase, and glutathione reductase activities. Collén and Davison (1999), using vegetative tissues of *Fucus* spp., also demonstrated that stress tolerance is variable among different cultivars of the same species. Using lipid peroxidation as a measure of oxidative stress following freezing, drought, and high intensity light stresses, all species tested produced equal amounts of

AOS, even though individual species varied in their susceptibility to lipid peroxidation. An association was also established between a species' stress tolerance and a species' protective mechanisms against AOS. Species' protective mechanisms can be expressed through the ability to acclimate to stresses such as temperature or drought; or different plant parts, tissues, or developmental stages having variable tolerance to high or low light intensity stress. This association of species variation with AOS concentration and in AOS tolerance occurs in conjunction with AOS-scavenging enzymes; i.e., the ratio of AOS:AOS-scavenging enzymes activities, but not with the variation of nonenzymatic antioxidant contents. Another difference in species protective mechanisms is the ability of some species to prevent AOS formation altogether rather than deactivate them once formed.

Molecular Modification

Kerdnaimongkol and Woodson (1999) down-regulated the tomato catalase gene (ASTOMCAT1) by antisense genetic modification. They showed that a twofold reduction in total catalase activity resulted in a twofold increase in H_2O_2 content, eventually causing tissue damage and death. Molecular modification of plants using sense and antisense technology to modify the activity of enzymes involved in the oxidative defense mechanism has been accomplished in a number of transgenic plants. Recent investigations into enzymatic antioxidants include ascorbate peroxidase (Torsethaugen et al., 1997), catalase (Kerdnaimongkol and Woodson, 1999), glutathione reductase (Aono et al., 1991; Broadbent et al., 1995; Brüggemann et al., 1999), glutathione S-transferase (EC 2.5.1.18)/glutathione peroxidase (EC 1.11.1.9) (Roxas et al., 1997), and superoxide dismutase (Bowler et al., 1991; McKersie et al., 1993; Perl et al., 1993; Sen Gupta et al., 1993). Utilization of genetic modification via transgenic fruit with altered enzymatic AOS scavenging abilities will help provide clearer insights into where and when these antioxidants affect senescence, and how their concentration may be heightened during fruit senescence.

Modified Atmospheres

Modified atmospheres (Gorny, 1997), either using controlled atmospheres or film packaging systems with fresh-cut (minimally processed) fruits and vegetables, are beneficial in reducing hydroxyl radical formation or controlling its activity once formed. Use of noble gases (argon, krypton, xenon, and neon) are twice as effective as nitrogen in displacing oxygen and are extremely effective in retarding enzymatic activity (Brody, 1997). Studies using nitrogen and noble gases in reducing phospholipase, phosphatidic phosphatases, and other enzymes in the autocatalytic lipid degradation pathway could be helpful in reducing ROS production in minimally processed fresh fruit and vegetables.

Cellular Localization

Vitor et al. (1999), using epicarp tissue of mandarin citrus fruits, showed that the starting reaction of peroxidation involves the removal of H^+ from membrane-bound phospholipid fatty acid δ-linoleic by OH·. The resulting linoleate radical initiates a chain reaction by binding with molecular oxygen to produce a linolenate peroxy radical that in turn subtracts another H^+ from δ-linoleic acid, forming more linolenate peroxide. The resulting loss of acyl lipids affects the planar conformation of the membrane lipid bilayer, thus causing leakage and cellular death. Vitor et al. (1999) also state that the role of nonenzymatic antioxidants (such as carotenes, luteins, and others) is that of protecting the membrane by shielding the phospholipid fatty acids against attack by $^1O_2$, thus limiting OH· formation. However, according to Vitor et al. (1999), nonenzymatic antioxidant concentration is insufficient to block all free-radical attacks and thus functions only to slow oxidative-stress-affected fruit senescence.

CONCLUSION

Oxidative stress in senescent fruit tissues can be systematic, whereby hydrogen is released from unsaturated free fatty acids—either cytoplasmic free fatty acids or membrane-derived fatty acids generated by lipolytic acyl hydrolase via the autocatalytic lipid degradation pathway—to produce lipid radicals. These lipid radicals can interact with molecular oxygen (O_2) to produce a radical chain reaction, generating lipid monohydroperoxides (Kappus, 1985). In the presence of metal catalysts, monohydroperoxides produce lipid alkoxy radicals. Lipid alkoxy radicals undergo cleavage of the C-C bonds to form unsaturated acid aldehydes and alkyl radicals, which propagate the radical chain reaction. Peroxyl radicals undergo cyclization and generate malondialdehyde as a secondary end product of polyunsaturated peroxidation (Hodges et al., 1999). It is important to note that three isolated double bonds of unsaturated fatty acids are required to produce malondialdehyde. However, the absence of malondialdehyde does not, in itself, suggest an absence of lipid peroxidation (Evans et al., 1999). Once oxidative stress of senescing fruits is activated, and with sufficient oxygen present and without the immediate presence of antioxidants to dispose of or transform AOS into nonradical oxygen species, leakage of the microsomal membranes increases. Eventually, oxidative stress will damage all cellular molecules such as proteins, saccharides, and nucleic acids, etc., resulting in cellular or organ (fruit) death.

ISSUES FOR FURTHER RESEARCH

Antioxidants

Either one or all three SODs (Cu/ZnSOD, cytoplasmic localized; MnSOD, mitochondrial localized; and FeSOD, chloroplast localized) need to be investigated for cultivar variability of sustained activity. Other antioxidants such as catalase, peroxidases, and gluthione reductase should be examined for fruit cultivar variability of sustained antioxidant activity. This research could lead to the selection of existing cultivars, if not the breeding of new fruit cultivars, with enhanced AOS scavenging activity and resultant extended postharvest marketable quality and shelf life. A recent study by Masia (1998) demonstrated that higher levels of catalase and superoxide dismutase activities in various apple cultivars are reflected in theses fruits having longer shelf lives.

Molecular Modification

Up-regulation of enzymatic and nonenzymatic antioxidant activities should be investigated along with simultaneous down-regulation of enzymes within the autocatalytic degradation pathway. Utilizing molecular technology to generate transformed fruit with sense and antisense orientation of enzymes, involved in oxidative defense, is fundamental in both the specific biochemical functioning as well as cellular localization of these antioxidants.

Modified Atmospheres

As molecular oxygen is mandatory for the production of AOS, reduction of O_2 through modified or controlled atmospheres and time-controlled studies could be helpful in determining when and where AOS initiate unchecked catabolism in fruits. Oxygen isotope labeling studies have been used to show that fatty acid hydroperoxide lyase can function as an acidic amino acid (Lewis acid) by cleaving fatty acid hydroperoxides heterolytically at the active site, in addition to their known functioning as a substrate-derived free-radical cleaving enzyme through the use of heme proteins (Suurmeijer et al., 2000). Utilizing oxygen isotope labeling studies in modified atmosphere trials on fruit tissue may be helpful in describing the role of O_2 and O_2 availability in stimulating lipoxygenase activity, and may be helpful in identifying the site(s) within the microsomal membranes or membrane conditions most susceptible to oxidative damage.

Cellular Localization

Further research is required related to biochemical interactions between oxidative stress, physiological damage (such as surface pitting, discoloration or staining), and cellular-associated microsomal membrane changes during the onset of senescence. This research may identify the subcellular compartments and processes that initiate the specific signaling cascades of oxidative stress.

Understanding the biochemical initiation mechanisms of oxidative stress at the membrane level during fruit senescence will be helpful in developing molecular assays that can define how the autocatalytic lipid degradation pathway changes from catabolism and recycling to catabolism only. Utilizing extensive cellular biochemical assays, with a variety of fruits and various cultivars throughout their ripening and senescence phases, should aid in answering this foremost question of when oxidative stress occurs and how it is best regulated.

REFERENCES

Aono, M., A. Kubo, H. Saji, T. Natori, K. Tanaka, and N. Kondo (1991). Resistance to active oxygen toxicity of transgenic *Nicotiana tabacum* that expresses the gene for gluthione reductase from *Escherichia coli*. *Plant and Cell Physiology* 32:691-697.

Ben-Amor, B., A. Latche, M. Bouzayen, J.C. Pech, and F. Romojoro (1999). Inhibition of ethylene biosynthesis by antisense ACC oxidase RNA prevents chilling injury in Charentais cantaloupe melons. *Plant, Cell and Environment* 22:1579-1586.

Bowler, C., L. Slooten, S. Vanderbraden, R. De Rycle, J. Botterman, C. Sybesma, M. Van Montagu, and D. Inzé (1991). Manganese SOD can reduce cellular damage mediated by oxygen radicals in transgenic plants. *European Molecular Biology Organization Journal* 10:1723-1732.

Bowler, C., M. Van Montagu, and D. Inzé (1992). Superoxide dismutase and stress tolerance. *Annual Review of Plant Physiology and Plant Molecular Biology* 43:83-116.

Brennan, T. and C. Frenkel (1977). Involvement of hydrogen peroxide in the regulation of senescence in pear. *Plant Physiology* 59:411-416.

Broadbent, P., G.P. Creissen, B. Kular, A.R. Wellburn, and P.M. Mullineaux (1995). Oxidative stress responses in transgenic tobacco containing altered levels of glutathione reductase activity. *Plant Journal* 8:247-255.

Brody, A. (1997). Modified atmosphere packaging: a future outlook. In *Proceedings of the 7th Controlled Atmosphere Research Conference,* J.R. Gorny (Ed.). University of California, Davis. 5:104-112.

Brüggemann, W., V. Beyel, M. Brodka, H. Poth, M. Weil, and J. Stockhaus (1999). Antioxidants and antioxidative enzymes in wild-type transgenic *Lycopersicon* genotypes of different chilling tolerance. *Plant Science* 140:145-154.

Buchanan-Wollaston, V. (1997). The molecular biology of leaf senescence. *Journal of Experimental Botany* 48:181-199.

Collén, J. and I.R. Davison (1999). Reactive oxygen metabolism in intertidal *Fucus* spp. (Phaeophyceae). *Journal of Phycology* 35:62-69.

Dörnenburg, H., and C. Davies (1999). The relationships between lipid oxidation and antioxidant content in postharvest vegetables. *Food Reviews International* 15:435-453.

Du, Z., and W.J. Bramlage (1995). Peroxidative activity of apple peel in relation to development of poststorage disorders. *HortScience* 30:205-209.

Evans, D.E., S.A. Briaro, and L.E. Williams (1991). Active calcium transport by plant cell membranes. *Journal of Experimental Botany* 42:285-303.

Evans, P.J., D. Gallsi, C. Mathieu, M.J. Hernandez, M. de Felipe, B. Halliwell, and A. Puppo (1999). Oxidative stress occurs during soybean nodule senescence. *Planta* 208:73-79.

Foyer, C.H., H. Lopez-Delgado, J.F. Dat, and I.M. Scott (1997). Hydrogen peroxide and glutathione-associated mechanisms of acclamatory stress tolerance and signaling. *Physiologia Plantarum* 100:241-254.

Gorny, J.R. (1997). A summary of Ca and MA requirements and recommendations for the storage of freshly-cut (minimally processed) fruits and vegetables. In *Proceedings of the 7th Controlled Atmosphere Research Conference*, J.R. Gorny (Ed.). University of California, Davis. 5:30-66.

Gross, J. (1991). *Pigments in vegetables, chlorophylls and carotenoids*. New York: AVI Book, Van Nostrand Reinold.

Hodges, D.M., J.M. DeLong, C.F. Forney, and R.P. Prange (1999). Improving the thiobarbituric acid reactive-substance assay for estimating lipid peroxidation in plant tissues containing anthocyanin and other interfering compounds. *Planta* 207:604-611.

Hultin, H.O. and M. Milner (1978). *Postharvest biology and biotechnology*. Westport, CT: Food and Nutrition Press.

Kader, A.A. (1992). Modified atmospheres during transport and storage. In *Postharvest technology of horticultural crops*, A.A. Kader (Ed.). Oakland, CA: University of California, Division of Agri. and Natural Resources, Pub. 3311, pp. 85-92.

Kappus, H. (1985). Lipid peroxidation: Mechanisms, analyses, enzymology and biological relevance. In *Oxidative stresses*, H. Siers (Ed.). New York: Academic Press, pp. 273-309.

Kays, S.J. (1991). Stress in harvested products. In *Postharvest physiology of perishable plant products*, S.J. Kays (Ed.). New York, NY: AVI Book, Van Nostrand Reinold, pp. 335-407.

Kerdnaimongkol, K. and W.R. Woodson (1999). Inhibition of catalase by antisense RNA increase susceptibility to oxidative stress and chilling injury in transgenic tomato plants. *Journal of the American Society for Horticultural Science* 124:330-336.

Leshem, Y.Y. (1988). Plant senescence processes and free radicals. *Free Radical Biology and Medicine* 5:39-49.

Leshem, Y.Y. (1992). *Plant membranes: A biophysical approach to structure, development and senescence.* Dordrecht, The Netherlands: Kluwer Academic Pub.

Lester, G.E. (1988). Comparison of 'honey dew' and netted muskmelon fruit tissues in relation to storage life. *HortScience* 23:180-182.

Lester, G.E. (1990). Lipoxygenase activity of hypodermal- and middle-mesocarp tissues from netted muskmelon fruit during maturation and storage. *Journal of the American Society for Horticultural Science* 115:612-615.

Lester, G.E. (1996). Calcium alters senescence rate of postharvest muskmelon fruit disks. *Postharvest Biology and Technology* 7:91-96.

Lester, G.E., V.M. Baizabal-Aguirre, L.E. Gonzalez de la Vara, and W. Michalke (1998). Calcium-stimulated protein kinase activity of the hypodermal-mesocarp plasma membrane from preharvest-mature and postharvest muskmelon. *Journal of Agricultural and Food Chemistry* 46:1242-1246.

Lester, G.E. and M.A. Grusak (1999). Postharvest application of calcium and magnesium to honeydew and netted muskmelons: Effect on tissue ion concentrations, quality and senescence. *Journal of the American Society for Horticultural Science* 124:545-552.

Lester, G.E. and B.D. Whitaker (1996). Gamma-ray-induced changes in hypodermal mesocarp tissue plasma membrane of pre- and post-harvest muskmelon. *Physiologia Plantarum* 98:265-270.

Lurie, S., J.D. Klein, and E. Fallik (1994). Cross protection of one stress by another: Strategies in postharvest fruit storage. *NATO ASI Ser. H Cell Biology* 86:201-212.

Masia, A. (1998). Superoxide dismutase and catalase activities in apple fruit during ripening and post-harvest and with special reference to ethylene. *Physiologia Plantarum* 104:668-672.

Mathieu, C., S. Moreau, P. Frendo, A. Puppo, and M. Davies (1998). Direct detection of radicals in intact soybean nodules: Presence of nitric oxide-leghemoglobin complexes. *Free Radical Biology and Medicine* 24:1242-1249.

McCullum, T.G., D.J. Huber, and D.J. Cantliffe (1989). Modification of polyuronides and hemicelluloses during muskmelon fruit softening. *Physiologia Plantarum* 76:303-308.

McKersie, B.D., Y. Chen, M. de Beus, S.R. Bowley, C. Bowler, D. Inzé, K. D'Halluin, and J. Botterman (1993). Superoxide dismutase enhances tolerance of freezing stress in transgenic alfalfa (*Medicago sativa* L.). *Plant Physiology* 103:1155-1163.

Meir, S., S. Philosoph-Hadas, G. Zauberman, Y. Fuchs, M. Akerman, and N. Aharoni (1991). Increased formation of fluorescent lipid-peroxidation products in avocado peels precedes other signs of ripening. *Journal of the American Society for Horticultural Science* 116:823-826.

Monk, L.S., K.V. Fagerstedt, and R.M.M. Crawford (1989). Oxygen toxicity and superoxide dismutase as an antioxidant in physiological stress. *Physiologia Plantarum* 76:456-459.

Noodén, L.D. (1988). The phenomena of senescence and aging. In *Senescence and aging in plants,* Noodén, L.D. and A.C. Leopold (Eds.). San Diego, CA: Academic Press, pp. 1-50.

Paliyath, G. and M.J. Droillard (1992). The mechanisms of membrane deterioration and disassembly during senescence. *Plant Physiology and Biochemistry* 30:789-812.

Pastori, G.M. and L.A. del Rio (1997). Natural senescence of pea leaves: An activated oxygen-mediated function for peroxisomes. *Plant Physiology* 113:405-412.

Perl, A., R. Perl-Treves, S. Galili, D. Aviv, E. Shalgi, S. Malkin, and E. Gulan (1993). Enhanced oxidative-stress defense in transgenic potato expression tomato Cu, Zn superoxide dismutase. *Theoretical Applied Genetics* 85:568-576.

Roxas, V.P., R.P. Smith Jr., S.R. Allen, and R.P. Allen (1997). Overproduction of glutathione S-transferase/glutathione peroxidase enhances the growth of transgenic tobacco seedlings during stress. *Nature Biotechnology* 15:988-991.

Ryall, L.A. and W. J. Lipton (1979). *Handling, transportation and storage of fruits and vegetables,* Second edition, Vol. 1, Westport, CT: AVI Publishing Co.

Sala, J.M. (1998). Involvement of oxidative stress in chilling injury in cold-stored mandarin fruit. *Postharvest Biology and Technology* 13:255-261.

Sen Gupta, A., J.L. Heimen, A.S. Holaday, J.J. Burke, and R.P. Allen (1993). Increased resistance to oxidative stress in transgenic plants that overexpress chloroplastic Cu/Zn superoxide dismutase. *Proceedings of the National Academy of Sciences USA* 90:1629-1633.

Shewfelt, R.L. and B.A. del Rosario (2000). The role of lipid peroxidation in storage disorders of fresh fruits and vegetables. *HortScience* 35:575-579.

Shewfelt, R.L. and S.E. Prussia (1993). *Postharvest handling: A systems approach.* San Diego, CA: Academic Press, Inc.

Suurmeijer, C.N.S.P., M. Perez-Gilabert, D-J. van Unen, H.T.W.M. van der Hijden, G.A. Veldink, and J.F.G. Vliegenthart (2000). Purification, stabilization and characterization of tomato fatty acid hydroperoxide lyase. *Phytochemistry* 53:177-185.

Sylvestre, I., J. Bureau, A. Tremolieres, and A. Paulin (1989). Changes in membrane phospholipids and galactolipids during the senescence of cut carnations. Connection with ethylene rise. *Plant Physiology and Biochemistry* 27:931-937.

Thompson, J.E., R.L. Legge, and R.F. Barber (1987). The role of free radicals in senescence and wounding. *New Phytology* 105:317-344.

Torsethaugen, G., L.H. Pitcher, B.A. Zilinskas, and E.J. Pell (1997). Overproduction of ascorbate peroxidase in the tobacco chloroplast does not provide protection against ozone. *Plant Physiology* 114:529-537.

Vick, B.A. and D.C. Zimmerman (1989). Metabolism of fatty acid hydroperoxides by *Chorella pyrenoidosa*. *Plant Physiology* 90:125-132.

Vitor, R.F., F.C. Lidon, and C.S. Carvalho (1999). Dark stained tissues of the epicarp of encore mandarin: Interactions with the production of hydroxyl radicals. *Free Radical Research* 31 (Supplement):S163-S169.

Wang, C.Y. (1995). Effect of temperature preconditioning on catalase, peroxidase, and superoxide dismutase in chilled zucchini. *Postharvest Biology and Technology* 5:67-76.

Yapa, P.A.J., T. Kawasaki, and H. Matsumoto (1986). Changes of some membrane-associated enzyme activities and degradation of membrane phospholipid in cucumber roots due to Ca^{2+} starvation. *Plant and Cell Physiology* 27:223-232.

Chapter 7

Antioxidants

Susan Lurie

INTRODUCTION

The chief toxicity of reactive oxygen species (ROS) resides in their ability to initiate cascade reactions that result in the production of the hydroxyl radical and other destructive species, such as lipid peroxides. These cascades are prevented by efficient operation of antioxidant defense systems. The term *antioxidant* can be considered to describe any compound capable of quenching ROS without itself undergoing conversion to a destructive radical. Organisms have evolved complex enzyme and nonenzyme defenses to minimize oxidative damage to macromolecules and cellular structures. They also possess repair systems for renewing some oxidative modifications and disposal systems for removing modified macromolecules that are not repaired.

Plants are protected against conditions that generate high levels of ROS by a complex antioxidant system which includes three general classes: (1) lipid-soluble, membrane-associated antioxidants (e.g., α-tocopherol and β-carotene); (2) water-soluble reductants (e.g., glutathione and ascorbate); and (3) enzymatic antioxidants (e.g., superoxide dismutase [SOD, EC 1.15.1.1], catalase [CAT, EC 1.11.1.6], and peroxidase [POD, EC 1.11.1.7]). Other naturally occurring antioxidants in plants are also receiving increasing attention. These include isoflavonoids, phenols, polyamines, and specific amino acids, particularly cysteine and methionine (Larson, 1988; Levine et al., 1996; Robards et al., 1999).

Harvested horticultural crops undergo postharvest stress conditions somewhat similar to stresses that plants experience in the field. These include wounding or bruising from the harvesting and sorting process, water stress from water loss in storage, temperature stress from cold storage conditions, and anaerobic stress from controlled or modified atmosphere storage. These stresses are superimposed on the normal ripening or senescence process of the fruit or vegetable, which also produce ROS in the tissue. Therefore, to minimize the potentially deleterious effects of storage conditions, ripening, and senescence, it is important to maintain the antioxidative processes in the tissue.

LIPID-SOLUBLE MEMBRANE-ASSOCIATED ANTIOXIDANTS

Tocopherols

The designation of the tocopherol class of compounds as vitamin E was first made by animal nutritionists following the observation that these compounds were the necessary factors in pregnant rats for fetus viability (Evans and Bishop, 1922). Vitamin E is now known to be necessary in animal cells for a multitude of antioxidant functions essential for their well-being. Tocopherol compounds are a group of closely related phenolic benzochroman derivatives having extensive ring alkylation. The most active of four major tocopherols is α-tocopherol, being one of the most important in vitro chain-breaking antioxidants tested (Neely et al., 1988).

Tocopherols are amphipathic molecules; the hydrophobic phytl tail is located in a membrane, associated with the acyl chains of fatty acids or their residues, whereas the polar chromanol head group lies at the membrane-cytosol interface where it can interact with other cytosolic molecules. It scavenges free radicals, by acting as either a chemical scavenger or a physical deactivator (quencher), thereby preventing the proliferation of oxidative chain reactions (Fryer, 1992). The fact that the tocopherols partition into the membranes of cells and organelles allows them to inhibit oxidative damage to membranes caused by stress or aging. Maintenance of high α-tocopherol levels in fruit tissue can help prevent injury from water or temperature stress by preventing membrane damage.

The principal function of α-tocopherol in all biological membranes is as a highly efficient, recyclable, chain-reaction terminator for the removal of polyunsaturated fatty acid radicals generated during lipid oxidation (Figure 7.1). The radical is transferred to the chromanol hydroxyl group of α-tocopherol. The radical form of vitamin E is recycled back to α-tocopherol by interaction at the membrane-cytosol interface with either ascorbate or thiols such as reduced glutathione. The glutathionyl radical or monodehydroascorbate radical formed during the regeneration process is recycled by either NADH or NADPH and their specific reductase enzymes (Asada and Takahashi, 1987). There is evidence that the ratio of vitamin E to vitamin C (tocopherol to ascorbate) changes during senescence (Kunert and Ederer, 1985). During leaf aging, α-tocopherol transiently increases in membranes, but ascorbate decreases and the consequence is lipid oxidation.

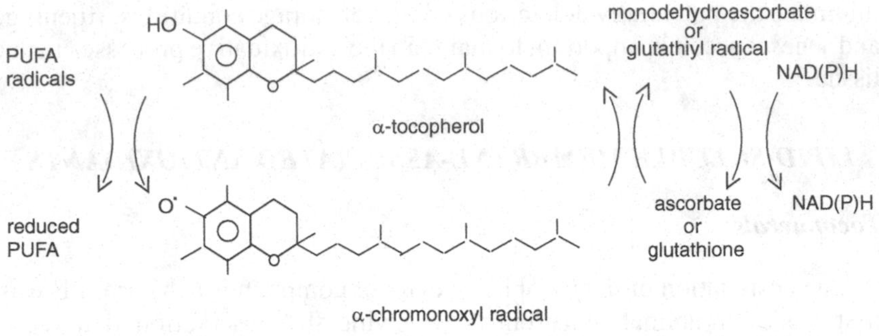

FIGURE 7.1. Chain terminating action of α-tocopherol and its regeneration.

Carotenoids

Although the principal recognized role of carotenoids is to act as photo-receptive antenna pigments for photosynthesis, gathering wavelengths of light that are not absorbed by chlorophyll, it has also been recognized that they, or at least β-carotene, also have a protective function against oxidative damage. In most plant tissues, carotenoids occur in the chloroplast membranes and help prevent oxidative damage from ROS generated by the absorption by chlorophyll of more light energy than can be transferred to the photosynthetic electron transport chain. Chloroplast membranes are rich in the polyunsaturated fatty acid linolenate which can easily become peroxidized. Carotenoids are very powerful quenchers of ROS and at relatively low concentration β-carotene can effectively protect membrane lipids from oxidation (Larson, 1988).

A number of fruits and vegetables are high in carotenoid concentration, including yellow squash, peppers, carrots, apricots, persimmons, and yellow plums. In this case, the carotenoids are concentrated in chromoplasts, not chloroplasts, and it is unclear how much they contribute to antioxidative protection of the plant tissue, although they may be important for nutritional health.

WATER-SOLUBLE ANTIOXIDANTS

Glutathione

Glutathione is an important antioxidant in all aerobic organisms and plays a crucial role in the defense against ROS arising as by-products of metabolism (Meister, 1983; Alscher, 1989; Noctor and Foyer, 1998). In addition, glutathione plays an important role in stress tolerance and has been implicated in the adaptation of plants to environmental stresses such as drought (Dhindsa, 1991), atmospheric pollution (Guri, 1983), and extremes of temperature (Esterbauer and Grill, 1978). Furthermore, its synthesis is induced in response to these stresses (Alscher, 1989; Noctor and Foyer, 1998).

Glutathione is synthesized by a two-step process. The first step, catalyzed by γ-glutamyl-cysteine synthetase, results in the production of γ-glytamyl-cysteine; the second, catalyzed by glutathione synthetase, produces glutathione through the addition of glycine (Rennenberg, 1982; Alscher, 1989). Glutathione synthesis has been shown to respond either directly or indirectly to hydrogen peroxide (Smith et al., 1984; Smith, 1985). Inhibition of catalase by aminotriazole led to leakage of hydrogen peroxide from the peroxisomes and to a stimulation of glutathione synthesis (Smith, 1985; May and Leaver, 1993).

Glutathione exists in both reduced (GSH) and oxidized (GSSG) forms in plant tissues (Rennenberg, 1982). The reduced form of glutathione reacts with ROS to prevent oxidation of thiol groups in enzymes and thus plays an important role as an antioxidant in the stabilization of many enyzmes. Once oxidized, the glutathione can be regenerated by the enzyme glutathione reductase (EC 1.6.4.2), the key enzyme in the ascorbate-glutathione cycle (Figure 7.2). A high ratio of GSH/GSSG seems to be necessary for the detoxification of ROS and for the adaptation of plants to environmental stresses including drought and temperature extremes. Stress conditions such as low temperature can often enhance glutathione reductase activity (Edwards et al., 1994). However, in fruits and vegetables of tropical or subtropical origin, chilling injury occurs when they are stored at low temperatures. Preconditioning zucchini squash at 15°C before 2°C storage enhanced glutathione reductase activity and maintained high ratio of GSH/GSSG while preventing chilling injury (Wang, 1995b).

Ascorbate

Ascorbate plays a major role in the prevention of peroxidative damage by scavenging ROS and, as a consequence, producing its own free radical, monodehydroascorbate. This is a relatively stable radical and reacts preferentially with itself, thus preventing the propagation of free-radical reactions (Bielski et al., 1981; Noctor and Foyer, 1998). Monodehydroascorbate can also nonenzymatically disproportionate to dehydroascorbate. Ascorbic acid also acts against ROS in concert with other antioxidants such as glutathione in the ascorbate-glutathione cycle, and this cycle can also interact with α- tocopherol. This ascorbate-glutathione cycle is found in chloroplasts to protect against

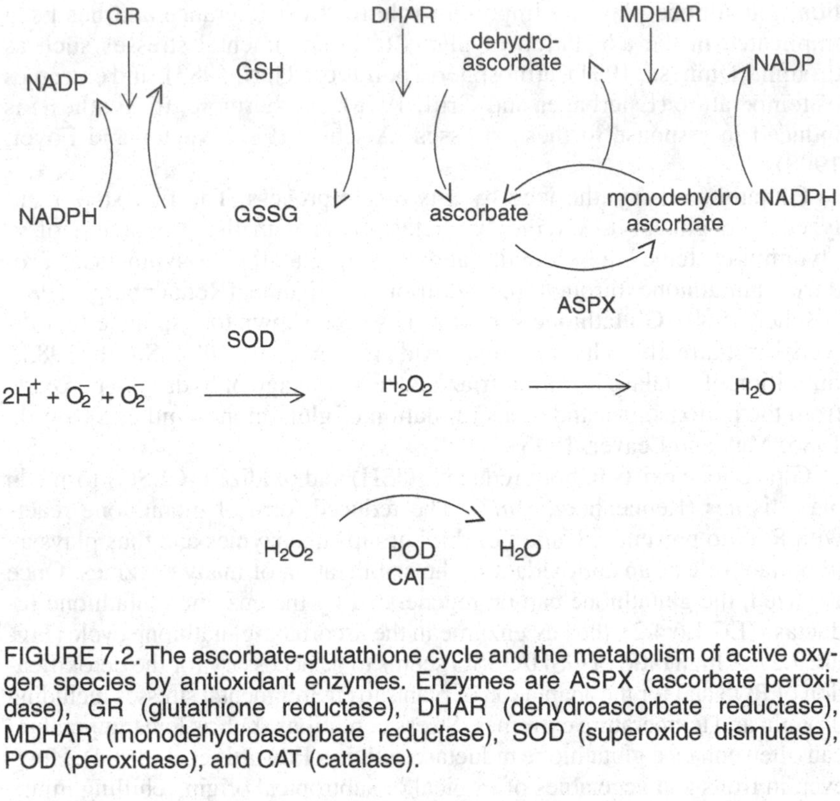

FIGURE 7.2. The ascorbate-glutathione cycle and the metabolism of active oxygen species by antioxidant enzymes. Enzymes are ASPX (ascorbate peroxidase), GR (glutathione reductase), DHAR (dehydroascorbate reductase), MDHAR (monodehydroascorbate reductase), SOD (superoxide dismutase), POD (peroxidase), and CAT (catalase).

ROS generated by photosynthesis (Bielawski and Joy, 1986; Kalt-Torres et al., 1984), but is also found in mitochondria (Jiménéz et al., 1997) and the cytoplasm (Mullineaux et al., 1996). In this cycle ascorbate undergoes continual oxidation and reduction. Ascorbate peroxidase (EC 1.11.1.11) catalyzes the reduction of H_2O_2 to water using the reducing power of ascorbate and, in the process, produces monodehydroascorbate (Loewus and Loewus, 1987). Monodehydroascorbate can spontaneously disproportionate by reacting with another monodehydroascorbate to form ascorbic acid and dehydroascorbate (Bielski et al., 1981), the latter being converted back to ascorbic acid by dehydroascorbate reductase (EC 1.8.5.1) using glutathione as the electron donor (Law et al., 1983). Dehydroascorbate can also react nonenzymatically with glutathione to form reduced ascorbate. However, monodehydroascorbate can also be reduced to ascorbic acid by a monodehydroascorbate reductase (EC 1.6.5.4) using NADH or NADPH as the electron donor (Hossain et al., 1984). By these mechanisms, ascorbate is maintained in its reduced form as a protectant against cellular oxidative degradation.

Ascorbic acid is synthesized in a number of steps from glucose (Loewus and Loewus, 1987). It is involved in many metabolic activities in addition to its role as an antioxidant. One of these roles is the biosynthesis of hydroxyproline rich proteins in the cell wall, in which ascorbate oxidase (EC 1.10.3.3) is involved . This is a cell-wall-located enzyme and utilizes oxygen rather than H_2O_2 to oxidize ascorbic acid (Lin and Varner, 1991). Interestingly, this enzyme increases in activity during fruit maturation and may have a role in cell expansion (for review, see Smirnoff, 1996).

Ascorbic acid is important not only during the life of plants and during storage or shelf life of fruits and vegetables, but also for human health. In this regard, ascorbic acid is a quality parameter of fruits and vegetables, and should be kept at an appropriate level. Ascorbic acid levels, however, tend to decrease during storage and processing of fruits and vegetables. In apples and pears stored in controlled atmosphere, ascorbic acid levels were lower than under regular air storage (Haffner et al., 1997; Veltman et al., 2000). In potatoes, tuber levels declined as storage was extended, to 15 percent of the original level (Keijbets and Ebbenhorst-Seller, 1990). In pears, a relationship was found beween ascorbate levels and susceptibility to browning (Veltman et al., 1999). Browning was initiated in a cutivar-dependent manner when ascorbate acid fell below a certain threshold, which varied between 2 and 6 mg per 100 g fresh weight (Veltman et al., 2000).

ENZYMATIC ANTIOXIDANT SYSTEMS

Superoxide Dismutase

Reactive oxygen species such as superoxide (O_2^-), H_2O_2, and hydroxyl radicals (·OH) are by-products of normal cellular metabolism. These active oxygen species result in the peroxidation of membrane lipids (Mead, 1976), breakage of DNA strands, and inactivation of enzymes (Brawn and Fridovich, 1981). Both chloroplasts and mitochondria are major sources of ROS production either under normal growth conditions or during exposure to various stresses. SODs are a group of metalloenzymes that protect cells from O_2^- radicals by catalyzing the dismutation of O_2^- to molecular O_2 and H_2O_2. Three classes of SOD activity have been identified that differ by the active site metal cofactors (Fe, Mn, or Cu/Zn). The primary sequences of FeSOD and MnSOD apoproteins are related, whereas Cu/ZnSOD is distinct. Cu/ZnSOD occurs mainly in the cytosol and chloroplast stroma of plants, whereas MnSOD occurs in the mitochondrial matrix, although a thylakoid-bound MnSOD has been reported to exist in some plants (Scandalios, 1993). FeSODs are generally found in prokaryotes, but have been found in some plants such as the Cruciferae in association with the chloroplasts (Bowler et al., 1994). In monocotyledonous plants, only choroplastic Cu/ZnSOD and mitochondrial MnSOD have been found (Wu et al., 1999). The SODs differ in their quantitative distribution in plant tissues in addition to their subcellular localization (Du and Bramlage, 1994).

SOD activity has been linked to physiological stresses such as low temperature, high intensity light, water stress, and oxidative stress (Bowler et al., 1992). SOD activity was inversely related with susceptibility to sunscald damage in tomato fruits (Rabinowitch and Sklan, 1981; Rabinowitch et al., 1982). SOD activity in tomato fruit is highest in the immature-green fruit, passes through a minimum level at the mature-green and breaker stages, and rises again at the pink stage of ripening. The change in SOD activity between immature-green and red-ripe is a 50 percent decrease, while the difference between mature-green and red-ripe is only 5 percent (Rabinowitch et al., 1982). This decrease in activity at the beginning of ripening has been found in cucumbers and peppers as well as between immature and mature fruit (Rabinowitch and Sklan, 1981). Apples differed greatly in their SOD activity at harvest from cultivar to cultivar, with 'Cortland' apples having 10-fold higher activity than 'Delicious' fruit (Du and Bramlage, 1994). However, there was little change in activity during 15 weeks of 0°C storage. Baker (1976) also reported that there were no major quantitative changes in total SOD activity between pre- and postclimacteric apple, banana, avocado, and tomato fruits. Therefore, SOD does not always respond to changes in physiological states of fruit. However, the differences in SOD activity among cultivars may predispose them to storage disorders. Ju et al. (1994) found that SOD and CAT activities in brown-heart-sensitive cultivars of pears were lower than in tolerant fruits, and peroxide accumulation during storage coincided with brown-heart development.

Catalase

CAT is a tetrameric iron porphyrin protein that catalyzes the dismutation of H_2O_2 to water and oxygen. No multicellular organism has been found that does not possess at least some CAT activity. Peroxisomes, for which CAT activity is a biological marker, are present in almost all eukaryotes (Subramani, 1993). In addition, it is associated to some degree with mitochondria in plants such as maize (Scandalios et al., 1980). Hydrogen peroxide is produced during many cellular processes. It is synthesized as a by-product of photorespiration, β-oxidation of fatty acids, and also as a consequence of biotic and abiotic stresses. Catalases, together with SOD and peroxidases, make up a defense system for the scavenging of O_2^- and hydroperoxides (Beyer and Fridovich, 1987).

As might be suspected from the multiple functions of CAT, plants contain multiple CAT isozymes, from two isozymes in castor bean (Ota et al., 1992) to as many as 12 in mustard (Drumm and Schopfer, 1974). Different isozymes are found in different plant tissues. For example, in *Arabidopsis*, six isozymes are detected in flowers and leaves and two are found in roots (Frugoli et al., 1996). Some of the genes in each plant appear to be regulated by the circadian clock at the level of mRNA abundance; others are not.

Catalase is also found in fruits. In citrus fruits, high CAT activity has been linked to resistance of the fruits to chilling injury (Sala, 1998; Sala and Lafuente, 1999). It was found that fruits of 'Clementine' and 'Clemenules' mandarins, which are chilling tolerant, have a more efficient antioxidant enzyme system than the chilling-sensitive 'Fortune' cultivar (Sala, 1998). The sensitive 'Fortune' cultivar can become tolerant to cold storage if heat treated, and the resistance was correlated with an increase and maintenance of CAT activity (Sala and Lafuente, 1999). Inhibitors of CAT activity prevented the heat-induced cold tolerance from developing, indicating that this is the critical antioxidant enzyme in this system. On the other hand, Wang (1995a) has shown that the tolerance to chilling of zucchini squash may be increased by conditioning the fruits at 15°C, and that this pretreatment reduced the chilling-induced decline in CAT, SOD, and POD. Therefore, cold storage may alter the equilibrium between free-radical production and defense mechanisms in favor of free-radical production, and maintaining the defense mechanisms may be critical for prevention of storage disorders.

Peroxidases

PODs are heme-containing enzymes ubiquitous in the plant kingdom and belonging to a superfamily which comprises Class I enzymes from mitochondria, chloroplasts, and bacteria; Class II from fungi; and Class III from higher plants. Plant peroxidases are glycoproteins characterized by the presence of oligosaccharide chains linked to the protein moiety and having effects on the stability of the enzymes (Hu and van Huystee, 1989). The enzyme is also stabilized by the presence of calcium ions that are involved in holding the confirmation of the active site (Barber et al., 1995). PODs have many substrates that act as hydrogen donors in the presence of H_2O_2. Due to the broad substrate specificity and the presence of many closely related isoforms, it has been difficult to assign a specific function to the isoform associated to a certain cellular compartment or tissue. PODs are involved in many growth-related processes, including cell wall extension, lignin biogenesis, and auxin catabolism. In addition, they are involved in stress-related processes such as wounding and disease resistance (Moerschbacher, 1992).

In the past decade, in studies to elucidate the function of PODs in plants, molecular biological approaches have been used to isolate, characterize, and study the expression of genes encoding POD isozymes in plants. More than 100 cDNAs for Class III peroxidases have been cloned and their expression studied in various plants (Esnault and Chibbar, 1997). The great majority of these isoforms are found in extracellular matrix, followed in abundance in the vacuole. The deduced amino acid sequences reveal similarity ranging from 30 to 80 percent among the cDNAs. Accumulation of POD transcripts, particularly those found extracellularly, has been observed during elicitation of defense reactions, after wounding or treatment with ethylene (Esnault and Chibbar, 1997).

The responsiveness to ethylene may be why POD activities increased during ripening (Da Silva et al., 1990). However, fruit ripening is generally viewed as a regulated senescence phenomenon where increased levels of ROS are involved. During ripening, in many cases, both enhanced POD as well as higher levels of peroxides are found (Brennan and Frenkel, 1977). It is unclear whether higher POD activity is present to try to decrease peroxide levels, or if, in spite of higher POD activity, the peroxide levels are high because of their generation during ripening-related processes.

PODs are involved in the responses to preconditioning treatments utilized to prevent chilling injury during storage. Holding zucchini squash for two days at 15°C before low-temperature storage enhanced ascorbate peroxidase activity as well as the activity of other enzymes in the ascorbate antioxidant system (Wang, 1996). High-temperature preconditioning of citrus fruit at 37°C also enhanced POD activity during storage at low temperature. This temperature conditioning prevented chilling injury in the fruits (Martinez-Tellez and Lafuente, 1997). In tomatoes, new isoforms of POD were found to be induced by high temperature treatment, which were correlated with prevention of chilling injury (Lurie, Laamim, et al., 1997).

Other Antioxidants

Fruits and vegetables contain many different antioxidant components. Most of the antioxidant capacity of a fruit or vegetable may be from compounds other than ascorbate, α-tocopherol, or β-carotene. Flavonoids (including flavones, isoflavones, flavonones, anthocyanins, catechin, and isocatechin) that are found in fruits and vegetables also demonstrated strong antioxidant activities (Bors and Saran, 1987; Bors et al., 1990; Hanasaki et al., 1994). Therefore, there have been a number of studies measuring total antioxidant capacity of fruits, vegetables, and juices by various assay methods

(Webman et al., 1989; Miller et al., 1995; Wang et al., 1996; Fogliano et al., 1999).

Flavonoids occur widely in the plant kingdom, and are present in all tissues of the plant. The metabolic pool of flavonoids is by no means static, but subject to turnover at very different rates depending on the plant organ and its stage of development. The concentration of flavonoids in plant cells often exceeds 1 mM, with concentrations from 3 to 10 mM present in some epidermal cells of certain fruit and vegetable peels (Robards et al., 1999). Early work by Clemetson and Andersen (1966) showed that many flavonoids and other plant phenolics protected ascorbic acid against oxidation. Other in vitro tests of the antioxidant efficiency of flavonoids used inhibition of oxidation of linolenic acid or β-carotene (Robards et al., 1999). The highest antioxidant activity was shown by free flavonols and the least by flavonoids having fewer than four hydroxyl groups (Larson, 1988). Several flavonoids were shown to be potent inhibitors of lipoxygenase, an enzyme involved in fruit ripening and senescence. Lipoxygenase converts polyunsaturated fatty acids to oxidized forms. Lipoxygenase activity increased during both ripening and senescence, and inhibition of this enzyme can delay these processes (Leshem, 1988). Most of the flavonoids are in vacuoles, although some are in chloroplasts and in cell walls. In fruit and vegetables, anthocyanins and flavonol glycosides are concentrated in the peel and exert protective effects (Caldwell et al., 1983).

Phenolic acids have also been implicated as active antioxidants. Caffeic acid, chlorogenic acid, and its isomer, caffeoylquinic acid, were isolated from sweet potato and demonstrated protective action in both the linoleic and the β-carotene test systems (Pratt, 1965). Chlorogenic acid was the most common phenol found in apples and can be used as a substrate by peroxidase to convert H_2O_2 to water (Lurie et al., 1989). However, as well as being antioxidants, phenols can be condensed to polyphenols by polyphenol oxidase or peroxidase, and these compounds can be involved in postharvest disorders of numerous fruits. The browning reaction in fruit flesh is due to this reaction, as is the darkening of sun-scalded apples that occurs in storage (Lurie et al., 1991). Oxidative skin darkening, such as scald in pomegranates and apples, has been tied to the formation of polyphenols in the peel tissue (Ben-Arie and Or, 1986; Lurie et al., 1989). Therefore, these antioxidants may have protective effects as a food dietary component, but may cause problems in fruit and vegetable storage.

ANTIOXIDANTS DURING POSTHARVEST STORAGE

The largest body of research conducted on changes in antioxidant activity of stored produce has probably been conducted on apples. This is because many cultivars suffer from a storage disorder called superficial scald, which causes blackening of the fruit peel. This disorder is caused by oxidative processes that occur only in low-temperature storage, and, therefore, is one of the many manifestations of chilling injury found in fruits and vegetables (Watkins et al., 1995). The commercial method of preventing the development of this disorder is to give a prestorage dip in synthetic antioxidants. This prevents the oxidation to conjugated trienes of α-farnesene, a compound accumulating in the apple cuticle during low-temperature storage, and concurrently prevents scald development.

Endogenous high levels of antioxidants at harvest have been found to be correlated with reduced susceptibility to scald (Meir and Bramlage, 1988; Barden and Bramlage, 1994). Correlation has been made between total antioxidants in apple peel during storage and conjugated trienes and scald at the end of storage (Gallerani et al., 1990; Barden and Bramlage, 1994). Although correlation was not good with individual antioxidants, and while lipid-soluble antioxidants were stable or increased during storage, the water-soluble antioxidants declined during storage as scald susceptibility increased.

Attempts have been made to identify specific antioxidants that may be involved in preventing scald. These include phenolic fatty acid esters (Whitaker, 1998) and free phenols (Abdallah et al., 1997; Ju and Bramlage, 2000). However, the symptom expression of scald has also been linked to oxidation of phenols (Ju et al., 1996). Polyphenol oxidase activity is elevated in peel having scald (Lurie et al., 1989) and has been found to increase more in scald-susceptible than scald-resistant cultivars (Kang and Seung, 1987). Therefore, the role of phenols in preventing or expressing scald is not yet clear (Piretti et al., 1994).

Superficial scald is just one example of physiological storage disorders that have been associated with the antioxidant capacity of fruit tissue, albeit one which has been intensively investigated. Storage scalds on other fruits have been found to be affected by peel antioxidants including pear (Oleszek et al., 1994), persimmons (Lee et al., 1993), pomegranates (Ben-Arie and Or, 1986), and citrus (Cohen et al., 1988). Internal flesh disorders can be also traced to antioxidant levels or activities of the antioxidant enzymes.

Some pear cultivars are susceptible to internal browning in storage. The onset of browning is initiated when tissue ascorbate levels fall below a certain threshold (Veltman et al., 2000). In addition to lower ascorbate, a decrease during storage of SOD, CAT, and glutathione peroxidase activities in pears was found in a cultivar susceptible to fresh browning (Vanoli et al., 1995). The maturity of the pears at harvest can influence the levels of the antioxidants and enzyme activities. With increasing maturity, ascorbate and glutathione decreased, as did SOD and CAT activities (Lentheric et al.,

1999). POD activity increased with increasing maturity. This may help explain the widely described influence of harvest date on the occurrence of physiological disorders in pears.

Internal browning in other fruits has also been connected to breakdowns in the tissue antioxidant defense systems. Ascorbate levels were low in 'Bracburn' apples afflicted with browning disorder (Burmeister and Roughan, 1997). Browning in litchi fruit was due to oxidative reactions and could be prevented by keeping the antioxidant system high (Jiang and Fu, 1999).

Internal quality and the rate of senescence of fruits and vegetables in storage have been linked to antioxidants. The antioxidants delay lipid peroxidation and concomitant increase in membrane leakage associated with senescence. A study of short storage life and long storage life muskmelons found that there were high levels of SOD and CAT in the long storage life melons, including SOD isozymes not present in the short storage life cultivar (Lucan and Baccou, 1998). Leafy vegetables and edible spices also showed a similar relationship between antioxidant levels and senescence in storage. Spinach treated with ethylene senesced rapidly and showed a decrease in antioxidant enzyme activities and water-soluble antioxidants compared to leaves stored in controlled atmosphere where senescence was delayed (Hodges and Forney, 2000). In fresh edible herbs, such as watercress, parsley, and basil, total reducing capacity was correlated with storage potential (Meir et al., 1995). Leaves with greater antioxidant activity senesced more slowly, including yellowing from chlorophyll loss. The rate of broccoli yellowing in storage was also related to the antioxidant enzyme activities of SOD and POD, with longer-storing cultivars having higher activity than a poor-storing cultivar (Toivonen and Sweeney, 1998). The storage system can also help to retain antioxidant quality. Broccoli stored in modified atmosphere retained its green color and did not lose ascorbic acid. POD activity remained low in modified-atmosphere broccoli but increased, along with a 80 percent decline in ascorbate levels, in senescing broccoli (Barth and Hong, 1996).

An often-neglected aspect of antioxidant systems in harvested commodities is their involvement in responses to pathogen attack. It is a truism that freshly harvested fruits are less susceptible to fungal attack than fruits after storage, even if ripening has not perceptibly occurred during storage. Fruits also become more susceptible to decay as they ripen. The increased susceptibility comes from the disappearance of antifungal compounds in the fruit tissue and a decrease in the ability to initiate defense reactions when presented with an attack. Some of these antifungal compounds are also antioxidants. Resveratrol is a natural antioxidant in grapes which is high in concentration in the immature berry, declines as the grape ripens, and declines still further in storage. The ability of *Botrytis cinerea* to invade different grape cultivars was well correlated with the level of resveratrol in the grapes (Sarig et al., 1997). Mature green tomatoes are refractory to infection by *B. cinarea* when freshly harvested, but this resistance to infection disappears after 24 to 48 hours, while ripening has not yet occurred. The increase in susceptibility was correlated to the disappearance of an mRNA coding for an anionic POD (Lurie, Fallik, et al., 1997).

A more complicated situation is found in avocado storage and ripening, where generally there are quiescent infections of *Colletotrichum gloeosporioides*. The fungus begins to develop when an antifungal diene present in the fruit falls below a certain level. The breakdown of the diene is due to the activity of lipoxygenase. Lipoxygenase activity rises only when the flavanol epicatechin decreases. The decrease of a flavanol antioxidant apparently activates lipoxygenase, which metabolizes the antifungal diene and allows decay to develop (Prusky and Keen, 1993). An external antioxidant added to the fruit during sorting before storage was able to maintain the diene level longer and prevent infection (Prusky and Keen, 1993).

In terms of nutritional value, the decrease in antioxidant enzymes during storage and, sometimes, the increase in POD during ripening or senescence is of less interest than the fate of the antioxidant compounds. Losses of these compounds in storage will vary by the type of fruit or vegetable, physical damage, storage temperature, and environment (Shewfelt, 1990). Most studies of antioxidant losses have examined ascorbic acid as being the most labile of the antioxidants. Generally, during storage dehydroascorbic acid increases at the expense of ascorbic acid and may be responsible for over 50 percent of the vitamin C measured in fresh market fruits and vegetables (Wills et al., 1984). Lipid-soluble antioxidants are generally more stable in storage than are water-soluble ones; for example, arotenoids and α-tocopherol are much more stable than ascorbic acid. Lipid-soluble antioxidants were found to increase during apple storage, while water-soluble antioxidants declined (Barden and Bramlage, 1994). Antioxidant levels in apples, as measured by absorbance of a hexane extract of cuticle at 200 nm, varied among cultivars. However, in all cases the level was higher in stored apples than in freshly harvested fruit (Curry, 1997). This increase in antioxidants during storage can be found in other commodities as well. During storage there was an increase in the antioxidant capacity of strawberries and raspberries due to an increase in anthocyanins in strawberries and in anthocyanins and total phenolics in raspberries (Kalt et al., 1999). In the same study, blueberries did not increase in antioxidant capacity during storage, but they began with threefold higher capacity than either strawberries or raspberries.

Therefore, it is necessary to examine each commodity in terms of what antioxidant compounds are present and how they change during storage. It is not possible to make a statement that will be true for all cases. In addition, it should be recognized that no matter how optimal the storage, the fresh commodity will deteriorate in quality once it is purchased. Indeed, it was found that the greatest losses in antioxidants in fruits and vegetables occurred during home or food service preparation and not in commercial handling and storage (Williams, 1996).

CONCLUSION AND DIRECTIONS FOR FUTURE RESEARCH

A well-functioning antioxidant system is necessary to protect against postharvest stresses, maintain fruit and vegetable quality in storage, and prevent postharvest storage disorders. The elucidation of the particular systems involved in stress responses of different commodities for the most part remains to be investigated. In recent years, a number of reports have correlated increases in one or more of the antioxidant enzymes with increased resistance to storage disorders. In work where several enzymes have been studied under the same stress conditions, differential responses have frequently been observed. The degree to which the activities of individual antioxidant enzymes are increased is variable and, in many cases, relatively minor. However, each of the antioxidant enzymes comprises a family of isoforms, and in vitro enzyme assays will not often register increased activity of particular isoforms. This drawback can be overcome by investigation of isoforms by activity gels, but this has rarely been done.

Molecular techniques can make a major contribution to this area of research. Specific probes for the genes of the different isoforms can help determine which are affected by the conditions imposed on the fruit or vegetable. However, although transcripts for specific isoforms can be quantified, increases in transcript abundance may not necessarily be accompanied by corresponding increases in enzyme activities. Therefore, the research should be a combination of determination of both gene expression and enzyme activity. Only after the specific genes have been targeted can breeding for improved storage be instigated.

REFERENCES

Abdallah, A.U., M.T. Gil, W. Biasi, and E.J. Mitcham (1997). Inhibition of superficial scald in apples by wounding: Changes in lipids and phenolics. *Postharvest Biology and Technology* 12: 203-212.

Alscher, R.G. (1989). Biosynthesis and antioxidant function of glutathione in plants. *Physiologia Plantarum* 77: 457-464.

Asada, K. and M. Takahashi (1987). Production and scavenging of active oxygen in photosynthesis. In *Photoinhibition,* eds. D.J. Kyle, C.B. Osmond, and C.J. Arntzen (Eds.). Amsterdam, Holland: Elsevier Publ., pp. 227-280.

Baker, J.E. (1976). Superoxide dismutase in ripening fruits. *Plant Physiology* 58: 644-647.

Barber, K.R., M. Maranon, G. Shaw, and R.B. van Huystee (1995). Structural influence of calcium on the cationic peanut peroxidase as determined by NMR spectroscopy. *European Journal of Biochemistry* 232: 825-833.

Barden, C.L. and W.J. Bramlage (1994). Relationships of antioxidants in apple peel to changes in α-farnesene and conjugated trienes during storage, and to superficial scald development after storage. *Postharvest Biology and Technology* 4: 23-33.

Barth, M. and Z. Hong (1996). Packaging design affects antioxidant vitamin retention and quality of broccoli florets druing postharvest storage. *Postharvest Biology and Technology* 8: 141-150.

Ben-Arie, R. and E. Or (1986). The development and control of hust scald on 'Wonderful' pomegranate fruit druing storage. *Journal of the American Society of Horticultural Science* 111: 395-399.

Beyer, W.F. and I. Fridovich (1987). Catalases—with and without heme. In *Oxygen Radicals in Biology and Medicine,* M.G. Simic, K.A. Taylor, J.F. Ward, and C. Von Sonntag (Eds.). New York: Plenum Press, pp. 651-661.

Bielawski, W. and K.N. Joy (1986). Properties of glutathione reductase from chloroplasts and roots of pea. *Phytochemistry* 25: 2261-2265.

Bielski, B.H.J., A.O. Allen, and H.A. Schwarz (1981). Mechanism of disproportionation of ascorbate radicals. *Journal of the American Chemical Society* 103: 3516-3518.

Bors, W. and M. Saran (1987). Radical scavenging by flavonoid antioxidants. *Free Radical Research Communication* 2: 289-294.

Bors, W., H. Werner, C. Michel, and M. Saran (1990). Flavonoids as antioxidants: Determination of radical scavenging efficiencies. *Methods in Enzymology* 186: 343-355.

Bowler, C., W. Van Camp, M. Van Montagu, and D. Inzé (1994). Superoxide dismutase in plants. *Critical Reviews in Plant Sciences* 13: 199-208.

Bowler, C., M. Van Montagu, and D. Inzé (1992). Superoxide dismutase and stress tolerance. *Annual Review of Plant Physiology and Plant Molecular Biology* 43: 83-116.

Brawn, K. and I. Fridovich (1981). DNA strand scission by enzymically generated oxygen radicals. *Archives of Biochemistry and Biophysics* 206: 414-419.

Brennan, T. and C. Frenkel (1977). Involvement of hydrogen peroxide in the regulation of senescence in pear. *Plant Physiology* 59: 411-415.

Burmeister, D.M. and S. Roughan (1997). Physiological and biochemical basis for the Braeburn browning disorder. *Controlled Atmosphere Research Conference, CA 97,* Vol. 2, E.J. Mitcham (Ed.). Davis, CA: University of California Press, pp. 13-18.

Caldwell, M.M., R. Robberecht, and S.D. Flint (1983). Internal filters: Prospects for UV-acclimation in higher plants. *Physiologia Plantarum* 58: 445-450.

Clemetson, C. and L. Andersen (1966). The protection afforded by plant phenols to ascorbate oxidation. *Annals of the New York Academy of Sciences* 123: 341-347.

Cohen, E., S. Lurie, and Y. Shalom (1988). Prevention of red blotch in degreened lemon fruit. *HortScience* 23: 864-865.

Curry, E.A. (1997). Effect of postharvest handling and storage on apple nutritional status using antioxidants as a model. *HortTechnology* 7: 240-243.

Da Silva, E., E. Lourenco, and V. Neves (1990). Soluble and bound peroxidases from papaya fruit. *Phytochemistry* 29: 1051-1056.

Dhindsa, R.S. (1991). Drought stress, enzymes of glutathione metabolism, oxidation injury and protein synthesis in *Tortula ruralis*. *Plant Physiology* 95: 648-651.

Drumm, H. and P. Schopfer (1974). Effect of phytochrome on development of catalase activity and isoenzyme pattern in mustard (*Sinapis alba* L.) seedlings. A reinvestigation. *Planta* 120: 13-30.

Du, Z. and W.J. Bramlage (1994). Superoxide dismutase activities in sensecing apple fruit (*Malus domestica* Borkh.). *Journal of Food Science* 59: 581-584.

Edwards, E.A., C. Enard, G.P. Creissen, and P.M. Mullineaux (1994). Synthesis and properties of glutathione reductase in stressed peas. *Planta* 192: 137-143.

Esnault, R. and R.N. Chibbar (1997). Peroxidases: At the gene expression level. *Plant Peroxidase Newsletter* 10: 7-14.

Esterbauer, H. and D. Grill (1978). Seasonal variation of glutathione and glutathione reductase in needles of *Picea abies*. *Plant Physiology* 61: 119-121.

Evans, H. and K. Bishop (1922). The relations between fertility and nutrition. I. The ovulation rhythm in the rat on a standard nutritional regime; II. The ovulation rhythm in the rat on inadequate nutritional regimes. *Journal of Metabolic Research* 1: 319-355.

Fogliano, V., V. Verde, G. Randazzo, and A. Ritieni (1999). Method for measuring antioxidant activity and its application to monitoring the antioxidant capacity of wines. *Journal of Agricultural and Food Chemistry* 47: 1035-1040.

Frugoli, J.A., H.H. Zhong, M.L. Nuccio, P. McCort, M.A. McPeek, T.L. Thomas, and C.R. McClung (1996). Catalase is encoded by a multigene family in *Arabidopsis thaliana* (L.) Heynh. *Plant Physiology* 112: 327-336.

Fryer, M.J. (1992). The antioxidant effects of Vitamin E (α-tocopherol). *Plant, Cell and Environment* 15: 381-392.

Gallerani, G., G.C. Pratella, and R.A. Budini (1990). The distribution and role of natural antioxidant substances in apple fruit affected by superficial scald. *Advances in Horticultural Science* 4: 144-146.

Guri, A. (1983). Variation in glutathione and ascorbic acid content among selected cultivars of *Phaseolus vulgaris* prior to and after exposure to oxone. *Canadian Journal of Plant Science* 63: 733-737.

Haffner, K., W.K. Jeksrud, and G. Tengesdal (1997). L-ascorbic acid contents and other quality criteria in apples (*Malus domestica* Borkh.) after storage in cold store and controlled atmosphere. In *Postharvest Horticulture*, E.J. Mitcham (Ed.). Series no. 16. Davis, CA: University of California Press, pp. 10-21.

Hanasaki, Y., S. Ogawa, and S. Fukui (1994). The correlation between active oxygen scavenging and antioxidative effects of flavonoids. *Free Radical Biology and Medicine* 16: 845-850.

Hodges, D.M. and C.F. Forney (2000). The effects of ethylene, depressed oxygen and elevated carbon dioxide on antioxidant profiles of senescing spinach leaves. *Journal of Experimental Botany* 51: 645-655.

Hossain, M.A., Y. Nakano, and K. Asada (1984). Monodehydroascorbate reductase in spinach chloroplasts and its participation in regeneration of ascorbate for scavenging hydrogen peroxide. *Plant and Cell Physiology* 25: 385-395.

Hu, C. and R.B. van Huystee (1989). Role of carbohydrate moieties in peanut peroxidases. *Biochemical Journal* 263: 129-135.

Jiang, Y.M. and J.R. Fu (1999). Biochemical and physiological changes involved in browning of litchi fruit. *Journal of Horticultural Science and Biotechnology* 74: 43-46.

Jiménez, A., J.A. Hernandez, L.A. del Rio, and F. Sevilla (1997). Evidence for the presence of the ascorbate-glutathione cycle in mitochondria and peroxisomes of pea leaves. *Plant Physiology* 114: 275-284.

Ju, Z. and W.J. Bramlage (2000). Cuticular phenolics and scald development in 'Delicious' apples. *Journal of the American Society of Horticultural Science* 125: 498-504.

Ju, Z.G., Y.B. Yuan, C.L. Liu, S.M. Zhan, and M.X. Wang (1996). Relationships among simple phenol, flavonoid and anthocyanin in apple fruit peel at harvest and scald susceptibility. *Postharvest Biology and Technology* 8: 83-93.

Ju, Z.G., Y.B. Yuan, C.L. Liou, S.M. Zhan, and S. Xin (1994). Effects of low temperature on H_2O_2 and heart browning of Chili and Yali pear (*Pirus bred Scheideri*, R.). *Science of Agriculture Sinica* 27:77-81.

Kalt, W., C. Forney, A. Martin, and R. Prior (1999). Antioxidant capacity, vitamin C, phenolics, and anthocyanins after fresh storage of small fruits. *Journal of Agricultural and Food Chemistry* 47: 4638-4644.

Kalt-Torres, W., J.J. Burke, and J.M. Anderson (1984). Chloroplast glutathione reductase: Purification and properties. *Physiologia Plantarum* 61: 271-278.

Kang, S.D. and K. Seung (1987). Nature and control of apple scald during storage. *Journal of the Korean Society of Horticultural Science* 28: 343-345.

Keijbets, M.J.H. and G. Ebbenhorst-Seller (1990). Loss of vitamin C (L-ascorbic acid) during long term cold storage of Dutch table potatoes. *Potato Research* 33: 125-130.

Kunert, K.J. and M. Ederer (1985). Leaf aging and lipid peroxidation: The role of antioxidants vitamin C and E. *Physiologia Plantarum* 65: 85-88.

Larson, R.A. (1988). The antioxidants of higher plants. *Phytochemistry* 27: 969-978.

Law, M.Y., S.A. Charles, and B. Halliwell (1983). Glutathione and ascorbic acid in spinach (*Spinacea oleracea*) chloroplasts. *Biochemical Journal* 210: 899-903.

Lee, S.K., I.S. Shin, and Y.M. Park (1993). Factors involved in skin browning of non-astringent 'Fuyu' persimmon . *Acta Horticultura* 343: 300-303.

Lentheric, I., E. Pinto, M. Vendress, and C. Larrigaudiere (1999). Harvest date affects the antioxidative systems in pear fruits. *Journal of Horticultural Science and Biotechnology* 74: 791-795.

Leshem, Y. (1988). Plant senescence processes and free radicals. *Free Radical Biology and Medicine* 5: 39-49.

Levine, R.L., L. Mosoni, B.S. Berlett, and E.R. Stadtman (1996). Methionine residues as endogenous antioxidants in proteins. *Proceedings of the National Academy of Sciences USA* 93: 15036-15040.

Lin, L.S. and J.E. Varner (1991). Expression of ascorbic acid oxidase in zucchini squash (*Cucurbita pepo* L.). *Plant Physiology* 96: 159-165.

Loewus, F.A. and M.W. Loewus (1987). Biosynthesis and metabolism of ascorbic acid in plants. *CRC Critical Reviews in Plant Sciences* 5: 101-119.

Lucan, D. and J.C. Baccou (1998). High levels of antioxidant enzymes correlate with delayed senescence in non netted muskmelon fruits. *Planta* 204: 377-382.

Lurie, S., E. Fallik, A. Handros, and R. Shapira (1997). The involvement of peroxidase in resistance to *Botrytis cinerea* in heat treated fruit. *Physiological and Molecular Plant Pathology* 50: 141-159.

Lurie, S., J.D. Klein, and R. Ben-Arie (1989). Physiological effects of DPA on apples. *Israeli Journal of Botany* 38: 199-208.

Lurie, S., M. Laamim, Z. Lapsker, and E. Fallik (1997). Heat treatments to decrease chilling injury in tomato fruit. Effects on lipids, pericarp lesions and fungal growth. *Physiologia Plantarum* 100: 297-302.

Lurie, S., E. Pesis, and R. Ben-Arie (1991). The darkening of sunscald on apples in storage is a non-enzymatic and non-oxidative process. *Postharvest Biology and Technology* 1: 119-125.

Martinez-Tellez, M. and M.T. Lafuente (1997). Effect of high temperature conditioning on ethylene, phenylalanine ammonia-lyase, peroxidase and polyphenol oxidase activities in flavedo of chilled 'Fortune' mandarin fruit. *Journal of Plant Physiology* 150: 674-678.

May, M.J. and C.J. Leaver (1993). Oxidative stimulation of glutathione synthesis in *Arabidopsis thaliana* suspension cultures. *Plant Physiology* 103: 621-627.

Mead, J.F. (1976). Free radical mechanisms of lipid damage and consequences for cellular membranes. In *Free Radicals in Biology*, Vol. 1., W.A. Pryor (Ed.). New York: Academic Press, pp. 51-68.

Meir, S. and W.J. Bramlage (1988). Antioxidant activity in 'Cortland' apple peel and susceptibility to superficial scald after storage. *Journal of the American Society of Horticultural Science* 113: 412-418.

Meir, S., J. Kanner, B. Akiri, and S. Philosoph-Hadas (1995). Determination and involvement of aqueous reducing compounds in oxidative defense systems of various senescing leaves. *Journal of Agricultural and Food Chemistry* 43: 1813-1819.

Meister, A. (1983). Selective modification of glutathione metabolism. *Science* 220: 472-477.

Miller, N.J., A.T. Diplock, and C.A. Rice-Evans (1995). Evaluation of the total antioxidant activity as a marker of the deterioration of apple juice on storage. *Journal of Agricultural and Food Chemistry* 43: 1794-1801.

Moerschbacher, B.M. (1992). Plant peroxidases: Involvement in response to pathogen. In *Plant Peroxidases 1980-1990*, C. Penel, T. Gaspar, and H. Greppin (Eds.). Geneva, Switzerland: University of Geneva Press, pp. 91-99.

Mullineaux, P., C. Enard, R. Hellens, and G. Creissen (1996). Characterization of a glutathione reductase gene and its genetic locus from pea (*Pisum sativum* L.). *Planta* 200: 186-194.

Neely, W., J. Martin, and S. Barker (1988). Products and relative reaction rates of the oxidation of tocopherols with singlet molecular oxygen. *Photochemistry and Photobiology* 48: 423-428.

Noctor, G. and C. Foyer (1998). Ascorbate and glutathione: Keeping active oxygen under control. *Annual Review of Plant Physiology Plant Molecular Biology* 49: 249-279.

Oleszek, W., M. Amiot, and S. Aubert (1994). Identification of some phenolics in pear fruit. *Journal of Agricultural and Food Chemistry* 42: 1261-1265.

Ota, Y., T. Ario, K. Hayashi, T. Nakagawa, T. Hattor, M. Maeshima, and T. Asahi (1992). Tissue-specific isoforms of catalase subunits in castor bean seedlings. *Plant and Cell Physiology* 33: 225-232.

Piretti, M.V., G. Gallerani, and G.C. Pratella (1994). Polyphenol fate and superficial scald in apple. *Postharvest Biology and Technology* 4: 213-224.

Pratt, D.E. (1965). Antioxidative properties of phenols. *Journal of Food Science* 30: 737-743.

Prusky, D. and N.T. Keen (1993). Involvement of preformed antifungal compounds in the resistance of subtropical fruits to fungal decay. *Plant Disease* 77: 114-119.

Rabinowitch, H.D. and D. Sklan (1981). Superoxide dismutase activity in ripening cucumber and pepper fruit. *Physiological Plantarum* 52: 380-384.

Rabinowitch, H.D., D. Sklan, and P. Budowski (1982). Photo-oxidative damage in the ripening tomato fruit: Protective role of superoxide dismutase. *Physiological Plantarum* 54: 369-374.

Rennenberg, H. (1982). Glutathione metabolism and possible biological roles in higher plants. *Phytochemistry* 21: 2778-2781.

Robards, K., P. Prenzler, G. Tucker, P. Swatsitang, and W. Glover (1999). Phenolic compounds and their role in oxidative processes in fruits. *Food Chemistry* 66: 401-436.

Sala, J.M. (1998). Involvement of oxidative stress in chilling injury in cold-stored mandarin fruits. *Postharvest Biology and Technology* 13: 255-261.

Sala, J.M. and M.T. Lafuente (1999). Catalase in the heat-induced chilling tolerance of cold-stored hybrid Fortune mandarin fruits. *Journal of Agricultural and Food Chemistry* 47: 2410-2414.

Sarig, P., Y. Zutkhi, A. Monjauze, N. Lisker, and R. Ben-Arie (1997). Phytoalexin elicitation in grape berries and their susceptibility to *Rhyzopus stolonifer*. *Physiological and Molecular Plant Pathology* 50: 337-347.

Scandalios, J.G. (1993). Oxygen stress and superoxide dismutases. *Plant Physiology* 101: 7-12.

Scandalios, J.G., W.F. Tong, and D.G. Roupakias (1980). *Cat3,* a third gene locus coding for a tissue-specific catalase in maize; genetics, intracellular location, and some biochemical properties. *Molecular General Genetics* 179: 33-41.

Shewfelt, R. (1990). Sources of variation in the nutrient content of agricultural commodities from the farm to the consumer. *Journal of Food Quality* 13: 37-54.

Smirnoff, N. (1996). The function and metabolism of ascorbic acid in plants. *Annals of Botany* 78:61-669.

Smith, I.K. (1985). Stimulation of glutathione synthesis in photorespiring plants by catalase inhibitors. *Plant Physiology* 79: 1044-1047.

Smith, I.K., A.C. Kendall, A.J. Keys, J.C. Turner, and P.J. Lea (1984). Increased levels of glutathione in a catalase-deficient mutant of barley (*Hordeum vulgare* L.). *Plant Science Letters* 37: 29-33.

Subramani, S. (1993). Protein import into peroxisomes and biogenesis of the organelle. *Annual Review of Cell Biology* 9: 445-478.

Toivonen, P.M.A. and M. Sweeney (1998). Difference in chlrophyll loss at 13°C for two broccoli (*Brassica oleracea* L.) cultivars associated with antioxidant enzyme activities. *Journal of Agricultural and Food Chemistry* 46: 20-24.

Vanoli, M., C. Visai, M. Zini, and R. Budini (1995). Enzymatic activities and internal browning of Passa Crassana pears. *Acta Horticultura* 379: 405-411.

Veltman, R.H., R.M. Kho, A.C.R. van Schaik, M.G. Sanders, and J. Oosterhaven (2000). Ascorbic acid and tissue browning in pears (*Pyrus communis* L. cvs Rocha and Conference) under controlled atmosphere conditions. *Postharvest Biology and Technology* 19: 129-137.

Veltman, R.H., M.G. Sanders, S.T. Persijn, H.W. Peppelenbos, and J. Oosterhaven (1999). Decreased ascorbic acid levels and brown core development in pears (*Pyrus communis* cv 'Conference'). *Physiologia Plantarum* 107:39-45.

Wang, C.Y. (1995a). Effect of temperature preconditioning on catalase, peroxidase, and superoxide dismutase in chilled zucchini squash. *Postharvest Biology and Technology* 5: 67-75.

Wang, C.Y. (1995b). Temperature preconditioning affects glutathione content and glutathione reductase activity in chilled zucchini squash. *Journal of Plant Physiology* 145: 148-152.

Wang, C.Y. (1996). Temperature preconditioning affects ascorbate antioxidant system in chilled zucchini squash. *Postharvest Biology and Technology* 8: 29-36.

Wang, H., G. Cao, and R.L. Prior (1996). Total antioxidant capacity of fruits. *Journal of Agricultural and Food Chemistry* 44: 701-705.

Watkins, C.B., W.J. Bramlage, and B.A. Cregoe (1995). Superficial scald of 'Granny Smith' apples is expressed as a typical chilling injury. *Journal of the American Society of Horticultural Science* 120: 88-94.

Webman, E.J., G. Edlin, and H.F. Mower (1989). Free radical scavenging activity of papaya juice. *International Journal of Radiation Biology* 55: 347-351.

Whitaker, B.D. (1998). Phenolic fatty-acid esters from the peel of 'Gala' apples and their possible role in resistance to superficial scald. *Postharvest Biology and Technology* 13: 1-10.

Williams, P. (1996). Vitamin retention in cook/chill and cook/hot-hold food services. *Journal of the American Dietary Association* 96: 490-498.

Wills, R., P. Wimalasiri, and H. Greenfeld (1984). Dehydroascorbic acid levels in fresh fruit and vegetables in relation to total vitamin C activity. *Journal of Agricultural and Food Chemistry* 32: 836-838.

Chapter 8

How Respiring Plant Cells Limit the Production of Active Oxygen Species

Albert C. Purvis

INTRODUCTION

Diatomic oxygen is the second most abundant component in the earth's atmosphere, comprising nearly 21 percent of the atmospheric gases. Oxygen is essential for efficient energy metabolism and most (85 to 95 percent) of the oxygen consumed by plant (and animal) tissues is reduced to water by the terminal oxidase of the respiratory electron transport chain located in the mitochondria. The remaining 5 to 15 percent of the oxygen consumed is reduced, either fully or partially, by other oxidases or incorporated into various organic molecules by oxygenases. Complete reduction of oxygen to water requires four electrons, and the oxidases involved bind the partially reduced oxygen intermediates until all four electrons have been transferred (Turrens, 1997). Monovalent (one electron) reduction of oxygen to superoxide (O_2^-) may be catalyzed by any one of several oxidases located in the endoplasmic reticulum, plasma membrane, and cell wall, or by the leakage of electrons from single-electron-reduced components in the electron transport chain directly to oxygen. Superoxide is rapidly dismutated to hydrogen peroxide (H_2O_2), either spontaneously or enzymatically through the action of superoxide dismutase. The more stable H_2O_2 readily penetrates organellar membranes and, in the presence of metal reductants such as Fe^{2+}, may be reduced to the highly reactive and degradative hydroxyl radical ($OH\cdot$). Superoxide, H_2O_2, and $OH\cdot$, collectively, are called active oxygen species (AOS). Singlet oxygen is also an AOS that is produced in the thylakoid membranes of stressed plants.

Superoxide and H_2O_2 produced by the oxidases associated with cell walls and plasma membranes are important in cell wall lignification, the detoxification of xenobiotics, and the hypersensitive response to pathogens. A role for O_2^- and H_2O_2 to serve as signals or messengers to increase the production of antioxidants and AOS scavenging enzymes in the tissues of many plants during acclimation or exposure to mild stress also has been documented. Active oxygen species, however, are also responsible for degradative reactions in cells, such as peroxidation of lipids, inactivation and cross-linking of proteins, and base degradation and mutation of DNA (McKersie and Leshem, 1994). Active oxygen species in plant (and animal) tissues are regulated, in part, by an elaborate system that includes AOS scavenging enzymes and lipid-soluble and water-soluble antioxidants (Scandalios, 1993), which are the subject of Chapter 7 in this book.

There is increasing evidence that (1) disruption of metabolic reactions or processes involving electron transfer in plant tissues exposed to any one of several environmental stresses results in increased production, or decreased scavenging of AOS, and that (2) injuries are manifested when the production of AOS increases to the point that the scavenging system is compromised, a situation known as oxidative stress (Scandalios, 1993). Because the mitochondrial electron transport chain is the major consumer of oxygen in plant cells, mitochondria are major producers of AOS when plants are exposed to various environmental stresses (Turrens, 1997). This chapter outlines five mechanisms that potentially reduce the production of AOS by the electron transport chain of mitochondria in stressed plant tissues.

SITES OF ACTIVE OXYGEN SPECIES PRODUCTION IN THE MITOCHONDRIAL ELECTRON TRANSPORT CHAIN

The respiratory electron transport chain located in the inner membrane of mitochondria is composed of four major enzyme complexes embedded in a phospholipid matrix along with a mobile Q pool (Figure 8.1). In addition, plant mitochondria have a NAD(P)H dehydrogenase located on the external surface of the inner mitochondrial membrane and also a rotenone-insensitive NAD(P)H dehydrogenase located on the matrix side of the inner membrane. Electrons derived from the oxidation of organic acids by the tricarboxylic acid (TCA) cycle enzymes located in the matrix of the mitochondria are transferred to the Q pool by complex I (NADH-Q reductase) and complex II (succinate-Q reductase) and the rotenone-insensitive NAD(P)H dehydrogenase, reducing Q to QH_2. Electrons removed during cytoplasmic oxidations are also transferred to the Q pool via the external NAD(P)H dehydrogenase. Complex III (cytochrome bc_1 complex) transfers electrons singly from QH_2 through cytochrome c to complex IV. Complex IV, the major consumer of oxygen in plant tissues, transfers four electrons simultaneously to oxygen reducing it to water. Plant mitochondria have an additional terminal oxidase, the alternative oxidase (AOX), which also reduces oxygen to water.

Intermembrane Space

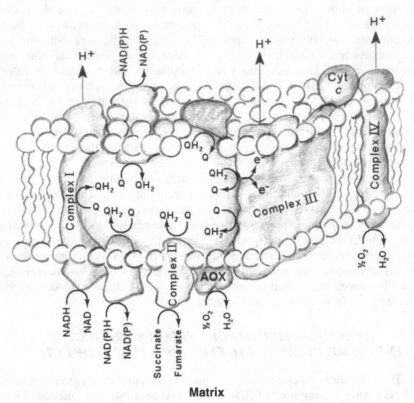

Matrix

FIGURE 8.1. Three-dimensional view of the mitochondrial inner membrane. The primary focus is on the mobile Q pool and the components that reduce Q to QH_2 and the components that transfer electrons from QH_2 to oxygen. The oxidizing components are oriented so that electrons can flow from the mobile Q pool through the components to oxygen. Details of the operation of the Q-cycle (complex III) are in the text.

Complex III, the largest of the protein components of the mitochondrial electron transport chain, consists of several components, including two Q-reaction centers, an iron-sulfur protein, a c-type heme (cytochrome c_1) and two b-type hemes (cytochrome b_{566} and cytochrome b_{562}). The b cytochromes are part of a transmembrane protein that forms an electrical circuit across the inner mitochondrial membrane (Trumpower, 1990). The overall electron pathway within complex III is described by the widely accepted mechanism known as the protonmotive Q-cycle (Mitchell, 1975). The main feature of the Q-cycle hypothesis is the presence of the two Q-reaction centers, the Q-oxidation center on the cytoplasmic side and the Q-reduction center on the matrix side of the inner membrane of the mitochondria (Figure 8.1). Ubiquinol binds to the Q-oxidation center and an electron is transferred first to the iron-sulfur protein, then to cytochrome c_1, cytochrome c, cytochrome oxidase, and ultimately to oxygen. The QH formed at the Q-oxidation center is oxidized by cytochrome b_{566}, which transfers the electron to cytochrome b_{562} near the Q-reduction center. Ubiquinone binds at the Q-reduction center and is reduced to QH by cytochrome b_{562}. The cycle is repeated with the net result that two electrons are transferred from two QH_2 molecules to cytochrome c at the Q-oxidation site and one Q molecule is reduced to QH_2 at the Q-reuction center. Only QH_2 and Q are mobile, i.e., move across the membrane (Mitchell, 1975); the unstable QHs are bound to the Q-oxidation and the Q-reduction centers. For a more detailed description of the operation of the Q-cycle, see Mitchell (1975), Trumpower (1990), and Brandt (1996).

Three proton gradients across the inner mitochondrial membrane are also established when electrons are transferred through complex I, complex III, and complex IV and are dissipated during the phosphorylation of ADP to ATP. Thus, respiratory electron transfer through the mitochondrial electron transport chain and the reduction of oxygen to water are tightly coupled with the synthesis of ATP. In addition, the various components of the membrane have to be properly oriented for electron transfer between components to occur. Stress-induced physical changes in membrane components, such as changes in the fluidity of membrane phospholipids at low temperatures (Lyons and Raison, 1970), as well as low turnover of ATP, result in disruption or restriction of electron transfer through the electron transport chain and can lead to the excessive production of AOS (Purvis and Shewfelt, 1993).

Sites identified in the mitochondrial electron transport chain where single electrons are transferred nonenzymatically to oxygen, reducing it to O_2^-, include the flavoprotein components of the internal and external dehydrogenases and the Q-oxidation and Q-reduction centers of complex III (Rich and Bonner, 1978). During state 3 coupled respiration electron flow through the electron transport chain is rapid and the unstable single-electron-reduced components are short lived. During state 4 respiration, however, electron flow through the electron transport chain is slow and the unstable single-reduced-components are long lived, thus enhancing their chance for the random collision with diatomic oxygen and reducing it to O_2^-.

Two situations favor the production of O_2^- by mitochondria: a high intracellular oxygen concentration and a highly reduced mitochondrial electron transport chain, as in state 4 respiration, when unstable single-electron-reduced components are long lived. At least five potential mechanisms are available or present in plant mitochondria to reduce the production of O_2^- during state 4 respiration or when electron flow through the respiratory chain is otherwise restricted or disrupted, as found in stressed tissues.

INTRACELLULAR OXYGEN CONCENTRATION

Superoxide production increases when the oxygen concentration in the mitochondria milieu increases (Boveris and Chance, 1973). Since O_2^- production by the electron transport chain results from the random collision of oxygen with the unstable single-electron-reduced components, an obvious way to reduce O_2^- production is to optimize the intracellular oxygen concentration. Living organisms consist mainly of water in which oxygen is dissolved, allowing it to cross membranes and move into cells and subcellular compartments and organelles. Solubility of oxygen in cell sap is lower than in water due to other solutes, but the oxygen in solution is in equilibrium with the oxygen in the atmosphere (Burton, 1982). At 25°C, the concentration of oxygen in air-saturated water is approximately 230 µM. Solubility increases as the temperature decreases. Millar et al. (1994) calculated K_m oxygen values of approximately 0.14 µM for cytochrome oxidase and 1.7 µM for the alternative oxidase, whereas the K_m oxygen of phenolase is 150 to 300 µM (Mapson and Burton, 1962). Thus, the terminal oxidases have an extremely high affinity for oxygen compared to other oxidases in organelles and cells. Skulachev (1997) has distinguished three regions of oxygen concentrations by comparing the rates of enzymatic and nonenzymatic oxygen reductions as a function of oxygen concentration. Below 0.5 µM, the oxygen concentration is too low to saturate the terminal oxidases. Respiration is limited by oxygen availability, and nonenzymatic oxygen consumption (O_2^- production) is negligible. Between 0.5 µM and 10 µM, the oxygen concentration is high enough to saturate the terminal oxidases and respiration rate is maximal, but nonenzymatic reactions are still very slow. Above 10 µM, the terminal oxidases are saturated and enzymatic oxygen reduction cannot be enhanced. Superoxide production increases to the point that AOS accumulate in measurable amounts. Thus, optimizing the intracellular concentration of oxygen will reduce the chances for random collision oxygen reduction to O_2^-.

THE ALTERNATIVE OXIDASE

Superoxide production increases when all of the electron carriers in the electron transport chain are highly reduced and oxygen is available. Unlike animal mitochondria, plant mitochondria possess a bifurcated electron transport chain and have an alternative terminal pathway in addition to the universal cytochrome pathway. The alternative pathway oxidizes QH_2 directly and is not coupled to the phosphorylation of ADP. The alternative pathway is comprised of a single dimer protein, the AOX, which is inducible in plant tissues by several stresses, including wounding, low temperature, osmotic, and drought (Wagner and Krab, 1995). It is also induced by inhibitors of the cytochrome pathway, such as carbon monoxide and antimycin A (Minagawa and Yoshimoto, 1987, 1989; Vanlerberghe and McIntosh, 1992; Wagner and Wagner, 1997). Alternative oxidase activity is regulated by the redox status of intermolecular sulfhydryl groups on the enzyme, by the intra- and extra-mitochondrial pyruvate concentration (Umbach et al., 1994), and by the concentration and redox status of the Q pool (Ribas-Carbo et al., 1995; Wagner and Wagner, 1997). Purvis and Shewfelt (1993) were the first to propose that the AOX reduced the production of AOS, since it oxidizes QH_2 directly without the formation of the unstable QHs and is active when the energy consumption of the cells is low. Subsequent studies in which AOX activity was varied by the use of inhibitors of the cytochrome and the alternative pathways, and uncouplers of the electron transport chain and ADP, strongly supported the hypothesis that O_2^- production is reduced when the AOX is active (Purvis et al., 1995; Purvis, 1997; Popov et al., 1997; Wagner and Purvis, 1998; Braidot et al., 1999). Recent studies with transgenic cultured tobacco cells overexpressing and underexpressing the AOX unequivocally confirm that the AOX lowers the production of AOS in plant cells (Maxwell et al., 1999; Parsons et al., 1999). Antisense suppression of the AOX resulted in tobacco cells that produced significantly higher levels of AOS compared with wild-type cells when cytochrome electron transport was inhibited with antimycin A (Maxwell et al., 1999). Overexpression of the AOX resulted in cells with lower production of AOS than in wild-type cells. Alternative pathway respiration increases dramatically in wild-type tobacco cells grown under phosphate limitation, but the antisense cells are unable to induce the AOX (Parsons et al., 1999). Thus, under phosphate limitation, the antisense cells displayed high in vivo rates of generation of AOS compared with wild-type cells, but this difference could be abolished by the addition of an uncoupler of mitochondrial oxidative phosphorylation. The AOX, which does not generate a protonmotive force across the membrane, thus reduces the redox potential of the electron transport chain and the production of AOS in plant mitochondria.

THE PLANT UNCOUPLING MITOCHONDRIAL PROTEIN

In addition to the AOX, some plant mitochondria also possess a proton electrochemical potential dissipating system. A plant uncoupling mitochondrial protein (PUMP) was first described for potato tuber mitochondria by

Vercesi et al. (1995). It is similar to the mitochondrial uncoupling protein linked to the transient thermogenesis in brown adipose tissue of mammals. The plant uncoupling mitochondrial protein is 32 kilodalton (kDa) and has recently been characterized in mitochondria isolated from potato tuber and tomato fruit tissues (Je ek et al., 1996, 1997, 1998; Jarmuszkiewicz et al., 1998; Kowaltowski et al., 1998; Sluse et al., 1998; Almeida et al., 1999). Antibodies raised against the potato PUMP cross-reacted with a 32 kDa PUMP band and its dimer in mitochondria of tomato, corn, and avocado, and mitochondrial fractions prepared from homogenates of banana, apple, peach, melons, orange, and other fruits (Je ek et al., 1998). Thus, PUMP appears to be widespread in the plant kingdom. PUMP is activated by linoleic acid and other fatty acids and is inhibited by various purine nucleotides, such as ATP, GDP, and GTP (Je ek et al., 1996, 1997; Kowaltowski et al., 1998).

Free fatty acids accumulate as a consequence of the activities of several lipases and hydrolases that increase especially during ripening and senescence (Paliyath et al., 1987). The protonated long-chain fatty acids readily cross the mitochondrial inner membrane where they can be oxidized by β-oxidation and the electrons fed to the mitochondrial electron transport chain. Anionic fatty acids are removed by the PUMP from the mitochondrial matrix back into the intermembrane space, resulting in fatty acid recycling and mitochondrial uncoupling (Je ek et al., 1996). Kowaltowski et al. (1998) observed that the addition of the PUMP activator and substrate, linoleic acid, to potato mitochondria significantly decreased respiration-supported H_2O_2 generation. The addition of ATP, which inhibits PUMP activity, and bovine serum albumin (BSA), which removes free fatty acids from the mitochondrial suspension, increased mitochondrial H_2O_2 generation. The uncoupler, carbonyl cyanide *m*-chlorophenylhydrazone, completely inhibited H_2O_2 generation in the presence of BSA and ATP, suggesting that the increase in mitochondrial H_2O_2 generation induced by the inhibition of PUMP depends on the effects of PUMP on mitochondrial membrane potential. Thus, the regulated uncoupling in plant mitochondria may be especially important in reducing AOS production during ripening and senescence when a rapid reduction in ATP production is needed.

Sluse et al. (1998) reported that in tomato fruit mitochondria the PUMP substrate, linoleic acid, inhibited the AOX-sustained respiration. They suggested that AOX and PUMP do not work simultaneously as energy-dissipating systems but sequentially during growth and development. The AOX functions as a redox potential dissipating system that does not form a protonmotive force across the membrane and is active when the demand for ATP is low, whereas PUMP is a proton electrochemical potential dissipating system active in the recycling of free fatty acids across the mitochondrial membrane (Jarmuszkiewicz et al., 1998).

Q-CYCLE INHIBITORS

It has long been known that electron flow through the cytochrome pathway is required in order to generate QH, on the basis that cyanide inhibits succinate-supported O_2^- formation by mitochondria (Rich and Bonner, 1978; Ksenzenko et al., 1983; Konstantinov et al., 1987). An essential feature of the cytochrome pathway is the Q-cycle. In the Q-cycle, QH_2 is oxidized to QH by the transfer of an electron to the iron-sulfur protein at the Q-oxidizing center and ultimately to oxygen, reducing it to water. A second site in the Q-cycle where QH is formed is at the Q-reduction center. At the Q-reduction center, cytochrome b_{562} alternately reduces Q and QH to form QH_2, which then returns to the membranous mobile Q pool (Dawson et al., 1993). When mitochondria are exposed to cyanide, an inhibitor of cytochrome oxidase, all of the components of the electron transport remain fully reduced, as opposed to state 4 respiration, and no unstable QHs are formed in the presence of cyanide.

Myxothiazol, an antibiotic from *Myxococcus fulvus,* is a potent inhibitor of mitochondrial respiration and may interact with the *b* cytochromes and the iron-sulfur protein of the bc_1 complex (von Jagow et al., 1984). It blocks the reduction of the iron-sulfur protein by QH_2 resulting in the complete oxidation of cytochrome c_1, cytochrome *c,* and cytochrome oxidase, unlike the effects of cyanide, while the *b* cytochromes remain entirely reduced. Thus the Q pool becomes increasingly reduced in the presence of myxothiazol (Van den Bergen et al., 1994; Millar et al., 1996). Another compound that apparently binds to the Q-oxidation center and acts similarly to myxothiazol is 5-*n*-undecyl-6-hydrox-4,7-diobenzothiazol (UHDBT) (von Jagaw et al., 1984). The effects of myxothiazol are consistent with its binding at the Q-oxidation center on the cytoplasmic side of the inner mitochondrial membrane and preventing the formation of QH. Myxothiazol prevented the formation of H_2O_2 by rat heart mitochondria (Turrens et al., 1985). Myxothiazol specifically inhibited O_2^- formation at complex III and reduced the rate of hydroperoxide formation in rat hepatocytes following their subjection to reductive stress (Dawson et al., 1993). During conditions of severe hypoxia or ischemia, oxidation-reduction components that are normally oxidized in the aerobic state become reduced. When oxygen is restored, this reductive stress may promote intracellular formation of AOS that attack lipids, thiols, and other cellular components, culminating in lethal cell injury. Skulachev (1996) suggests that continuous oscillation between aerobic and anaerobic conditions is especially dangerous for cells because aeration after anoxia appears to be very favorable for one-electron reductions.

In contrast to myxothiazol and UHDBT, antimycin is a complex III inhibitor that promotes the formation of AOS in mitochondria of several species (Boveris and Chance, 1973; Turrens and Boveris, 1980; Ksenzenko et al., 1983; Konstantinov et al., 1987; Minagawa and Yoshimoto, 1987; Purvis et al., 1995; Maxwell et al., 1999; Parsons et al., 1999). There is evidence of competition between Q and antimycin for the antimycin-binding site as revealed by spectroscopic analyses which suggests that these compounds act at or very close to cytochrome b_{562} (Tyler, 1992). Antimycin may act by displacing the QH from the Q-reduction center, thus short-circuiting the Q-cycle and causing the rapid dismutation of QH to form Q and QH_2 (Tyler, 1992). The displaced QH, however, could also be oxidized by oxygen producing O_2^-. The b cytochromes become fully reduced in the presence of antimycin, while the iron-sulfur protein, cytochrome c_1, cytochrome c, and cytochrome oxidase are fully oxidized. Thus, the oxidation of QH_2 at the Q-oxidation center can continue, but since the b cytochromes are fully reduced, the QH reduces molecular oxygen to O_2^-. In the presence of both antimycin and myxothiazol, oxidation of QH_2 to QH is prevented, and the production of O_2^- is reduced (A. C. Purvis, unpublished data). The compound 2-n-heptyl-4-hydroxyquinoline N-oxide (HOQNO), similar to antimycin, also appears to bind at the Q-reduction center (Tyler, 1992).

THE Q POOL

Ubiquinone, present in large molar excess over the other respiratory components (Søballe and Poole, 2000), plays a central role in the mitochondrial electron transport chain. The Q pool is reduced by the various dehydrogenases and is oxidized by the cytochrome and the alternative pathways. The completely reduced and oxidized forms of lipid-soluble Q apparently move freely through the hydrophobic part of the inner mitochondrial membrane. The concentration of the Q pool increased about twofold in green bell peppers stored at low temperature and in *Petunia hybrida* cell suspensions cultured in the presence of antimycin A, an inhibitor of the cytochrome pathway (Wagner and Purvis, 1998; Wagner and Wagner, 1997). Furthermore, a large part of the Q pool appeared to be "unoxidizable" in that 30 to 40 percent of the pool remained reduced in the absence of respiratory substrates. Although the concentration and redox status of the Q pool may alter the production of O_2^- by regulating AOX activity in plant mitochondria (Purvis and Shewfelt, 1993; Purvis et al., 1995; Purvis, 1997; Popov et al., 1997; Maxwell et al., 1999; Parsons et al., 1999), QH_2 may also directly scavenge AOS (Beyer, 1992; Søballe and Poole, 2000). Ubiquinol inhibited the NADPH- and ADP-Fe^{2+}-supported peroxidation of lipid catalyzed by rat liver microsomes (Beyer, 1992). The concentration of the Q pool increased in several tissues of rats during prolonged exposure to cold and enforced and voluntary endurance training (Beyer et al., 1962, 1984). The rate of O_2^- production by rapidly respiring wild-type *Eschericia coli* membranes was twofold higher than in the slowly respiring membranes from a *ubi*CA knockout mutant (Søballe and Poole, 2000). However, large amounts of O_2^- accumulated in the Ubi$^-$ membranes compared to wild-type membranes, which possess O_2^--scavenging QH_2. Thus, the concentration and redox status of the Q pool can protect the inner mitochondrial membrane from damage inflicted by AOS not only by regulating alternative oxidase activity and thereby reducing the production of AOS but also by scavenging AOS at the sites of its formation.

CONCLUSION

Many sites produce AOS in respiring plant (and animal) cells. This chapter dealt primarily with only one of those sites, O_2^- production by complex III of the mitochondrial electron transport chain. McLennan and Degli Espositi (2000) concluded that complex I and complex II normally generate most of the AOS produced by respiring rat liver mitochondria, but that the contribution of complex III to AOS production can become relevant during certain conditions of cellular stress. Stresses are imposed on horticultural commodities, especially during harvesting, handling, and storage, which alter their metabolism. This chapter focused on five potential mechanisms, one of which is unique to plant mitochondria, the AOX, which can reduce O_2^- production by complex III.

ISSUES AND TOPICS FOR FUTURE RESEARCH

Regulating the production of AOS may be especially important during the postharvest handling of fresh fruits and vegetables. Decreasing the intracellular oxygen concentration and reducing the production of AOS currently forms a part of the basis for controlled and modified atmosphere storage of fresh fruits and vegetables. The overexpression of AOX in plant mitochondria has been found to reduce the production of AOS and lipid peroxidation in plant cells, but whether fresh fruits and vegetables from plants with mitochondria modified to overexpress AOX resist stresses imposed during postharvest handling has yet to be studied. It is also possible to modify plants to have mitochondria with enhanced PUMP activity. Treatments that increase the Q concentration in the mitochondrial inner membrane can moderate AOS production, but few studies with such treatments have been made. Although compounds are available that inhibit the production of QHs and the production of AOS, their usefulness in postharvest handling of fresh fruits and vegetables is not readily apparent.

REFERENCES

Almeida, A. M., W. Jarmuszkiewicz, H. Khomsi, P. Arruda, A. E. Vercesi, and F. E. Sluse (1999). Cyanide-resistant, ATP-synthesis-sustained, and uncoupling-protein-sustained respiration during postharvest ripening of tomato fruit. *Plant Physiology* 119: 1323-1329.

Beyer, R. E. (1992). An analysis of the role of coenzyme Q in free radical generation and as an antioxidant. *Biochemistry and Cell Biology* 70: 390-403.

Beyer, R. E., P. G. Morales-Corral, B. J. Ramp, K. R. Kreitman, M. J. Falzon, S. Y. S. Rhee, T. W. Kuhn, M. Stein, M. J. Rosenwasser, and K. J. Cartwright (1984). Elevation of tissue coenzyme Q (ubiquinone) and cytochrome *c* concentrations by endurance exercise in the rat. *Archives of Biochemistry and Biophysics* 234: 323-329.

Beyer, R. E., W. M. Noble, and T. J. Hirschfeld (1962). Coenzyme Q (ubiquinone) levels of tissues of rats during acclimation to cold. *Canadian Journal of Biochemistry and Physiology* 40: 511-518.

Boveris, A. and B. Chance (1973). The mitochondrial generation of hydrogen peroxide. *Biochemistry Journal* 134: 707-716.

Braidot, E., E. Petrussa, A. Vianello, and F. Macrì (1999). Hydrogen peroxide generation by higher plant mitochondria oxidizing complex I or complex II substrates. *Federation of European Biochemical Societies Letters* 451: 347-350.

Brandt, U. (1996). Energy conservation by bifurcated electron-transfer in the cytochrome-bc_1 complex. *Biochimica et Biophysica Acta* 1275: 41-46.

Burton, W. G. (1982). The physiological implications of structure: Exchange of gases. In *Post-Harvest Physiology of Food Crops*. London and New York: Longman, pp. 69-96.

Dawson, T. L., G. J. Gores, A.-L. Nieminen, B. Herman, and J. J. Lemasters (1993). Mitochondria as a source of reactive oxygen species during reductive stress in rat hepatocytes. *American Journal of Physiology* 264 (*Cell Physiology* 33): C961-C967.

Jarmuszkiewicz, W., A. M. Almeida, C. M. Sluse-Goffart, F. E. Sluse, and A. E. Vercesi (1998). Linoleic acid-induced activity of plant uncoupling mitochondrial protein in purified tomato fruit mitochondria during resting, phosphorylating, and progressively uncoupled respiration. *Journal of Biological Chemistry* 273: 34882-34886.

Je ek, P., A. D. T. Costa, and A. E. Vercesi (1996). Evidence for anion-translocating plant uncoupling mitochondrial protein in potato mitochondria. *Journal of Biological Chemistry* 271: 32743-32748.

Je ek, P., A. D. T. Costa, and A. E. Vercesi (1997). Reconstituted plant uncoupling mitochondrial protein allows for proton translocation via fatty acid cycling mechanism. *Journal of Biological Chemistry* 272: 24272-24278.

Je ek, P., H. Engstová, M. áèková, A. E. Vercesi, A. D. T. Costa, P. Arruda, and K. D. Garlid (1998). Fatty acid cycling mechanism and mitochondrial uncoupling proteins. *Biochimica et Biophysica Acta* 1365: 319-327.

Konstantinov, A. A., A. V. Peskin, E. Y. Popova, G. B. Khomutov, and E. K. Ruuge (1987). Superoxide generation by the respiratory chain of tumor mitochondria. *Biochimica et Biophysica Acta* 894: 1-10.

Kowaltowski, A. J., A. D. T. Costa, and A. E. Vercesi (1998). Activation of the potato plant uncoupling mitochondria protein inhibit reactive oxygen species generation by the respiratory chain. *Federation of European Biochemical Societies Letters* 425: 213-216.

Ksenzenko, M., A. A. Konstantinov, G. B. Khomutov, A. N. Tikhonov, and E. K. Ruuge (1983). Effect of electron transfer inhibitors on superoxide generation in the cytochrome bc_1 site of the mitochondrial respiratory chain. *Federation of European Biochemical Societies Letters* 155: 19-24.

Lyons, J. M. and J. K. Raison (1970). Oxidative activity of mitochondria isolated from plant tissues sensitive and resistant to chilling injury. *Plant Physiology* 45: 386-389.

Mapson, L. W. and W. G. Burton (1962). The terminal oxidases of the potato tuber. *Biochemistry Journal* 82: 19-5.

Maxwell, D. P., Y. Wang, and L. McIntosh (1999). The alternative oxidase lowers mitochondrial reactive oxygen production in plant cells. *Proceedings of the National Academy of Sciences USA* 99: 8271-8276.

McKersie, B. D. and Y. Y. Leshem (1994). Oxidative stress. In *Stress and Stress Coping in Cultivated Plants.* Dordrecht, The Netherlands: Kluwer Academic Publishers, pp. 15-54.

McLennan, H. R. and M. Degli Espositi (2000). The contribution of mitochondrial respiratory complexes to the production of reactive oxygen species. *Journal of Bioenergetics and Biomembranes* 32: 153-162.

Millar, A. H., F. J. Bergersen, and D. A. Day (1994). Oxygen affinity of terminal oxidases in soybean mitochondria. *Plant Physiology and Biochemistry* 32: 847-852.

Millar, A. H., M. H. N. Hoefnagel, D. A. Day and J. T. Wiskich (1996). Specificity of the organic acid activation of alternative oxidase in plant mitochondria. *Plant Physiology* 111: 613-618.

Minagawa, N. and A. Yoshimoto (1987). The induction of cyanide-resistant respiration in the absence of respiratory inhibitors in *Hansenula anomala. Agricultural and Biological Chemistry* 51: 2263-2265.

Minagawa, N. and A. Yoshimoto (1989). The induction of cyanide-resistant respiration by carbon monoxide in *Hansenula anomala. Agricultural and Biological Chemistry* 53: 2025-2026.

Mitchell, P. (1975). The protonmotive Q cycle: A general formulation. *Federation of European Biochemical Societies Letters* 59: 137-139.

Paliyath, G., D. V. Lynch, and J. E. Thompson (1987). Regulation of membrane phospholipid catabolism in senescing carnation flowers. *Physiologia Plantarum* 71: 503-511.

Parsons, H. L., J. Y. H. Yip, and G. C. Vanlerberghe (1999). Increased respiratory restriction during phosphate-limited growth in transgenic tobacco cells lacking alternative oxidase. *Plant Physiology* 121: 1309-1320.

Popov, V. N., R. A. Simonian, V. P. Skulachev, and A. A. Starkov (1997). Inhibition of the alternative oxidase stimulates H_2O_2 production in plant mitochondria. *Federation of European Biochemical Societies Letters* 415: 87-90.

Purvis, A. C. (1997). Role of the alternative oxidase in limiting superoxide production by plant mitochondria. *Physiologia Plantarum* 100: 165-170.

Purvis, A. C. and R. L. Shewfelt (1993). Does the alternative pathway ameliorate chilling injury in sensitive plant tissues? *Physiologia Plantarum* 88: 712-718.

Purvis, A. C., R. L. Shewfelt, and J. W. Gegogeine (1995). Superoxide production by mitochondria isolated from green bell pepper fruit. *Physiologia Plantarum* 94: 743-749.

Ribas-Carbo, M., J. T. Wiskich, J. A. Berry, and J. N. Siedow (1995). Ubiquinone redox behavior in plant mitochondria during electron transport. *Archives of Biochemistry and Biophysics* 317: 156-160.

Rich, P. R. and W. D. Bonner Jr. (1978). The sites of superoxide anion generation in higher plant mitochondria. *Archives of Biochemistry and Biophysics* 188: 206-213.

Scandalios, J. G. (1993). Oxygen stress and superoxide dismutases. *Plant Physiology* 101: 7-12.

Skulachev, V. P. (1996). Role of uncoupled and non-coupled oxidations in maintenance of safely low levels of oxygen and its one-electron reductants. *Quarterly Reviews of Biophysics* 29: 169-202.

Skulachev, V. P. (1997). Membrane-linked systems preventing superoxide formation. *Bioscience Reports* 17: 347-366.

Sluse, F. E., A. M. Almeida, W. Jarmuszkiewicz, and A. E. Vercesi (1998). Free fatty acids regulate the uncoupling protein and alternative oxidase activities in plant mitochondria. *Federation of European Biochemical Societies Letters* 433: 237-240.

Søballe, B. and R. K. Poole (2000). Ubiquinone limits oxidative stress in *Escherichia coli*. *Microbiology* 146: 787-796.

Trumpower, B. L. (1990). The protonmotive Q cycle. *Journal of Biological Chemistry* 265: 11409-11412.

Turrens, J. F. (1997). Superoxide production by the mitochondrial respiratory chain. *Bioscience Reports* 17: 3-8.

Turrens, J. F., A. Alexandre, and A. L. Lehninger (1985). Ubisemiquinone is the electron donor for superoxide formation by complex III of heart mitochondria. *Archives of Biochemistry and Biophysics* 237: 408-414.

Turrens, J. F. and A. Boveris (1980). Generation of superoxide anion by the NADH dehydrogenase of bovine heart mitochondria. *Biochemical Journal* 191: 421-427.

Tyler, D. D. (1992). Respiratory enzyme systems of mitochondria. In *The Mitochondrion in Health and Disease*. New York:VCH Publishers, Inc., pp. 270-351.

Umbach, A. L., J. T. Wiskich, and J. N. Siedow (1994). Regulation of alternative oxidase kinetics by pyruvate and intermolecular disulfide bond redox status in soybean seedling mitochondria. *Federation of European Biochemical Societies Letters* 348: 181-184.

Van den Bergen, C. W. M., A. M. Wagner, and A. L. Moore (1994). The relationship between electron flux and the redox poise of the quinone pool in plant mitochondria. *European Journal of Biochemistry* 226: 1071-1078.

Vanlerberghe, G. C. and L. McIntosh (1992). Coordinate regulation of cytochrome and alternative pathway respiration in tobacco. *Plant Physiology* 100: 1846-1851.

Vercesi, A. E., I. S. Martins, M. A. P. Silva, H. M. F. Lelte, I. M. Cuccovia, and H. Chalmovich (1995). PUMPing plants. *Nature* 375: 24.

von Jagow, G., P. O. Ljungdahl, P. Graf, T. Ohnishi, and B. L. Trumpower (1984). An inhibitor of mitochondrial respiration which binds to cytochrome *b* and displaces quinone from the iron-sulfur protein of the cytochrome bc_1 complex. *Journal of Biological Chemistry* 259: 6318-6326.

Wagner, A. M. and K. Krab (1995). The alternative respiration pathway in plants: Role and regulation. *Physiologia Plantarum* 95: 318-325.

Wagner, A. M. and A. C. Purvis (1998). Production of reactive oxygen species in plant mitochondria. A dual role for ubiquinone? In *Plant Mitochondria: From Gene to Function,* I. M. Møller, P. Gardeström, K. Glimelius, and E. Glaser (Eds.). Leiden, The Netherlands: Backhuys Publishers, pp. 537-541.

Wagner, A. M. and M. J. Wagner (1997). Changes in mitochondrial respiratory chain components of petunia cells during culture in the presence of antimycin A. *Plant Physiology* 115: 617-622.

Chapter 9

Physiological Effects of Oxidative Stress in Relation to Ethylene in Postharvest Produce

Andrea Masia

INTRODUCTION

Oxidative Stress

Oxidative stress occurs when the rate of formation of prooxidant molecules exceeds the rate of their removal by an organism's antioxidant defense mechanisms (Hippeli and Elstner, 1996; Levine, 1999). An antioxidant is any compound capable of quenching reactive oxygen species (ROS) without undergoing conversion to a destructive radical. Aerobic organisms primarily utilize dioxygen (O_2) as a hydrogen and electron acceptor to generate water (Elstner, 1982; Levine, 1999). Oxygen is also incorporated into organic molecules through several biochemical pathways and reactions, such as enzyme catalysis (e.g., RuBP oxygenase, hydroxylases, phenoloxidases), nonenzymatic chemical reactions (reduction of oxygen and incorporation of reduced species), and physical (photodynamic) activation (Elstner, 1982). This complexity in oxygen incorporation is derived from the ground state situation of dioxygen, with its two unpaired electrons with parallel spins (triplet) in the outer electronic shell, and the consequent low reactivity toward other molecules presenting the more common (singlet) ground state with paired electrons. Triplet oxygen, owing to the spin restriction principle which requires that electrons entering the orbital must have an opposite spin, can react with electron donor molecules only one e^- at a time, i.e., react to radical species or unpaired e^- (Leshem, Halevy, Frenkel, and Frimer, 1986). The O_2 molecule can then accept four e^- (as well as four H^+), with the production of two molecules of water (Levine, 1999). The overall reaction of O_2 reduction is:

$$O_2 + 4\ e^- + 4H^+ \rightarrow 2H_2O$$

During the four linked univalent reductions of dioxygen, different active oxygen species (AOS) are generated. The four-step reaction chain requires initiation for the first endothermic one, while all the subsequent exothermic steps occur spontaneously, either catalyzed or uncatalyzed (Elstner, 1982).

$$\begin{array}{ccccccccc}
 & e^- & & e^- & & e^- & & e^- & \\
 & \downarrow & & \downarrow & & \downarrow & & \downarrow & \\
O_2 & \rightarrow & O_2{\cdot}^- & \rightarrow & HO_2{\cdot} & \rightarrow & H_2O_2 & \rightarrow & OH{\cdot} + H_2O \rightarrow 2H_2O \\
 & \uparrow & & \uparrow & & \uparrow & & \uparrow & \\
 & H^+ & & H^+ & & H^+ & & H^+ &
\end{array}$$

Unfortunately, the chemistry of oxygen is also related to the production AOS, namely oxygen free radicals plus hydrogen peroxide (H_2O_2). The term *free radical* describes any molecule that has unpaired electrons in its outer electronic orbitals, which begets another radical when it reacts with a nonradical, and is normally used for relatively stable radicals whose longevity is enough to exist separately and react with other molecules. The most common free radical in the atmosphere is the oxygen molecule itself, which with the two unpaired electrons in the outer shell constitutes a biradical molecule. Fortunately, its reactivity at the ground state is decreased considerably by the spin restriction principle. This restriction no longer exists in the reduced oxygen species, as they have only one unpaired electron in the outer electron shell. The superoxide radical ($O_2{\cdot}^-$) is negatively charged, cannot cross membranes, has a very short life, and, if not enzymatically catalyzed, reacts only with a limited number of biomolecules (Levine, 1999). However, if rapid and immediate scavenging of $O_2{\cdot}^-$ at the site of its generation is not performed, it can react with proteins, nucleic acids, carbohydrates, and lipids, in particular polyunsaturated fatty acids (PUFAs) (Leshem, Halevy, Frenkel, and Frimer, 1986). Superoxide is rapidly converted to membrane-diffusable H_2O_2 by superoxide dismutase (SOD; EC 1.15.1.1), ascorbate, thiols, or spontaneous dismutation (Knorzer et al., 1996; Foyer et al., 1997; Yamasaki et al., 1997; Levine, 1999). In thylakoids, H_2O_2 is photoproduced via the spontaneous disproportionation of $O{\cdot}^-$ but not directly through the two-electron reduction of O_2 (Asada, 1999). The toxicity of hydrogen peroxide itself is relatively weak in comparison with that of other AOS, but in the presence of $O_2{\cdot}^-$ it can generate the reactive hydroxyl radical ($OH{\cdot}$), one of the most reactive molecules known. This radical is formed very rapidly in the presence of metal ions such as Fe^{2+} or Cu^{2+}, (metal-catalyzed Haber-Weiss reaction), and it can react with almost any biomolecule, from proteins to nucleic acids and lipids (Scandalios, 1993; Shen et al., 1997; Asada, 1999).

Hydrogen peroxide represents a crucial crossing in the mechanism of oxidative stress, and its constraint inside a physiological range represents a crucial struggle in cell life. It is, therefore, not so strange that a plethora of different mechanisms is utilized by cells to scavenge the excess of H_2O_2. Many different enzymes are involved in this task; ascorbate peroxidase (APX; EC 1.11.1.11) and related enzymes in chloroplasts, catalase (CAT; EC 1.11.1.6) in peroxisomes, glyoxisomes, and mitochondria, and the cytoplasm-associated anionic peroxidase (POX; EC 1.11.1.7) are considered the primary ones (Asada, 1992; Zheng and van Huystee, 1992; Foyer et al., 1994; Knorzer et al., 1996; Yamasaki et al., 1997; Noctor and Foyer, 1998).

Hydrogen peroxide, a membrane-permeable molecule, has been found to act as a diffusable intercellular signal involved in many different stress factors, such as chilling (Prasad et al., 1994), heat (Dat et al., 1998), and pathogen defense (Levine, 1999). It induces a number of genes and proteins involved in stress defense, such as CAT, POX, glutathione S-transferase (GST; EC 2.5.1.18), SOD, pathogenesis-related proteins, and alternative oxidase (Prasad et al., 1994; Foyer et al., 1997; Morita et al., 1999). It seems likely that H_2O_2 accumulation is involved in cytosolic APX regulation in some, if not all, stress conditions (Morita et al., 1999). These and other results support the hypothesis that common redox signals are involved in the induction of acclimatory responses to both abiotic and biotic stresses.

Postharvest Aging As the Final Step in the Ripening-Senescence Syndrome

Senescence in plants is strictly related to the species and organ being considered. In annuals, flowers, fruits, leaves, and finally roots all senesce and die, usually in this order. In deciduous trees, only flowers, fruits, and leaves generally senesce and abscise, and death may be due to infection or other environmental causes rather than to true senescence (Thimann, 1987). In fruits, at least the so-called climacteric ones, the first stages of senescence are related to ripening, the phase usually associated with the development of optimal eating quality and representing the final stages of maturation. The subsequent senescence processes, or aging, are normally slowed down for long periods by low-temperature storage treatments. The last stage, decomposition, seems to be largely a function of secondary infections (Thimann, 1987; Ludford, 1995).

Oxidative Stress Involvement in the Aging Mechanism

Fruit ripening and senescence may be regarded as an oxidative process involving marked alterations in fruit metabolism and in the activity of many enzymatic classes, including those related to the regulation of AOS. Active oxygen species are continuously produced in all aerobic biological systems, and their generation is usually maintained at levels compatible with a cell's life by the antioxidant machinery of the cell (Figure 9.1). Therefore, mild oxidative stress (OS) is a normal facet of a cell's life cycle (Foyer et al., 1994; Noctor and Foyer, 1998). Only under stressful conditions and/or in peculiar stages of cell/organ development, such as fruit senescence, are the antioxidant systems often overwhelmed by AOS generated within the cell (Sen Gupta et al., 1993; Hippeli and Elstner, 1996; Foyer et al., 1997). The effect of AOS on cells and tissues depends upon their concentration and longevity and can induce either antioxidant responses, apoptosis (programmed cell death) or tissue necrosis. When OS is strong and prolonged, it induces senescence of the stressed compartments, along with partial loosening of the regulatory machinery (Hippeli and Elstner, 1996; Foyer et al., 1997). Lipid peroxidation may be regarded as one of the earliest detectable processes, occurring during fruit ripening (Meir et al., 1991). The accumulation of peroxidized lipids in senescing membranes, mainly due to their decreased blebbing capacity, contributes to loss of membrane function (Hudak et al., 1995). At the end of the oxidative process, aldehydes and lipofuscins are chemically produced from peroxidized lipids and protein amino groups (Meir et al., 1991; Hippeli and Elstner, 1996). These molecules contribute to cell necrosis and, in addition to phenoloxidases and peroxidase products (i.e., o-quinones from o-dihydroxy phenols), are responsible for browning effects (Hippeli and Elstner, 1996).

Do the enzymatic processes that generate AOS control the rate of aging? The answer to this question seems to be affirmative. In ripening, the activity of AOS-producing enzymes such as xanthine oxidase and glycolate oxidase increases. Hydrogen peroxide levels are also increased by the concomitant decrease in catalase activity. Moreover, activity of a specific cell wall peroxidase, strictly related to fruit ripening, is stimulated (Leshem, Halevy, Frenkel, and Frimer, 1986; Levine, 1999). Active oxygen species can also be generated in nonenzymatic reactions. In fact, many different systems are able to act as electron donors, including components of biological e⁻ transport systems (in chloroplasts, mitochondria, microsomes), reduced metal ions, and solvated e⁻ produced by radiation (Elstner, 1982).

Lipid Peroxidation and Membrane Permeability

Any condition in which cellular redox homeostasis is disrupted can be defined as oxidative stress (Alscher et al., 1997). The processes of fruit ripening and senescence may be linked to oxidative phenomena (Brennan and Frenkel, 1977; Masia, 1998). Several lines of evidence indicate that perox-

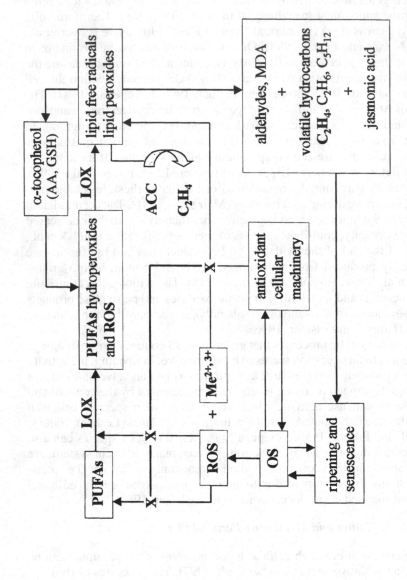

FIGURE 9.1. Mechanisms of lipid peroxidation in membrane destabilization during ripening and senescence. Details are provided in text.

idation of membrane lipids is deeply involved in ripening and senescence of plant tissues (Hudak et al., 1995; Izzo et al., 1995; Shewfelt and Purvis, 1995). Levels of H_2O_2 and lipid hydroperoxides also increase in ripening fruits (Meir et al., 1991; Hudak et al., 1995). The ability of membranes to eliminate undesired cellular catabolites by the blebbing process decreases with the progression of senescence; this decrease may be due in part to the decrease in bulk membrane fluidity that is characteristic of senescence (Hudak et al., 1995). Senescence and aging are reflected in a progressive increase in cell membrane permeability and a concomitant loss in membrane function (Izzo et al., 1995). Different environmental stress factors such as chilling, water stress, mechanical injury, and air pollutants are also able to induce, at the cellular level, similar results (Hippeli and Elstner, 1996; Levine, 1999). Membrane destabilization has been imputed to lipid peroxidation, due to an overproduction of AOS (Du and Bramlage, 1995).

Is any role played by ethylene in the lipoxygenase-controlled lipid peroxidation pathway? A strict correlation seems to exist between the enzyme and the hormone, especially during stressful conditions (Figure 9.1). The initial products of lipoxygenase (LOX; EC 1.13.11.12) action are the fatty acid hydroperoxides. Decomposition of these products give rise to aldehydes such as malonyldialdehyde (MDA) and low molecular weight hydrocarbons such as ethane, hexenals, pentane, and ethylene (C_2H_4) (Leshem, Halevy, and Frenkel, 1986). Moreover, enhanced ethylene synthesis in stressed in vivo tissues requires both promoted synthesis of ACC (1-aminocyclopropane-1-carboxylic acid) and easier chemical conversion of ACC to ethylene derived from a LOX-mediated increase of lipoperoxides (Kacperska and Kubacka-Zebalska, 1989), although this role for LOX-mediated in vivo source of ethylene has subsequently been refuted (Siedow, 1991).

Lipid peroxidation can be the result of two different oxidative routes. The interaction of AOS, qua superoxide, and H_2O_2 with iron species (Fe^{2+}, Fe^{3+}) in membranes leads to the formation of lipid free radicals (Shewfelt and Purvis, 1995). Membrane lipids are likely peroxidized by the hydroxyl radical generated in the membrane in close proximity to a PUFA. The resulting lipid peroxyl radical (LOO·) is rapidly converted to a lipid hydroperoxide (LOOH) by interaction with other fatty acids, α-tocopherol, and/or other hydrogen donors. Lipid hydroperoxides are also directly produced by the lipoxygenase-mediated oxidation of free fatty acids derived from the concomitant action of hydrolytic enzymes on membrane lipids. Under normal conditions in the cell, minimal peroxidation of membrane lipids and proteins occurs due to the efficiency of defense mechanisms. Under oxidative stress conditions, the cell antioxidant enzymatic machinery is unable to control the upsurge of AOS and of lipid hydroperoxides, the result being a peroxidative attack (Figure 9.1). Recently, a new pathway, activated under stress induced by excess H_2O_2, has been proposed, with the induction of glutathione S-transferase and glutathione peroxidases (Eshdat et al., 1997).

A comprehensive model for lipid peroxidation (Shewfelt and Purvis, 1995) has been sketched in recent years as follows: AOS are continuously produced as by-products of natural metabolic processes in the cell (LOX activity included), in particular superoxide and H_2O_2, and can interact with membranes in the presence of iron (e.g., Fe^{2+}, Fe^{3+}) or other transition metal ions (Shen et al., 1997), leading to the formation of lipid free radicals. Lipid hydroperoxides are also formed in membranes by the action of LOX on free fatty acids. For each of these sources of lipid peroxidation in the cell membrane, a different defense mechanism is at work. Active oxygen species are degraded by antioxidant enzymes such as SOD, CAT, APX, and the like, and antioxidant compounds such as α-tocopherol, β-carotene, and ascorbic acid (AA), while LOX activity is controlled by compartmentation and the concomitant presence of PUFAs in the (enzymatic attack hindered) esterified form. Membrane defense systems are usually able to protect membrane lipids from OS damage induced by continuous AOS production, and any peroxidized lipids and proteins are quickly repaired. Under stress conditions, the levels of AOS increase, often stimulating the synthesis of defense molecules and antioxidant enzymes such as SOD, CAT, APX, and GPX (Elstner, 1982; Scandalios, 1993; Noctor and Foyer, 1998; Levine, 1999). If the OS is severe enough and/or prolonged in duration, however, the level of prooxidants can exceed the defense capacity of the cell, resulting in damage mainly at the membrane level with the peroxidation of PUFAs, which is the most kinetically favored reaction (Shewfelt and Purvis, 1995). Polyunsaturated fatty acid peroxidation can, to a certain extent, act to protect against the direct attack of proteins and nucleic acids by AOS. As lipid free radicals are converted to lipid hydroperoxides, α-tocopherol concentration decreases, particularly as a consequence of the depletion of AA, glutathione (GSH), and other cell reductants. When the tocopherol levels in the membrane reach a concentration that is too low to protect the lipids, free-radical chain propagation reactions occur at a much faster rate than repairing mechanisms, resulting in modification of the membrane's physical properties. The subsequent direct peroxidation of proteins, derived primarily from the action of lipid free radicals, leads to decreased enzymatic activity, metabolic imbalances, cellular dysfunction, and tissue disorders (Shewfelt and Purvis, 1995).

Also known as vitamin E, α-tocopherol is a lipophilic molecule with a high reducing capacity, and it is the most important antioxidant in lipophilic media such as membranes, which are a preferential target for attack by free radicals (Fryer, 1992). It is also an effective radical scavenger for both singlet oxygen and organic peroxyl. The principal role of α-tocopherol in all biological membranes is to act as a highly efficient and recyclable chain reaction terminator for the removal of PUFA radical species generated during lipid peroxidation. The oxidized vitamin E radical is then recycled back to α-tocopherol by interaction with either ascorbate or GSH. It also contributes to membrane stabilization by incorporation into membranes with the phytyl chain, and seems to significantly influence membrane fluidity (Fryer, 1992; Shewfelt and Purvis, 1995).

Similar to α-tocopherol, ascorbic acid can scavenge hydroxyl radicals, singlet oxygen, and superoxide. Ascorbate also acts as a reductant in the regeneration of α-tocopherol and in the zeaxanthin cycle (Foyer, 1993). Ascorbate preserves the activities of the enzymes containing prosthetic transition metal ions. Ascorbate is present at high concentrations in many compartments of the cell, for example, in chloroplasts it is about 25 mM. It is the most important antioxidant molecule in plants, with a paramount role in the removal of H_2O_2 through the glutathione-ascorbate cycle (Foyer et al., 1994; Noctor and Foyer, 1998).

Glutathione is a tripeptide bearing a thiol group and has a pivotal role in antioxidant defense. It acts as a disulphide reductant, protecting thiols on enzymes; regenerates ascorbate from dehydroascorbate (DHA) via the enzyme dehydroascorbate reductase (DHAR; EC 1.8.5.1); and reacts with singlet oxygen and hydroxyl radicals. In normal conditions it is found mostly in the tripeptide reduced form, while, as a consequence of severe stress, GSH explicates its antioxidant activity and forms the oxidized glutathione disulphide (GSSG) molecule that is reduced back to GSH by glutathione reductase (GR; EC 1.6.4.2). An excess of GSSG in the cell can inactivate enzymes by the formation of mixed disulphides (Foyer et al., 1997).

Mechanism of Senescence and Aging

In avocado fruit, lipid peroxidation, which involves free-radical formation, precedes ripening (Meir et al., 1991). The process of senescence during ripening is started by oxidative deterioration of cell membranes and precedes the other characteristic biochemical and enzymatic processes. In fact, the accumulation of lipid peroxidation products starts before the onset of climacteric ethylene and respiration and the decrease in fruit firmness. At the onset of ripening, rapid membrane changes increase membrane permeability, inducing a resultant decompartmentation of cellular components, an increase in catabolic processes, thus initiating fruit senescence. In peach fruit, the greatest changes in lipids composition occur at the end of the SIII growth phase, corresponding to the onset of the climacteric (Izzo et al., 1995). During ripening, senescence, and abiotic stresses, an increase in the free sterol/phospholipid molar ratio has been related to a decrease in membrane fluidity. In ripening peach fruit, a significant increase in the double bond index was found in all lipid classes, indicating a progressive increase in desaturases activity. Microsomal desaturases are activated during senescence, with the subsequent formation of polyunsaturated molecular species prone to lipolytic hydrolysis, disorganization of membranes, an increase in catabolic processes and subsequent fruit aging and senescence. Senescence development in plant tissues is generally paralleled by a huge increase in toxic AOS, a likely result of rapidly modified enzymatic activities that are genetically determined and controlled (Leshem, Halevy, Frenkel, and Frimer, 1986). Oxidative processes may have a deep impact on hormonal levels and activities, and this is particularly true for ethylene. Furthermore, oxidative processes may have a general effect on the transition from the reduced state of nonsenescing cells to the highly oxidized state of aging cells. Ripening can, therefore, be seen as a protracted form of senescence during which the OS increases progressively, mainly as a consequence of a declining activity of key enzymes responsible for the antioxidant cellular machinery. In the climacteric saskatoon fruit, ripening is paralleled by a progressive rise in OS that dramatically affects the upsurge in polar lipids (Rogiers et al., 1998). An increased production of LOOH, derived from the higher LOX activity, in conjunction with a decrease in SOD and CAT activities and a concomitant increase in respiration, all contribute to an upsurge of cellular oxidative status. The surge in LOOH and in AOS eventually induces higher activities of POX, GR, and GST during the later stages of ripening. Fruit maturation and ripening, along with a steady decrease in respiration, are all senescence phenomena and are closely linked to ethylene physiology (Ludford, 1995). In climacteric fruit, a pronounced increase in the production of C_2H_4 precedes the surge in respiration while the inhibition of ethylene synthesis and/or perception inhibit ripening. Exogenous applications of C_2H_4 promote ripening in such different crops as apple, banana, blackberry, cherry, grape, pepper, and tomato. Moreover, the inhibition of C_2H_4 production, either by limiting the formation of the ethylene synthesis enzymes or by stimulating ACC degradation, is effective in inhibiting ripening.

ETHYLENE PHYSIOLOGY IN POSTHARVEST

Ethylene Synthesis

It is commonly accepted that in flowering plants the most prevalent pathway of ethylene biosynthesis proceeds from methionine, through S-adenosylmethyionine (SAM, AdoMet) and 1-aminocyclopropane-1-carboxylic acid to ethylene (Figure 9.2), and that the two key enzymes involved are ACC synthase (ACS) and ACC oxidase (ACO), formerly known as ethyl-

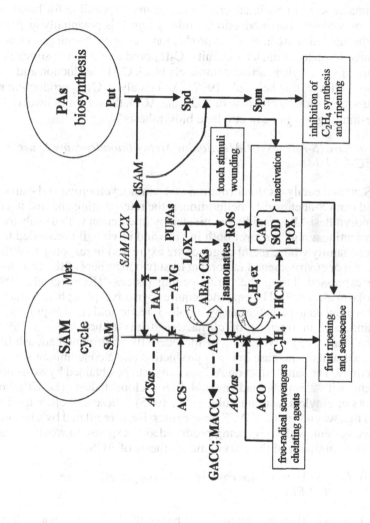

FIGURE 9.2. Schematic illustration of the main factors, reactions, hormones, and enzymes involved in the ripening syndrome. Abbreviations: ACOas, ACO antisense; ACSas, ACS antisense; dSAM, decarboxylated SAM; put, putrescine; SAM DCX, SAM decarboxylase; Spd, spermidine; Spm, spermine. Details are provided in text.

ene-forming enzyme (EFE) (Yang and Hoffman, 1984). It has been hypothesized that two different systems regulating the production of ethylene are functioning in higher plants. System I, operating in vegetative tissues and in both climacteric and nonclimacteric fruits, seems responsible for basal and wound-induced ethylene production, while system II is potentially responsible for the rapid increase in ethylene production during the ripening of climacteric fruits. In mature climacteric fruits, C_2H_4 production is autostimulated, and inhibitors of ethylene action completely block C_2H_4 production and fruit ripening (Lelievre, Latche, et al., 1997). Autocatalytic C_2H_4 production requires up-regulation by ethylene of ACS and ACO, the enzymes involved in the rate-limiting reactions in ethylene biosynthesis.

ACC Synthase (S-Adenosyl-L-Methionine Methylthioadenosine Lyase, ACS; EC 4.4.1.14)

ACS, which catalyzes the formation of 1-aminocyclopropane-1-carboxylic acid from S-adenosyl-L-methionine, is the rate-limiting enzyme in ethylene biosynthesis. It is rapidly inactivated by interaction with its substrate, thus presenting a high turnover both in vitro and in vivo. It is encoded by a multigene family whose member genes are expressed in response to different internal or environmental factors; at least two members of the gene family are expressed during the fruit-ripening process (Kubo et al., 1995; Mathooko et al., 1999). The enzyme requires pyridoxal phosphate for activity and is induced in plant tissues by indole-3-acetic acid (IAA) application, wounding, and in ripening of climacteric fruits (Abeles et al., 1992a; McKeon et al., 1995). Wounding of mesocarp tissue of winter squash fruit induces a dramatic increase in C_2H_4 production and a concomitant surge in ACS activity. The same effect on ACS activity can be obtained by exogenous treatments with 2,5-norbornadiene and diazocyclopentadiene (DACP), both inhibitors of ethylene action (Kubo et al., 1995). These and other findings suggest that wound-induced ACS gene expression is regulated by a negative feedback system, with C_2H_4 being produced as a response to wounding and acting as an inhibitor of the enzymatic synthesis of ACS.

ACC Oxidase (1-Aminocyclopropane-1-Carboxylate Oxidase, ACO, Formerly EFE)

ACO, an oxidative enzyme and member of the Fe(II)-dependent dioxygenase family, converts ACC to ethylene. As its activity was difficult to demonstrate in vitro, the first assays were performed in vivo (McKeon et al., 1995). The validation for authentic ACO activity relies on stereospecificity of the substrate, its dependence on O_2, the inhibition by Co^{2+} in the range of 10 to 100 µM, and the competitive inhibition by α-aminoisobutyrate. All of the subsequent enzyme preparations for in vitro activity required Fe^{2+} and ascorbate in the extraction media for activity. After enzyme purification, ACO was recognized as a monomer with a molecular mass of 35 kDa, which catalyzed the following reaction:

$$Fe^{2+}, CO_2$$

$$ACC(C_4NO_2H_7) + O_2 + AA \longrightarrow C_2H_4 + CO_2 + HCN + DHA + 2H_2O$$

ACO

Given that ACO is an exclusively cytosolic enzyme not associated with membranes, its reaction products are released in the cytoplasm (Grossmann, 1996; John, 1997). Initially, the enzyme was considered to be a constitutive one (Yang and Hoffman, 1984), but it is now clear that it can be induced by stress (Morgan and Drew, 1997) and during certain developmental stages such as fruit ripening (Yang and Hoffman, 1984). In mesocarp tissue of winter squash fruit, ACO activity is induced very rapidly by wounding and parallels the increase in the rate of ethylene production (Hyodo et al., 1993). ACO synthesis is controlled by a positive feedback system, being stimulated by ethylene (Kubo et al., 1995; Atta-Aly et al., 2000).

Regulation of ACO and ACS may represent the main control system of the rate of ethylene production in fruit. However, other steps in the ethylene biosynthetic pathway play a role. In apple, huge percentages of ACC synthesized in the skin of immature fruits can be stored in the vacuoles as inactive malonyl ACC (MACC) (Lelievre, Latche, et al., 1997). Recently, another diversion in the ACC pathway has been described in crude extracts of tomato, with ACC conjugation to a 1(γ-L-glutamylamino) derivative (GACC) in the presence of glutathione.

In the presence of particular cofactors, other enzymes besides ACO can oxidize ACC to ethylene. These enzymes include IAA oxidase, peroxidases, lipoxygenase, and H_2O_2-generating oxidases (Yang and Hoffman, 1984).

Effects of Ethylene

Ethylene is a plant hormone that regulates many physiological aspects of plant growth, development, maturation, and senescence (Yang and Hoffman, 1984; Abeles et al., 1992a; Ponnampalam et al., 1993; Oetiker and Yang, 1995; Picton et al., 1995; Beaulieu et al., 1998; Mathooko et al., 1998). Ethylene controls the ripening of climacteric fruits by coordinating the activation of many genes that cause changes in color, texture, aroma, and flavor (Oetiker and Yang, 1995). A clearer picture of its role in fruit ripening has recently been obtained by the characterization of genes encoding the main ethylene biosynthetic enzymes ACS and ACO and the isolation of genes involved in the ethylene signal transduction pathway, such as those encoding ethylene receptors (Lelievre, Latche, et al., 1997). The use of transgenic plants, with altered levels in C_2H_4 production or sensitivity, highlighted that autocatalytic ethylene production requires an up-regulation by C_2H_4 of ACS and ACO, and that ACO gene expression is developmentally controlled and precedes the expression of ACS. Many different experimental data suggest a model for ACS and ACO gene expression during the transition to autocatalytic ethylene production in climacteric fruit.

The rate of C_2H_4 evolution is normally used as a measure of the physiological degree of ripening in climacteric fruits and/or as an indicator of plant stress levels (Beaulieu et al., 1998). In fact, preclimacteric fruits produce low levels of ethylene; at the onset of the climacterium there is a surge in C_2H_4 production and its regulation becomes autocatalytic (Oetiker and Yang, 1995). Exogenous ethylene treatment induces only ACO synthesis in preclimacteric fruits such as tomato, cantaloupe, avocado, and apple. The increase in ACO occurs before the climacteric rise in ethylene production. The effects of ethylene treatments are strictly related to the fruit ripening stage; in fact, in immature climacteric fruits, exogenous C_2H_4 treatments induce a negative feedback regulation on ethylene production while in mature climacteric ones an autostimulation is obtained (Lelievre, Latche, et al., 1997). A recent study (Atta-Aly et al., 2000) clarifies the role of exogenous C_2H_4 treatments in tomato (climacteric) and strawberry (nonclimacteric) fruits at different stages of ripening. Ethylene application to strawberry tissues, excised from green to half-colored fruits, induces a short-term increase in C_2H_4 production; this is rapidly followed by a sharp reduction at the control level and paralleled by a marked reduction in ACC concentration. The same treatment on immature tomato fruit induces a negative C_2H_4 feedback mechanism, with a marked decrease in ACC concentration and C_2H_4 production. An opposite trend was observed working with fruits at the pink stage (positive C_2H_4 feedback mechanism). In all treated tissues, exogenous C_2H_4 significantly induced ACO activity. The proposed hypothesis suggests that a negative C_2H_4 feedback mechanism may be responsible for the nonclimacteric behavior of strawberry fruit and immature tomato fruit, while the climacteric behavior of tomato during ripening has to be imputed to a shift from a negative C_2H_4 feedback to a positive C_2H_4 feedback mechanism. Moreover, ACC formation may be the limiting step for both C_2H_4 feedback mechanisms, since exogenous C_2H_4 treatment induces ACO activity regardless of fruit species and physiological stage of maturation.

The term *stress ethylene* was coined for enhanced biosynthesis of this gas in association with environmental or biological stresses experienced by plants (Morgan and Drew, 1997; Beaulieu et al., 1998). Many different steps in ethylene biosynthesis and action can be altered by stress imposition (Morgan and Drew, 1997). For instance, wounding in mesocarp tissue of winter squash fruit induces a rapid increase in the rate of ethylene production with a parallel rapid induction of ACO activity (Hyodo et al., 1993). The conversion of SAM to ACC is the key reaction that controls the production of stress ethylene (Yang and Hoffman, 1984). In fact, the application of aminoethoxyvinylglycine (AVG), an inhibitor of ACS, effectively eliminates the increase in ACC formation and the production of stress ethylene.

During climacteric fruit ripening, two different stages for ethylene production and effects can be identified (Oetiker and Yang, 1995). At stage 1, ethylene is produced at low levels (*system1* ethylene), its main role to induce ACO transcript expression and accelerate stage 1. The tissue is neither competent to express system 2 ACS genes required for system 2 ethylene production nor to ripen in response to exogenous ethylene treatments. At stage 2, autocatalytic ethylene production (*system2* ethylene) is triggered by the expression of ACS genes, leading to ripening.

In nonclimacteric fruit, although the ripening process is not so strictly related to ethylene, at a determined stage of fruit development endogenous ethylene may be involved in the regulation of some ripening parameters. In citrus fruit, ethylene antagonists are able to block the degreening of flavedo, indicating a role of ethylene in the process (Goldschmidt et al., 1993). In the same species, exogenous ethylene treatments stimulate both chlorophyll degradation and carotenoid biosynthesis. Other pigments that increase during ripening, such as anthocyanins, can be stimulated by exogenous ethylene, depending on the species involved (Lelievre, Latche, et al., 1997). Nonclimacteric fruits, with the exception of cherries, show an increased rate of respiration in response to an exogenous C_2H_4 treatment—although this ethylene-stimulated respiration ceases with the removal of the exogenous hormone.

Ethylene is also an important factor for the induction of defense responses against biotic stresses such as pathogen attack (Ohtsubo et al., 1999). It activates many enzymatic activities, such as chitinase, peroxidase, phenylalanine ammonia-lyase, and polyphenol oxidase, and stimulates the production of defense-related molecules such as phenolics, lignins, and suberins. Ethylene production during hypersensitive reaction is restricted at the level of ACO activity. An important side effect is present in the biosynthetic pathway of C_2H_4: the oxidation of ACC to ethylene, catalyzed by ACO, leads to cyanoformic acid, which is converted to CO_2 and HCN (Grossmann, 1996). This pathway seems to be the main source of endogenous cyanide in many species; cyanide derived from C_2H_4 biosynthesis seems to be involved in the development of tissue necrosis following the hypersensitive response in plants. HCN is a strong phytotoxic agent and can inhibit several different classes of metalloenzymes. Many of the cyanide-sensitive enzymes are involved in protection from oxidative stress, including CAT, SOD, and POX. The capacity of HCN to mimic ethylene action in inducing climacteric respiration in fruits (Solomos and Laties, 1974) led to the hypothesis that cyanide is a positive regulator of ethylene synthesis, particularly in climacteric tissues (Pirrung and Brauman, 1987). In the respiration climacteric, the excess in cyanide produced in concomitance with ethylene may be scavenged by the cyanide-insensitive respiration, which can act as a protective mechanism (Lelievre, Latche, et al., 1997). Detoxification of HCN in higher plants is under the control of β-cyanoalanine synthase (β-CAS, EC 4.4.1.9), an enzyme mainly localized in the mitochondria. HCN removal outside these organelles is probably less efficient, and highly cyanide-sensitive enzymes such as Cu/Zn/SOD, CAT, and rubisco can, therefore, be inactivated (Grossmann, 1996).

Regulation of Ethylene Biosynthesis

The regulation of C_2H_4 biosynthesis is greatly influenced by many different environmental factors to which fruits and vegetables are subject during production. Moreover, other stress factors are added during harvest and transportation (Ludford, 1995) and by the environmental conditions imposed during postharvest storage and ripening (Ludford, 1995; Lelievre, Latche, et al., 1997; Morgan and Drew, 1997). The effects of each environmental factor on ethylene evolution are mainly related to the species and/or organ (Yang and Hoffman, 1984; Kubo et al., 1990; Lelievre, Tichit, et al., 1997) as well as to the cultivar (Larrigaudiere et al., 1997; Masia, 1998). For instance, touch stimuli during harvest and transportation may induce ethylene production from ACC in various plant species (Tatsuki and Mori, 1999). Manipulation of tomato fruits determines a marked increase in the levels of two isogenes for ACS (LE-ACS6 and LE-ACS1A mRNA transcripts), with a transient expression. The same results appear after fruit wounding. In recent years, the utilization of transgenic plants with reduced ethylene production, new ethylene antagonists, and the identification of the Nr tomato mutant as an ethylene receptor mutant have enabled a more precise analysis of the relationships between C_2H_4 metabolism and fruit ripening (Gray et al., 1994). Working in tomato with ACO-antisense fruit and with nonripening nor mutants, it was possible to demonstrate a developmental control of ACC synthesis (Lelievre, Latche, et al., 1997). Further experiments with ACS-antisense tomatoes and with overexpressing-ACC deaminase gene tomatoes have indicated that ACO gene expression is developmentally controlled and that it precedes the expression of ACS.

Active Oxygen Species

Many molecules deeply influence C_2H_4 production. Among them are an uncoupler of oxidative phosphorylation (2,4-dinitrophenol), free-radical scavengers (ascorbic acid, n-propyl gallate, Na-benzoate), chelating agents (sodium azide, EDTA, diethyldithiocarbamate, citric acid), and sulfhydryl-protecting reagents (cysteine, dithioerythritol, glutathione, $CoCl_2$). Multifunctional agents are apparently the most effective in inhibiting C_2H_4 production. The conversion of exogenous ACC to ethylene is inhibited in different tissues by application of various radical scavengers, suggesting that the reaction is mediated by a free-radical reaction (Kacperska and Kubacka-Zebalska, 1989; Ponnampalam et al., 1993). In tomato fruit tissue, the most effective compound is n-propyl gallate, probably due to its being a poly-functional agent. In addition to its role as a free-radical scavenger and a LOX activity inhibitor, it can react with Fe^{2+} ions to form dark compounds that decrease Fe^{2+} availability to act as a cofactor of ACO. Chelating agents are also capable of inhibiting the conversion of ACC to C_2H_4, with sodium azide being the most effective, probably because of its ability to inhibit metal-containing enzymes, as well as being a chaotropic agent and a free-radical scavenger. Another inhibiting compound is glutathione; its effect is mainly due to the interaction with SH groups present in ACO, besides its action as a free-radical scavenger (Ponnampalam et al., 1993).

Heavy metals stimulate ethylene production in treated cells and also affect activities of several enzymes such as POX, CAT, SOD, and LOX (Vangronsveld et al., 1993). Metal ions such as Cu, Fe, and Hg, which in solution perform one electron oxidoreduction reactions, directly induce free-radical formation (Abeles et al., 1992b), while nonredox active metals, such as Zn, produce ROS by mediation of lipoxygenase activity. The final effect is a LOX-mediated oxidation of PUFAs (Vangronsveld et al., 1993). Active oxygen species and hydroperoxides, generated by the action of LOX, stimulate the conversion of ACC into ethylene (Kacperska and Kubacka-Zebalska, 1989; Vangronsveld et al., 1993; Weckx et al., 1993). In mechanically wounded tissues, C_2H_4 can be the final result of fatty acid peroxidation (Yang and Hoffman, 1984). Although ethylene seems to have a consolidating effect on cell responses to heavy metal stress, it is not the triggering factor (Vangronsveld et al., 1993).

Hormones and Other Metabolites

Many different hormones are more and less directly involved in the regulation of ethylene production (Yang and Hoffman, 1984; Abeles et al., 1992b; Ludford, 1995; McKeon et al., 1995). Ethylene production rates are influenced by ethylene itself and other plant hormones (Figure 9.2). Depending on the tissue, this hormone can either promote ethylene production (autocatalysis) or inhibit it (autoinhibition). During the autocatalytic phase in ripening fruits, C_2H_4 at first affects ethylene production by promoting the conversion of ACC to ethylene; the massive increase in ACC synthesis takes place subsequently (Liu et al., 1985; McKeon et al., 1995).

In citrus leaf and tomato fruit, abscisic acid (ABA) appears to promote ethylene production primarily by stimulation of ACC synthesis. This effect can be increased by different forms of wounding (Riov et al., 1990). In many fruits (e.g., pear, avocado, citrus, grape, and cherry) the level of free ABA is constant during maturation but increases at the onset of ripening (Ludford, 1995). Auxin promotes ethylene production by inducing the synthesis of ACS, resulting in higher levels of ACC (Yang and Hoffman, 1984). The increase in ethylene production parallels the increase in ACC, and treatments with AVG block the IAA-induced ethylene increase. High levels of endogenous cytokinins delay fruit ripening, and levels may decline as ripening proceeds (Ludford, 1995). On the other hand, cytokinins stimulate ethylene production only in conjunction with other treatments that increase ethylene synthesis (McKeon et al., 1995). Cytokinins alone are unable to affect ACC levels, and their effect on ACS must be through some other factor affecting the level of induction.

Senescence in many plant organs, both in situ and upon their excision from the plant, is correlated with a decline in polyamine (PA) titer (Galston and Kaur-Sawhney, 1995). Polyamines are endogenous antisenescence agents, and free PA levels decline during fruit development in avocado, pear, and tomato (Ludford, 1995). It has been demonstrated that controlled atmosphere storage of apples reduces the rate of postharvest softening and also maintains a higher level of PAs. A possible working model involves binding of PAs to membranes, with the prevention of lipid peroxidation, and a quenching action on free radicals. Moreover, PA levels are associated with Ca^{2+} efflux from membranes, an effect which may imply the blocking of the senescence sequence (Leshem, Halevy, and Frenkel, 1986). Besides their action as free-radical scavengers, polyamines seem to react at the level of ACS and ACC (Ponnampalam et al., 1993; Ludford, 1995; Masia et al., 1998). A clear metabolic link exists between the pathways for biosynthesis of ethylene and polyamines (Picton et al., 1995). Ethylene and PAs share part of the biosynthetic pathway at the level of AdoMet (Figure 9.2), so that under certain conditions their biosynthesis can be restricted by competition for AdoMet, resulting in mutual inhibition (McKeon et al., 1995). Ethylene and PAs have opposite effects on senescence, particularly with excised plant parts (Galston and Kaur-Sawhney, 1995). The apparent inverse relation between PAs and senescence has been tested in ripening fruits; some tomato varieties, characterized by retarded senescence and an extended shelf life, contain very high levels of endogenous putrescine—which increase rather than showing the usual decline during ripening. It has been posited that PAs may constitute part of the fruit-associated ripening inhibitor system and that an increased synthesis of ethylene may be part of a mechanism to reduce the level of PAs at the onset of ripening (Picton et al., 1995). Nevertheless, PAs may have their own independent action on the ripening process; in fact, inhibition of both ripening and ethylene production is obtained only at high concentrations, while PAs at lower ones can affect only ripening, with no effect on ethylene formation (Galston and Kaur-Sawhney, 1995).

A strict correlation exists among jasmonates, ethylene, and LOX activity (Figure 9.1); in fact, PUFA hydroperoxides may be the substrates for LOX with the formation of further free radicals (Leshem, Halevy, and Frenkel, 1986). The lipid peroxides may subsequently decompose to aldehydes such as MDA and volatile hydrocarbons such as ethane, pentane, and ethylene. A further LOX catabolite that has been found is jasmonic acid (JA). An increase in ACO activity in preclimacteric apples is induced by methyl jasmonate (JA-Me) treatments, which also stimulate C_2H_4 production and increase ACC content (Saniewski, 1995). The hormone, however, has an opposite effect on ACC levels and ethylene production in climacteric and postclimacteric apples. Quite similar results were obtained on apple with *n*-propyldihydrojasmonate (Masia et al., 1998). Moreover, the effect on ACO activity is closely related to the cultivar under treatment and to its different levels in endogenous phenolic compounds (Ludford, 1995).

Postharvest Storage

Low Temperature. Usually low temperatures are applied to fruit and vegetables to increase their storability, although often in some species the treatment can result in an increased rate of ethylene biosynthesis (Larrigaudiere et al., 1997; Lelievre, Latche, et al., 1997). Winter pears need a cold treatment after harvest for the induction of autocatalytic ethylene production and ripening (Ludford, 1995; Lelievre, Latche, et al., 1997). In cultivar Passe Crassane, exposure to chilling conditions induces an accumulation of ACO and ACS transcripts and of ACO protein, all of which are required for the ethylene autocatalytic burst and ripening (Lelievre, Latche, et al., 1997). Without chilling exposure, exogenous ethylene is unable to induce ACS gene expression (Lelievre, Tichit, et al., 1997). Different mechanisms are represented by fruits without a chilling requirement, such as apples and kiwifruit. Low temperatures in this case can promote a more synchronous and homogeneous ripening and hasten the competence for autocatalytic ethylene synthesis (Ludford, 1995; Lelievre, Latche, et al., 1997).

Many fruits and vegetables of tropical and subtropical origin are particularly prone to chilling damages caused by nonfreezing low-temperature storage (Ben-Amor et al., 1999). The rapid decrease in temperature very often results in a chilling-induced OS with a surge in H_2O_2 cellular levels and the subsequent stimulation of antioxidant enzyme activity (Prasad et al., 1994; Morita et al., 1999). Chilling induces an oxidative stress in plant tissues subjected to low temperatures (Prasad et al., 1994; Alscher et al., 1997; Levine, 1999), with modifications in membrane fluidity (Murata and Los, 1997), electron leakage (Prasad, 1996), and induction of POX and IAA oxidase activities (Omran, 1980). Cold-temperature stress also induces LOX mRNA accumulation (Porta et al., 1999). The emerging model for chilling injury starts from a modification in membrane permeability due to oxidative stress with subsequent membrane lipid peroxidation, followed by a cascade of biochemical reactions that lead to cell death and tissue deterioration (Ben-Amor et al., 1999).

In transgenic cantaloupe melons, the inhibition of ethylene biosynthesis via the action of an antisense ACO gene can prevent this chilling injury principally through a reduced, chilling-dependent deterioration of the membrane (Ben-Amor et al., 1999). Other factors that are associated with fruit chilling tolerance are a higher activity in antioxidant enzymes, such as CAT, SOD, and POX, and a decreased production of stress metabolites such as ethanol and acetaldehyde (AcA). By contrast, in wild-type fruits C_2H_4 is responsible, in association with low temperatures, for metabolic dysfunctioning with the accumulation of ethanol and AcA in tissues, an increase in membrane deterioration, and a reduced capacity of the tissues to remove ROS caused by a decrease in activity of POX, CAT, and SOD. Chilling tolerance seems to be imputed mainly to CAT activity: in fact, overexpression of CAT in fruits results in higher chilling resistance (Ben-Amor et al., 1999). In the cantaloupe melon expressing an antisense ACO gene, the activity of CAT increases continuously during low-temperature storage while it steadily decreases in chilling-sensitive wild-type fruit. Heavy metal stress can also induce an increase in CAT activity in response to an enhanced AOS production (Weckx et al., 1993).

High Temperature. Heat shock (HS) may produce OS in plant cells (Bowler et al., 1992), and H_2O_2 accumulation during the treatment was recently reported (Foyer et al., 1997). A rapid temperature increase of as little as 10°C elicits a dramatic change in a cell's protein synthesis. This heat-shock protein (HSP) synthesis is common to all living organisms and many different classes of HSP are induced by the treatment (Guy, 1999). Among them a central role in sensing the rapid rise in temperature is played by HSP70s. The physiological response of cells and organs to high-temperature treatment is mainly related to small HSPs, α-crystallin family, which represent the major family of HSP induced by heat stress in plants.

Relatively high temperature treatments have been used in fruits and vegetables after harvest and before cold storage in order to, for instance, reduce chilling sensitivity, delay ripening, and control pathogenic levels (Lelievre, Latche, et al., 1997; Sabehat et al., 1998). The technique is based on physiological and metabolic modification induced in crops by high-temperature treatment. It has been demonstrated that the conversion of ACC to ethylene is inhibited when the temperature reaches 35°C or more (Yu et al., 1980). In tomato fruit, ACO HSP transcripts and protein levels are reduced after three days at 38°C (Lurie et al., 1996). HS treatments decrease ethylene production and reduce chilling injury in the subsequent low-temperature storage in cucumber (Ben-Amor et al., 1999). Mature-green tomatoes held at 38°C can be transferred to 2°C for several weeks without developing symptoms of chilling injury, and the high-temperature treatment induces the synthesis of small HSPs in tomato fruit (Kadyrzhanova et al., 1998; Sabehat et al., 1998). A close correlation seems to exist between HSP expression and chilling tolerance, supporting a role for HSPs in protection against chilling injury (Sabehat et al., 1998). This linkage between HS treatment and HSPs, resulting in an induced chilling tolerance, is presented by such horticultural crops as avocado, cucumber, pepper, and tomato, suggesting that this cross-tolerance may be representative of a general response.

Controlled Atmosphere, Oxygen, and Carbon Dioxide. The variations in gas composition present in modified and controlled atmosphere (CA) cold storage results in dramatic effects on crop postharvest life that are mainly due to marked modification of ethylene biosynthesis and metabolism (Beaudry, 1999). The effects are mainly related to O_2 and CO_2 changes (Lelievre, Latche, et al., 1997; Beaudry, 1999), but may also depend on the presence of trace elements in the air such as C_2H_4, C_3H_6, other hydrocarbons, peroxyacylnitrates (PANs), nitric oxides, etc. (Fergusson, 1985; Rao et al., 1995; Hippeli and Elstner, 1996). Other perturbing sources of internal origin are volatiles of low molecular mass such as acetaldehyde and ethanol (Beaulieu et al., 1998; Pesis et al., 1998), methanethiol and ammonia (Toivonen, 1997), α-farnesene and related oxidation metabolites (Wang and Dilley, 2000a,b). Responses to modified atmosphere often depend on plant species, organ type, and/or developmental stage, and can be beneficial or undesirable (Kubo et al., 1990; Beaudry, 1999). Chilling injury from low-temperature storage results in significant increases in putrescine levels in many different fruit and vegetable species (Galston and Kaur-Sawhney, 1995). The reduction of chilling injury in squash, Chinese cabbage, and apples by CA storage is coupled with increased PA titers. Polyamines can help to preserve membrane integrity, resulting in an increased viability during chilling. A marked effect of CA storage is also obtained at the level of antioxidant cellular machinery. Quite different senescing patterns were obtained in detached spinach leaves stored for 35 days at 10°C in ambient air, air plus 10 ppm ethylene, or CA (Hodges and Forney, 2000). Of all the antioxidant enzymes and molecules tested, significant declines were shown only in levels of APX, CAT, and AA in air and CA stored tissues with a sharp increase in lipid peroxidation at day 35 for air-treated ones only. On the other hand, all the antioxidants, with the exception of SOD, rapidly and continuously decreased in ambient air plus ethylene treated leaves. Common to all treatments is the decrease in AA and APX and CAT activities, suggesting a main role of H_2O_2 in the pattern and intensity of postharvest senescence in spinach.

It is well known that low O_2 atmospheres are able to inhibit C_2H_4 production in many different climacteric fruits such as apples, bananas, and pears (Abeles et al., 1992b). The same result is reported for nonclimacteric strawberries and for lettuce. Spinach leaves stored in 0.8 percent O_2 show a lower respiration rate coupled with a better quality in comparison to air-stored ones (Hodges and Forney, 2000). Levels of O_2 between 1 and 3 percent can reduce ethylene production by 50 percent in stored apples (Lelievre, Latche, et al., 1997). The decrease in ethylene production is to be imputed to a reduced ACO activity due to the fact that O_2, along with CO_2, is a cosubstrate for the enzyme. Low O_2 levels in stored apples are believed to block the signal transduction pathway involved in the response to C_2H_4 perception at the receptor site and hence prevent induction of autocatalytic C_2H_4 production (Gorny and Kader, 1996). The effect of low O_2 levels on C_2H_4 production is mainly related to an avoidance mechanism that impedes the rise of internal ethylene to a level sufficient to trigger its autocatalytic production. In apples, this result is cultivar dependent and in 'Cox' a 5 percent CO_2 pretreatment has to be applied to the fruits before low O_2 storage in order to maintain C_2H_4 fruit production under the critical triggering value (Stow et al., 2000). Low-temperature storage, combined with low O_2 levels, inhibits both ACS and ACO activities and increases the levels of their corresponding mRNAs, with a resultant delay in the onset of autocatalytic ethylene production (Gorny and Kader, 1996, 1997). However, not all the effects are C_2H_4 related; in fact, low O_2 levels have been shown to reduce the expression of ethylene-independent fruit ripening-related genes (Kanellis et al., 1991). In avocado fruit, O_2 concentrations in the 2.5 to 5.5 percent range reduced the accumulation of cellulase and polygalacturonase proteins and the expression of their isoenzymes. The same O_2 values induced the opposite effect on alcohol dehydrogenase. Very low O_2 levels (1-2 kPa O_2 corresponding to the crisis zone, and 2-4 kPa to the elastic stress zone) radically modify primary metabolic pathways such as glycolysis, fermentation, and aerobic respiration (Beaudry, 1999). The O_2 threshold for the induction of anaerobic respiration is species and cultivar dependent; for instance, in apple, 'Red Delicious' shows a 0.7 kPa O_2 partial pressure threshold limit, 'Red Rome' 0.8 kPa, while 'McIntosh' a 1.9 kPa value (Toivonen, 1997). Besides ethylene metabolism and the respiratory pathway, other secondary metabolic processes that are associated with the quality of the treated produce, and involving pigments, phenolics, and volatiles, are also affected. Obviously, low O_2 levels have a direct effect on limiting OS and AOS production, with the consequent decrease and/or delay in senescence evolution.

The effects of superatmospheric O_2 levels were recently reviewed by Kader and Ben-Yehoshua (2000). Very high O_2 levels were utilized in many different types of produce mainly to evaluate the effects on postharvest physiology and quality of fruits and vegetables. Depending on the commodity and its maturity and ripeness stage, the exposure to very high O_2 concentrations may stimulate, have no effect, or reduce rates of respiration and ethylene production. These effects are also dependent on O_2 concentration, storage period, temperature, and CO_2 and C_2H_4 levels.

Increased CO_2 levels affect primary biochemical metabolism (processes of glycolysis, fermentation, and aerobic respiration), as well as many secondary metabolism responses (ethylene, pigment, phenolic, cell wall, and volatile compounds metabolism). Responses depend on species, organ, and the latter's developmental stage (Beaudry, 1999). Cold storage of preclimacteric apples in enriched CO_2 (5 percent and 20 percent) atmospheres directly impedes C_2H_4 biosynthesis by reducing ACO activity and, indirectly, by blocking up-regulation of the genes encoding ACS and ACO (Gorny and Kader, 1996). The main effect seems related to ACS activity and expression (Gorny and Kader, 1997). The response is quite different when the treatment is made at room temperature (25°C), with the exposure of many kinds of fruits and vegetables to 60 percent CO_2. This treatment inhibited ethylene production in peach, broccoli, and apples, but promoted it in banana, lettuce, cucumber, tomato, and eggplant (Kubo et al., 1990). More recently, the molecular basis for inhibition of wound-induced ethylene biosynthesis by elevated concentrations of CO_2 has been studied in winter squash fruit (Kubo et al., 1995). The intact fruit can produce only trace amounts of ethylene, but mesocarp tissue cubes produce a huge amount of C_2H_4. Carbon dioxide suppresses wound-induced ethylene production and ACO activity at all concentrations utilized, starting from 5 percent and reaching 80 percent. This inhibition is due to suppression of the expression of both ACS and ACO at the mRNA level. In cold apple CA storage, ACO activity in vitro is enhanced by CO_2, with an optimum concentration at 2 percent (John, 1997).

However, at higher levels CO_2 acts as a competitive inhibitor of ethylene action, displacing the hormone at the level of the receptor binding site, and prevents the autocatalytic phase (Gorny and Kader, 1996; Lelievre, Latche, et al., 1997). In fact, at CO_2 levels of 5 percent or higher induction of ACS and ACO activities is reduced. Effects of different CO_2 levels were evaluated by in vivo ACO activity assays working with kiwifruit discs supplied with ACC (Rothan and Nicolas, 1994). From the experimental data it was apparent that when the ACC concentration was greater than 55 µM, high levels of CO_2 would stimulate C_2H_4 synthesis, while at ACC below this value high levels of CO_2 would inhibit C_2H_4 synthesis. Thus, the effect of CO_2 in either stimulating or inhibiting C_2H_4 synthesis is strictly related to the internal ACC tissue content (John, 1997). Elevated levels of CO_2 (combined or not with reduced O_2 levels) impede ethylene biosynthesis in a dual way, directly by reducing ACO catalytic ability to convert ACC to C_2H_4 and indirectly by blocking up-regulation of the genes encoding ACS and ACO (Gorny and Kader, 1996). However, the main effect of reduced O_2 and/or elevated CO_2 atmospheres in impeded C_2H_4 biosynthesis is to delay and suppress expression of ACS at the transcriptional level and to reduce the abundance of active ACO protein.

Extremely high levels of CO_2 (80 percent), combined with 20 percent O_2 and high (room) temperatures, have an oxidative effect on cucumber, which is deemed CO_2 stress (Mathooko et al., 1998, 1999). Carbon dioxide stimulates ethylene production through the enhancement of both ACS and ACO activities. The utilization of different inhibitors such as cordycepin, an inhibitor at the transcription level; cycloheximide, at translation level; dibucaine and 6-dimethylaminopurine, both inhibitors of protein kinases; and cantharidin, an inhibitor of phosphatases, demonstrates that although ACS and ACO induction requires de novo protein synthesis, it does not require new mRNA synthesis (Mathooko et al., 1998). Moreover, the experiments have shown that protein phosphorylation and dephosphorylation are involved in the carbon dioxide signaling cascade leading to induction of ACS and not of ACO (Mathooko et al., 1998). This reversible protein phosphorylation, although regulating CO_2 stress-induced ethylene biosynthesis in most cases, seems to have, under strong stress conditions, little or no role in mediating the CO_2 stress leading to the accumulation of ACS transcripts (Mathooko et al., 1999). In fact, transcription of ACS genes may not be the only factor regulating the production of ACC, and regulatory mechanisms at the posttranscriptional level may be equally important. This condition occurs when the accumulation of ACS transcripts is not correlated with ethylene production and/or ACS activity. The increase in ACS activity and ACC accumulation in cucumber fruit under CO_2 stress, leading to increased ethylene production, is mainly the result of an accumulation of the transcripts for the *CS-ACS1* gene, a primary response gene which is insensitive to protein synthesis inhibition but sensitive to wounding.

Volatiles. The effects of O_2 and CO_2 are not only related to ethylene and respiratory pathways but also affect other metabolic processes as well, greatly influencing fruit and vegetable quality. Among these processes is the metabolism of pigment, phenolic, and volatile compounds. Chlorophyll degradation is a desirable process for many climacteric fruits but should be avoided in green vegetables (Beaudry, 1999). It seems likely that low O_2 and elevated CO_2 can control chlorophyll loss in many green tissues acting at the level of C_2H_4 perception. The accumulation of brown pigments in mechanically stressed surfaces of many fruits and vegetables can be retarded by reduced levels of O_2. Brown pigments are formed mainly as the result of polyphenoloxidase (PPO) action on plant phenolics, with the formation of oxidation products that successively react with one another and with amino acids and proteins to form these pigments. The formation of volatile esters contributing to the aromas in many fruits is also strongly reduced by low O_2 concentration acting both on inhibition of C_2H_4 action and on oxidative processes, including respiration (Beaudry, 1999; Wang and Dilley, 2000a,b).

Acetaldehyde (AcA) and ethanol are two products of anaerobic respiration in fruits. They accumulate during ripening and are capable of retarding senescence and inhibiting ethylene production in plants (Pesis et al., 1998). Acetaldehyde inhibits fruit softening in peaches and polygalacturonase activity in tomato, while in grapes it can inhibit ethylene production with or without addition of ACC (Beaulieu et al., 1997, 1998; Pesis et al., 1998). Acetaldehyde vapor treatment inhibits avocado softening, as do low O_2 treatment or ethanol vapor (Pesis et al., 1998). It has been demonstrated that AcA is the agent responsible for ethanol-induced ripening inhibition in tomato fruit (Beaulieu et al., 1997). It is well known, in fact, that EtOH acts via AcA. Using a specific alcohol dehydrogenase (ADH) inhibitor, 4-methylpyrazole, it has been shown that AcA is the active agent inducing the effect of EtOH. In avocado fruit, AcA vapor inhibits ACO activity both in vivo and in vitro, probably through interaction with the enzyme's lysine residues. In tomato and apple, low concentrations in acetaldehyde vapor, in the presence of ascorbate and O_2, may stimulate nonenzymatic ACC cleavage with C_2H_4 production by either ascorbate-dependent or direct oxidation of ACC (Beaulieu et al., 1998). Another product of anaerobic respiration in broccoli is methanethiol, which can also be generated by disrupted tissues of cabbage (Toivonen, 1997). Methanethiol can inhibit cytocrome *c* oxidase and CAT activities in both in vivo and in vitro systems; decreased CAT activity results in an increased susceptibility of tissues to OS, although high levels of CO_2 can inhibit the production of this molecule.

FUTURE DIRECTIONS

The increasing understanding of fruit and vegetable postharvest physiology is rapidly changing the technology applied to the cold storage of horticultural crops, especially over prolonged periods. For each species and cultivar, however, many different aspects have to be cleared up for an optimal preservation of the commodities. In fact, produce quality during postharvest storage is dependent upon and linked to a plethora of different conditions and parameters.

Although it is widely accepted that postharvest behavior of different produce is largely influenced by the preharvest life cycle of fruits and vegetables, this linkage is quite difficult to assess. A better knowledge of the effects of the different cultural techniques during growth of horticultural crops on postharvest physiology could better clarify why storage disorders appear only in certain growth conditions.

The proper harvesting date also plays a role in produce postharvest behavior, and in many cases the maturation indices used for the choice of the optimal harvest time for each species and cultivar are insufficient (Masia, 1998). Future research has to focus on the search for new physiological ripening indices, mainly linked to the level of the crop's antioxidant machinery, for an optimization of commercial harvest practices (Masia, 1998; Rogiers et al., 1998). This can be achieved only by the evaluation, for each crop, of the patterns in antioxidant enzymes and protective metabolites during preharvest, harvest, and postharvest. Results will be addressed to the choice of the appropriate harvest date and the best storage conditions and parameters.

Nutritional and health aspects of produce are becoming increasingly important to consumers, indicating that more studies should be addressed to the variations during lengthy storage in the content of antioxidants such as ascorbic acid, α-tocopherol, glutathione, carotenes, and polyphenols. Another critical aspect concerns the control of pests and diseases during produce storage by chemical agents. Prestorage application of physical methods and the utilization of natural metabolites as substitutes to xenobiotic treatments during long-term conservation, combined with CA, may provide the answer.

The plant hormone ethylene plays a pivotal and crucial role in ripening and senescence processes, and its role in produce postharvest physiology has been studied in depth in its technological, biochemical, and molecular aspects (Abeles et al., 1992a,b; Oetiker and Yang, 1995; Lelievre, Latche, et al., 1997; Morgan and Drew, 1997). More recently, some C_2H_4 feedback mechanisms related to the ripening physiological stage in non- and climacteric fruits have also been examined (Lelievre, Latche, et al., 1997; Atta-Aly et al., 2000). In the C_2H_4-linked physiology of ripening-senescence syndrome, much has been done, but much remains to be elucidated, especially regarding C_2H_4 signal transduction pathways and their targets (Lelievre, Latche, et al., 1997). Yet our knowledge is still scanty if we consider the related oxidative stress mechanisms that are deeply involved in the process.

Ethylene biosynthesis and its effects are greatly influenced by the different environmental conditions utilized during postharvest storage. Controlled atmosphere is commonly and widely used in long-term produce storage and the effects of O_2, CO_2, temperature, and ethylene on different crops have been periodically reviewed (Abeles et al., 1992a; Lougheed, 1987; Kubo et al., 1990; Beaudry, 1999; Kader and Ben-Yehoshua, 2000). However, only in very few cases have the linkages among ethylene, CA or cold storage, OS, and their effects on produce storage been investigated (Masia, 1998; Rogiers et al., 1998; Hodges and Forney, 2000).

Variations in each of the CA parameters can induce rapid changes in the quality of fruit and vegetables, with a profound effect on aroma and off-flavor evolution. Special care has to be applied in the removal from CA of organic volatiles, some of which have positive effects on produce quality during storage (Toivonen, 1997). More experiments, focused on the effects of different O_2 and CO_2 pressures at the physiological, biochemical, and molecular level, are required. The development of new, selective, gas-permeable wrapping films as a substitute for produce CA storage may represent a functional alternative.

Future research should mainly focus on the relationships existing between ethylene physiology in ripening and senescence and the development and effects of oxidative stress on the evolution of crop antioxidant systems during cold storage under different techniques.

REFERENCES

Abeles, F.B., P.W. Morgan, and M.E. Saltveit (1992a). Fruit ripening, abscission, and postharvest disorders. In *Ethylene in Plant Biology,* F.B. Abeles, P.W. Morgan, and M.E. Saltveit (Eds.). San Diego, New York: Academic Press, Inc., pp. 182-221.

Abeles, F.B., P.W. Morgan, and M.E. Saltveit (1992b). Regulation of ethylene production by internal, environmental, and stress factors. In *Ethylene in Plant Biology,* F.B. Abeles, P.W. Morgan, and M.E. Saltveit (Eds.). San Diego, New York: Academic Press, Inc., pp. 56-119.

Alscher, R.G., J.L. Donahue, and C.L. Cramer (1997). Reactive oxygen species and antioxidants: Relationships in green cells. *Physiologia Plantarum* 100: 224-233.

Asada, K. (1992). Ascorbate peroxidase—A hydrogen peroxide-scavenging enzyme in plants. *Physiologia Plantarum* 85: 235-241.

Asada, K. (1999). The water-water cycle in chloroplasts: Scavenging of active oxygens and dissipation of excess photons. *Annual Review of Plant Physiology and Plant Molecular Biology* 50: 601-639.

Atta-Aly, M.A., J.K. Brecht, and D.J. Huber (2000). Ethylene feedback mechanisms in tomato and strawberry fruit tissues in relation to fruit ripening and climacteric patterns. *Postharvest Biology and Technology* 20: 151-162.

Beaudry, R.M. (1999). Effect of O_2 and CO_2 partial pressure on selected phenomena affecting fruit and vegetable quality. *Postharvest Biology and Technology* 15: 293-303.

Beaulieu, J.C., G. Peiser, and M.E. Saltveit (1997). Acetaldeyde is a causal agent responsible for ethanol-induced ripening inhibition in tomato fruit. *Plant Physiology* 113: 431-439.

Beaulieu, J.C., E. Pesis, and M.E. Saltveit (1998). AA or basic pH causes in vitro and nonenzymatic cleavage of ACC to ethylene. *Journal of the American Society for Horticultural Science* 123: 675-680.

Ben-Amor, M., B. Flores, A. Latche, M. Bouzayen, J.C. Pech, and F. Romojaro (1999). Inhibition of ethylene biosynthesis by antisense ACC oxidase RNA prevents chilling injury in Charentais cantaloupe melons. *Plant Cell and Environment* 22: 1579-1586.

Bowler, C., M. Van Montagu, and D. Inze (1992). Superoxide dismutase and stress tolerance. *Annual Review of Plant Biology and Plant Molecular Biology* 43: 83-116.

Brennan, T. and C. Frenkel (1977). Involvement of hydrogen peroxide in the regulation of senescence in pear. *Plant Physiology* 59: 411-416.

Dat, J.F., H. Lopez-Delgado, C.H. Foyer, and I.M. Scott (1998). Parallel changes in H_2O_2 and catalase during thermotolerance induced by salicylic acid or heat acclimation in mustard seedlings. *Plant Physiology* 116: 1351-1357.

Du, Z. and W.J. Bramlage (1995). Peroxidative activity of apple peel in relation to development of poststorage disorders. *Hortscience* 30: 205-209.

Elstner, E.F. (1982). Oxygen activation and oxygen toxicity. *Annual Review of Plant Physiology* 33: 73-96.

Eshdat, Y., D. Holland, Z. Faltin, and G. Ben-Hayyim (1997). Plant gluthathione peroxidases. *Physiologia Plantarum* 100: 234-240.

Fergusson, J.E. (1985). Air pollutants. In *Inorganic Chemistry and the Earth: Chemical Resources, Their Extraction, Use and Environmental Impact*, O. Hutzinger and S. Safe (Eds.). Oxford, United Kingdom: Pergamon Press, pp. 212-256.

Foyer, C.H. (1993). Ascorbic acid. In *Antioxidants in Higher Plants*, R.G. Alscher and J.L. Hess (Eds.). Boca Raton, FL: CRC Press, pp. 31-58.

Foyer, C.H., P. Descourvieres, and K.J. Kunert (1994). Protection against oxygen radicals: An important defense mechanism studied in transgenic plants. *Plant Cell and Environment* 17: 507-523.

Foyer, C.H., H. Lopez-Delgado, J.F. Dat, and I.M. Scott (1997). Hydrogen peroxide- and glutathione-associated mechanisms of acclimatory stress tolerance and signaling. *Physiologia Plantarum* 100: 241-254.

Fryer, M.J. (1992). The antioxidant effects of thylackoid vitamin E (α-tocopherol). *Plant, Cell and Environment* 15: 381-392.

Galston, A.W. and R. Kaur-Sawhney (1995). Polyamines as endogenous growth regulators. In *Plant Hormones: Physiology, Biochemistry and Molecular Biology*, P.J. Davies (Ed.). Dordrecht, The Netherlands: Kluwer Academic Publishers, pp. 158-178.

Goldschmidt, E.E., M. Huberman, and R. Goren (1993). Probing the roles of endogenous ethylene in the degreening of citrus fruit with ethylene antagonists. *Plant Growth Regulation* 12: 325-329.

Gorny, J.R. and A.A. Kader (1996). Controlled-atmosphere suppression of ACC synthase and ACC oxidase in 'Golden Delicious' apple during long-term cold storage. *Journal of the American Society for Horticultural Science* 121: 751-755.

Gorny, J.R. and A.A. Kader (1997). Low oxygen and elevated carbon dioxide atmospheres inhibit ethylene biosynthesis in preclimacteric and climacteric apple fruit. *Journal of the American Society for Horticultural Science* 122: 542-546.

Gray, J.E., S. Picton, J.J. Giovannoni, and D. Grierson (1994). The use of transgenic and naturally occurring mutants to understand and manipulate tomato fruit ripening. *Plant Cell and Environment* 17: 557-571.

Grossmann, K. (1996). A role for cyanide, derived from ethylene biosynthesis, in the development of stress symptoms. *Physiologia Plantarum* 97: 772-775.

Guy, C. (1999). The influence of temperature extremes on gene expression, genomic structure, and the evolution of induced tolerance in plants. In *Plant Responses to Environmental Stresses: From Phytohormones to Genome Reorganization*, H. R. Lerner (Ed.). New York: Marcel Dekker Inc., pp. 497-548.

Hippeli, S. and E.F. Elstner (1996). Mechanisms of oxygen activation during plant stress: Biochemical effects of air pollutants. *Journal of Plant Physiology* 148: 249-257.

Hodges, D.M. and C.F. Forney (2000). The effects of ethylene, depressed oxygen and elevated carbon dioxide on antioxidant profiles of senescing spinach leaves. *Journal of Experimental Botany* 51: 645-655.

Hudak, K., K. Yao, and J.E. Thompson (1995). Release of fluorescent peroxidized lipids from membranes in senescing tissue by blebbing of lipid-protein particles. *HortScience* 30: 209-213.

Hyodo, H., C. Hashimoto, S. Morozumi, W. Hu, and K. Tanaka (1993). Characterization and induction of the activity of 1-aminocyclopropane-1-carboxylate oxidase in the wounded mesocarp tissue of *Cucurbita maxima*. *Plant and Cell Physiology* 34: 667-671.

Izzo, R., A. Scartazza, A. Masia, L. Galleschi, M.F. Quartacci, and F. Navari-Izzo (1995). Lipid evolution during development and ripening of peach fruits. *Phytochemistry* 39: 1329-1334.

John, P. (1997). Ethylene biosynthesis: The role of 1-aminocyclopropane-1-carboxylate (ACC) oxidase, and its possible evolutionary origin. *Physiologia Plantarum* 100: 583-592.

Kacperska, A. and M. Kubacka-Zebalska (1989). Formation of stress ethylene depends both on ACC synthesis and on the activity of free radical-generating system. *Physiologia Plantarum* 77: 231-237.

Kader, A.A. and S. Ben-Yehoshua (2000). Effects of superatmospheric oxygen levels on postharvest physiology and quality of fresh fruits and vegetables. *Postharvest Biology and Technology* 20: 1-13.

Kadyrzhanova, D.K., K.E. Vlachonasios, P. Ververidis, and D.R. Dilley (1998). Molecular cloning of a novel heat-induced chilling tolerance related cDNA in tomato fruit by use of messenger-RNA differential display. *Plant Molecular Biology* 36: 885-895.

Kanellis, A.K., T. Solomos, and K.A. Roubelakis-Angelakis (1991). Suppression of cellulase and polygalacturonase and induction of alcohol dehydrogenase isoenzymes in avocado fruit mesocarp subjected to low oxygen stress. *Plant Physiology* 96: 269-274.

Knorzer, O.C., J. Durner, and P. Boger (1996). Alteration in the antioxidative system of suspension-cultured soybean cells *(Glycine max)* induced by oxidative stress. *Physiologia Plantarum* 97: 388-396.

Kubo, Y., A. Inaba, and R. Nakamura (1990). Respiration and C_2H_4 production in various harvested crops held in CO_2-enriched atmospheres. *Journal of the American Society for Horticultural Science* 115: 975-978.

Kubo, Y., Y. Yamashita, T. Ono, F.M. Mathooko, H. Imaseki, and A. Inaba (1995). Regulation by CO_2 of wound-induced ACS gene expression in mesocarp tissue of winter squash fruit. *Acta Horticulturae* 394: 219-226.

Larrigaudiere, C., J. Graell, J. Salas, and M. Vendrell (1997). Cultivar differences in the influence of a short period of cold storage on ethylene biosynthesis in apples. *Postharvest Biology and Technology* 10: 21-27.

Lelievre, J.M., A. Latche, B. Jones, M. Bouzayen, and J.C. Pech (1997). Ethylene and fruit ripening. *Physiologia Plantarum* 101: 727-739.

Lelievre, J.M., L. Tichit, P. Dao, L. Fillion, Y.W. Nam, J.C. Pech, and A. Latche (1997). Effect of chilling on the expression of ethylene biosynthetic genes in Passe-Crassane pear (*Pyrus communis* L.) fruits. *Plant Molecular Biology* 33: 847-855.

Leshem, Y.Y., A.H. Halevy, and C. Frenkel (1986). Membranes and senescence. In *Processes and Control of Plant Senescence,* Y.Y. Leshem, A.H. Halevy, and C. Frenkel (Eds.). Amsterdam, Holland: Elsevier. pp 54-83.

Leshem, Y.Y., A.H. Halevy, C. Frenkel, and A.A. Frimer (1986). Oxidative processes in biological systems and their role in plant senescence. In *Processes and Control of Plant Senescence,* Y.Y. Leshem, A.H. Halevy, and C. Frenkel (Eds.). Amsterdam, Holland: Elsevier. pp. 84-99.

Levine, A. (1999). Oxidative stress as a regulator of environmental responses in plants. In *Plant Responses to Environmental Stresses. From Phytohormones to Genome Reorganization,* H.R. Lerner (Ed.). New York: Marcel Dekker Inc., pp. 247-264.

Liu, Y., N.E. Hoffman, and S.F. Yang (1985). Promotion by ethylene of the capability to convert 1-aminocyclopropane-1-carboxylic acid to ethylene in preclimacteric tomato and cantaloupe fruit. *Plant Physiology* 77: 407-411.

Lougheed, E.C. (1987). Interactions of oxygen, carbon dioxide, temperature, and ethylene that may induce injuries in vegetables. *HortScience* 22: 791-794.

Ludford, P. (1995). Postharvest hormone changes in vegetables and fruit. In *Plant Hormones: Physiology, Biochemistry and Molecular Biology*, P.J. Davies (Ed.). Dordrecht, The Netherlands: Kluwer Academic Publishers, pp. 725-750.

Lurie, S., A. Handros, E. Fallik, and R. Shapira (1996). Reversible inhibition of tomato fruit gene expression at high temperatures. Effects on tomato fruit ripening. *Plant Physiology* 110: 1207-1214.

Masia, A. (1998). Superoxide dismutase and catalase activities in apple fruit during ripening and post-harvest and with special reference to ethylene. *Physiologia Plantarum* 104: 668-672.

Masia, A., M. Ventura, H. Gemma, and S. Sansavini (1998). Effect of some plant growth regulator treatments on apple fruit ripening. *Plant Growth Regulation* 25: 127-134.

Mathooko, F.M., A. Inaba, and R. Nakamura (1998). Characterization of carbon dioxide stress-induced ethylene biosynthesis in cucumber (*Cucumis sativus* L.) fruit. *Plant and Cell Physiology* 39: 285-293.

Mathooko, F.M., M.W. Mwaniki, A. Nakatsuka, S. Shiomi, Y. Kubo, A. Inaba, and R. Nakamura (1999). Expression characteristics of *CS-ACS1, CS-ACS2* and *CS-ACS3*, three members of the 1-aminocyclopropane-1-carboxylate synthase gene family in cucumber (*Cucumis sativus* L.) fruit under carbon dioxide stress. *Plant and Cell Physiology* 40: 164-172.

McKeon, T.A., J.C. Fernandez-Maculet, and S.F. Yang (1995). Biosynthesis and metabolism of ethylene. In *Plant Hormones: Physiology, Biochemistry and Molecular Biology*, P.J. Davies (Ed.). Dordrech, The Netherlands: Kluwer Academic Publishers, pp. 118-139.

Meir, S., S. Philosoph-Hadas, G. Zauberman, Y. Fuchs, M. Akerman, and N. Aharon (1991). Increased formation of fluorescent lipis-peroxidation products in avocado peels precedes other signs of ripening. *Journal of the American Society for Horticultural Science* 116: 823-826.

Morgan, P.W. and M.C. Drew (1997). Ethylene and plant responses to stress. *Physiologia Plantarum* 100: 620-630.

Morita, S., H. Kaminaka, T. Masumura, and K. Tanaka (1999). Induction of rice cytosolic ascorbate peroxidase mRNA by oxidative stress: The involvement of hydrogen peroxide in oxidative stress signalling. *Plant and Cell Physiology* 40: 417-422.

Murata, N. and D.A. Los (1997). Membrane fluidity and temperature perception. *Plant Physiology* 115: 875-879.

Noctor, G. and C.H. Foyer (1998). Ascorbate and glutathione: Keeping active oxygen under control. *Annual Review of Plant Physiology and Plant Molecular Biology* 49: 249-279.

Oetiker, J.H. and S.F. Yang (1995). The role of ethylene in fruit ripening. *Acta Horticulturae* 398:167-178.

Ohtsubo, N., I. Mitsuhara, M. Koga, S. Seo, and Y. Ohashi (1999). Ethylene promotes the necrotic lesion formation and basic PR genes expression in TMV-infected tobacco. *Plant and Cell Physiology* 40: 808-817.

Omran, R.G. (1980). Peroxide levels and the activities of catalase, peroxidase, and indoleacetic acid oxidase during and after chilling cucumber seedlings. *Plant Physiology* 65: 407-408.

Pesis, E., D. Faiman, and S. Dori (1998). Postharvest effects of acetaldehyde vapor on ripening-related enzyme activity in avocado fruit. *Postharvest Biology and Technology* 13: 245-253.

Picton, S., J.E. Gray, and D. Grierson (1995). Ethylene genes and fruit ripening. In *Plant Hormones. Physiology, Biochemistry and Molecular Biology*, P.J. Davies (Ed.). Dordrecht, The Netherlands: Kluwer Academic Press, pp. 372-394.

Pirrung, M.C. and J.I. Brauman (1987). Involvement of cyanide in the regulation of ethylene biosynthesis. *Plant Physiology and Biochemistry* 25: 55-61.

Ponnampalam, R., J. Arul, J. Makhlouf, and F. Castaigne (1993). Inhibition of 1-aminocyclopropane-1-carboxylic acid-dependent ethylene production in tomato-tissue discs by biochemical treatments. *Postharvest Biology and Technology* 2: 291-300.

Porta, H., P. Rueda-Benitez, F. Campos, J.M. Colmenero-Flores, J.M. Colorado, M.J. Carmona, A.A. Covarrubias, and M. Rocha-Sosa (1999). Analysis of lipoxygenase mRNA accumulation in the common bean (*Phaseolus vulgaris* L.) during development and under stress conditions. *Plant and Cell Physiology* 40: 850-858.

Prasad, T.K. (1996). Mechanism of chilling-induced oxidative stress injury and tolerance in developing maize seedlings: Changes in antioxidant system, oxidation of proteins and lipids, and protease activities. *Plant Journal* 10: 1017-1026.

Prasad, T.K., M.D. Anderson, B.A. Martin, and C.R. Stewart (1994). Evidence for chilling-induced oxidative stress in maize seedlings and a regulatory role for hydrogen peroxide. *Plant Cell* 6: 65-74.

Rao, M.V., B.A. Hale, and D.P. Ormrod (1995). Amelioration of ozone-induced oxidative damage in wheat plants grown under high carbon dioxide. Role of antioxidant enzymes. *Plant Physiology* 109: 421-432.

Riov, J., E. Dagan, R. Goren, and S.F. Yang (1990). Characterization of abscisic acid induced ethylene production in citrus leaf and tomato fruit tissues. *Plant Physiology* 92: 48-53.

Rogiers, S.Y., G.N. Mohan Kumar, and N.R. Knowles (1998). Maturation and ripening of fruit *Amelanchier alnifolia* Nutt. are accompanied by increasing oxidative stress. *Annals of Botany* 81: 203-211.

Rothan, C. and J. Nicolas (1994). High CO_2 levels reduce ethylene production in kiwifruit. *Physiologia Plantarum* 92: 1-8.

Sabehat, A., S. Lurie, and D. Weiss (1998). Expression of small heat-shock proteins at low temperatures. *Plant Physiology* 117: 851-658.

Saniewski, M. (1995). Methyl jasmonate in relation to ethylene production and other physiological processes in selected horticultural crops. *Acta Horticulturae* 394: 85-98.

Scandalios, J.G. (1993). Oxygen stress and superoxide dismutases. *Plant Physiology* 101: 7-12.

Sen Gupta, A., R.P. Webb, A.S. Holaday, and R.D. Allen (1993). Overexpression of superoxide dismutase protects plants from oxidative stress. *Plant Physiology* 103: 1067-1073.

Shen, B., R.G. Jensen, and H.J. Bohnert (1997). Mannitol protects against oxidation by hydroxyl radicals. *Plant Physiology* 115: 527-532.

Shewfelt, R.L. and A.C. Purvis (1995). Toward a comprehensive model for lipid peroxidation in plant tissue disorders. *HortScience* 30: 213-218.

Siedow, J.N. (1991). Plant lipoxygenase: Structure and function. *Annual Review of Plant Physiology and Plant Molecular Biology* 42: 145-188.

Solomos, T. and G.G. Laties (1974). Similarities between the actions of ethylene and cyanide in initiating the climacteric and fruit ripening of avocados. *Plant Physiology* 54: 506-511.

Stow, J.R., C.J. Dover, and P.M. Genge (2000). Control of ethylene biosynthesis and softening in "Cox's Orange Pippin" apples during low-ethylene, low-oxygen storage. *Postharvest Biology and Technology* 18: 215-225.

Tatsuki, M. and H. Mori (1999). Rapid and transient expression of 1-amino-cyclopropane-1-carboxylate synthase isogenes by touch and wound stimuli in tomato. *Plant and Cell Physiology* 40: 709-715.

Thimann, K.V. (1987). Plant senescence: A proposed integration of the constituent processes. In *Plant Senescence: Its Biochemistry and Physiology,* W.W. Thomson, E.A. Nothnagel, and R.C. Huffaker (Eds.). Rockville, MD: American Society of Plant Physiologists, pp. 1-19.

Toivonen, P.M.A. (1997). Non-ethylene, non-respiratory volatiles in harvested fruits and vegetables: Their occurrence, biological activity and control. *Postharvest Biology and Technology* 12: 109-125.

Vangronsveld, J., J. Weckx, M. Kubaka-Zebalska, and H. Clijsters (1993). Heavy metal induction of ethylene production and stress enzymes: II. Is ethylene involved in the signal transduction from stress perception to stress responses? In *Cellular and Molecular Aspects of the Plant Hormone Ethylene,* J.C. Pech, A. Latche, and C. Balague (Eds.). Dordrecht, Boston, London: Kluwer Academic Publishers, pp. 240-246.

Wang, Z.Y. and D.R. Dilley (2000a). Hypobaric storage removes scald-related volatiles during the low temperature induction of superficial scald in apples. *Postharvest Biology and Technology* 18: 191-199.

Wang, Z.Y. and D.R. Dilley (2000b). Initial low oxygen stress controls superficial scald of apples. *Postharvest Biology and Technology* 18: 201-213.

Weckx, J., J. Vangronsveld, and H. Clijsters (1993). Heavy metal induction of ethylene production and stress enzymes. I. Kinetics of the responses. In *Cellular and Molecular Aspects of the Plant Hormone Ethylene,* J.C. Pech, A. Latche, and C. Balague (Eds.). Dordrecht, Boston, London: Kluwer Academic Publishers, pp. 238-239.

Yamasaki, H., Y. Sakihama, and N. Ikehara (1997). Flavonoid-peroxidase reaction as a detoxification mechanism of plant cells against H_2O_2. *Plant Physiology* 115: 1405-1412.

Yang, S.F. and N.E. Hoffman (1984). Ethylene biosynthesis and its regulation in higher plants. *Annual Review of Plant Physiology* 35: 155-189.

Yu, Y.B., D.O. Adams, and S.F. Yang (1980). Inhibition of ethylene production by 2,4-dinitrophenol and high temperature. *Plant Physiology* 66: 286-290.

Zheng, X. and R.B. van Huystee (1992). Anionic peroxidase catalysed ascorbic acid and IAA oxidation in the presence of hydrogen peroxide: A defense system against peroxidative stress in peanut plant. *Phytochemistry* 31: 1895-1898.

Chapter 10

Genetic Variation and Prospects for Genetic Engineering of Horticultural Crops for Resistance to Oxidative Stress Induced by Postharvest Conditions

Christopher B. Watkins
Mulpuri V. Rao

INTRODUCTION

A successful horticultural product in today's market environment must possess attributes for quality retention during the storage, transport, and shelf life steps necessary for marketing of that product. Limitations to the commercial expansion of many horticultural products reside in their postharvest storability; postharvest research is concerned largely with delaying senescence and ripening processes in harvested produce to meet marketing requirements. Successful utilization of postharvest technologies such as cold storage, controlled atmosphere, and, to a lesser extent, modified atmosphere, underpins the international marketplace by allowing year-round availability of many produce items, especially in the developed world. Nevertheless, clear limitations exist to the storability of harvested produce, and much variability within any product type exists in response to postharvest treatments. The importance of this variability has increased with the utilization of technologies—e.g., ultra low O_2 storage—that induce stress close to the biological limits for a given product.

Critical preharvest factors affecting postharvest behavior include those associated with cultural practices and climate. Most research has focused on cultural practices, such as rootstock/scion age, soil management, nutrition, training and pruning practices, crop loads, product size, and growth regulators. Climatic factors including temperature, light, and soil water have been considered to a lesser extent because they are often more difficult to control experimentally. Extensive literature is available on the influence of cultural and climatic factors: Bramlage (1993) highlights the large number of preharvest variables that contribute to the diversity of postharvest responses. The role of cultivar (or breeding line), although tacitly critical to all postharvest responses of commodities, has been relatively less studied.

The loss in the quality and the quantity of horticultural produce during postharvest periods is the net result of pathogenic destruction and of cells undergoing ripening and senescence processes, some of which are manifested as physiological disorders. The latter two processes involve both anabolic and catabolic changes under specific genetic regulation; fruit ripening is especially well studied since changes in expression of genes encoding enzymes involved in ethylene production, color development, and softening have provided opportunities to attempt genetic manipulation of these processes (Picton et al., 1995). Ripening and senescence are, however, associated with a number of degradative changes to proteins, chlorophyll, nucleic acids, and membranes (Kays, 1997). The disintegration of cellular membranes leads to their dysfunction, presumably causing cell disruption, discoloration, and death. Circumstantial evidence suggests that increased production of active oxygen species (AOS) causes peroxidation of lipids and fatty acids (Parkin et al., 1989) and damage to cellular components (Elstner, 1991; McKersie, 1991). Also, it has been postulated that during cold storage, a combination of senescence and low temperature may generate excess AOS which overwhelm the defense and repair mechanisms and result in disorder development (Wang, 1993).

The objective of this chapter is to review the genetic variation in horticultural products for resistance to oxidative stresses induced by postharvest conditions. Most postharvest studies have focused on single cultivars or comparisons of different species, and are largely restricted to senescence and crop responses to chilling-injury-inducing storage temperatures. Relatively little information is available regarding comparison of oxidative metabolism among various cultivars or selections. In other plant systems, rapid progress is being made in manipulation of oxidative metabolism using genetic technology, with the objectives of modifying tolerances of plants to stresses such as low temperature, dehydration, and salt. Although no information on genetic modification of antioxidative enzymes in postharvest systems is available, results with other plant systems are reviewed as they may provide insight for future use in manipulating fruit and vegetables to extend their storability.

AOS METABOLISM

Partial reduction of molecular oxygen, leading to oxygen-radical generation, is an unfortunate consequence of aerobic life. Production of AOS such as the superoxide radical ($\cdot O_2^-$) and H_2O_2 in plant cells is constitutive, but their production is often enhanced under the influence of environmental stress factors (Halliwell and Gutteridge, 1989). Under normal conditions, plants are equipped with antioxidant defenses that are sufficient to metabolize these otherwise lethal AOS. However, changes in environmental conditions can cause excess AOS production and/or reduce defense capacity, overwhelming the system and necessitating additional defenses (Scandalios, 1997). By themselves, both $\cdot O_2^-$ and H_2O_2 are relatively unreactive, but they can react with transition metals via Fenton and/or Haber-Weiss reactions to form highly reactive species such as the hydroxyl radicals $\cdot OH$ (Halliwell and Gutteridge, 1989). Unless metabolized, these reactive oxygen species react indiscriminately to cause oxidative damage to membrane lipids, proteins, and DNA leading to cellular dysfunction and ultimately inducing cell death (Jabs, 1999; Rao et al., 2000). Since neither plant nor animal cells are equipped to metabolize the highly reactive $\cdot OH$ radicals, formation of these radicals is prevented by efficiently removing the less reactive forms ($\cdot O_2^-$, H_2O_2).

Plant cells are equipped with several enzymatic and nonenzymatic-based antioxidant defense systems that keep levels of these deleterious species to a minimum. Superoxide dismutase (SOD; EC 1.15.1.1) catalyzes the dismutation of $\cdot O_2^-$ to H_2O_2 and molecular oxygen (Scandalios, 1997). H_2O_2 is metabolized to H_2O by a wide array of antioxidant enzymes such as catalase (CAT; EC 1.11.1.6), peroxidase (POX; EC 1.11.1.7), ascorbate peroxidase (APX; EC 1.11.1.11), and glutathione reductase (GR; EC 1.6.4.2) (Figure 10.1). Both APX and GR metabolize H_2O_2 by participating in a pathway usually referred to as the Halliwell-Asada or ascorbate-glutathione cycle (Asada, 1992; Creissen et al., 1994). The concerted action of these antioxidative enzymes not only removes AOS, but also maintains a high redox state of ascorbate (ASA/DHA) and glutathione (GSH/GSSG). The redox state of glutathione-ascorbate appears important in the regeneration of membrane-bound carotenoids and α-tocopherol (Hess, 1993) and inactivation of Cu/ZnSOD (Hérouart, Van Montagu, et al., 1993). Furthermore, glutathione peroxidase (GPX), a major H_2O_2 scavenger in animal cells, also exists in plants (Holland et al., 1993). Although the subcellular localization of GPX is not known, it is believed to remove H_2O_2 by oxidizing glutathione. Likewise, glutathione-S-transferase (GST), which conjugates peroxidized metabolites, was also shown to possess GPX activity (Roxas et al., 1997). Thus, SOD is intertwined with antioxidative enzymes and/or antioxidant metabolites in what is likely a highly optimal balance that reduces the risk of $\cdot OH$ radical formation.

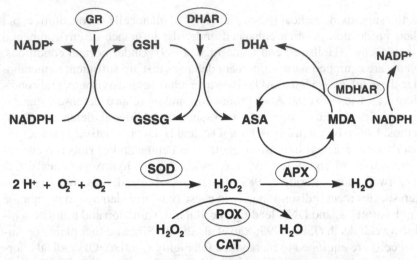

FIGURE 10.1. Schematic diagram illustrating the metabolism of active oxygen species by enzymes possessing antioxidative properties. (ASA, reduced ascorbate; APX, ascorbate peroxidase; CAT, catalase; DHA, oxidized ascorbate; DHAR, dehydroascorbate reductase; GR, glutathione reductase; GSH, reduced glutathione; GSSG, oxidized glutathione; MDHAR, monodehydroascorbate reductase; POX, peroxidase; SOD, superoxide dismutase.) GPX (not shown) removes H_2O_2 by oxidizing GSH.

SENESCENCE, RIPENING, AND AOS METABOLISM

Senescence, a slow form of cell death, is the terminal phase of development in the life of a whole plant or organ including leaves, flowers, and fruits (Buchanan-Wollaston, 1997; Kays, 1997; Nam, 1997; Quirino et al., 2000). It influences various levels of plant growth and development and is triggered by various internal age-dependent factors and can be induced prematurely by the changes in external environmental conditions. Various environmental stress factors are known to induce premature senescence (Quirino et al., 2000; reviewed by Rao et al., 2000), as does postharvest storage of plant organs at room temperature.

Cellular senescence is characterized by dramatic changes in the major organelles, particularly the plastids of mesophyll cells and in the parenchyma of fruit pericarp tissue (John et al., 1995; Buchanan-Wollaston, 1997). Further, senescence is also manifested by the changes in lipid metabolism and integrity, particularly membrane lipids and degradation of nucleic acids (Green, 1994; Thompson et al., 1998). Historically, progression of senescence in plant organs is associated with excess production of AOS and carbon-limiting conditions that generate energy by promoting the respiration of organic acid skeletons of amino acids (Buchanan-Wollaston, 1997). In addition, recent studies also suggest the involvement of several proteases, ribonucleases, and phytohormones such as ethylene and cytokinins in regulating senescence (Green, 1994; Gan and Amasino, 1995; Buchanan-Wollaston, 1997; Quirino et al., 2000). In several fruits, ripening is associated with increased respiration and ethylene production, and these changes may also be associated with an increase in AOS production (Brennan and Frenkel, 1977; Frenkel and Eskin, 1977; Warm and Laties, 1982).

Cellular oxidative metabolism is a central feature of senescence at the cellular level (Quirino et al., 2000; Rao et al., 2000). The shift in the balance of production of AOS and their catabolism by internal antioxidant enzymes and the subsequent changes in cellular redox state are believed to regulate senescent cell death (Rao et al., 2000). Various factors, both internal and external, are known to increase the production of AOS and initiate a battery of changes in cellular physiology, including lipid peroxidation, that ultimately induce cell death or senescence in plant cells. Several studies have investigated changes in AOS and associated enzymes in influencing both senescence and ripening processes in plant cells. Peroxide levels increased during the climacteric in tomato and pear fruit (Brennan and Frenkel, 1977; Frenkel and Eskin, 1977; Warm and Laties, 1982). In pear fruit, the infusion of glycolic acid or xanthine (substrates for H_2O_2 production), application of CAT inhibitors that maintained higher H_2O_2 levels, and application of exogenous ethylene, increased H_2O_2 levels and accelerated ripening (Brennan and Frenkel, 1977). Brennan et al. (1979) observed increased activities of xanthine oxidase and POX in pulp of ripening pears, but differences in CAT activity were affected by whether data were expressed on a fresh weight or protein content basis. Glycolate oxidase activity increased during the ripening of avocado, pear, and tomato fruit in the pulp, but declined in pear peel, and was not detectable in avocado peel. Baker (1976) found no major quantitative changes in SOD activity during ripening of apple, avocado, banana, or tomato fruit. However, Rabinowitch and Sklan (1980, 1981) detected increased SOD activities during ripening of tomato, cucumber, and peppers. Conner and Ng (1999) found that lipid peroxidation increased markedly during ripening of melon fruit, while α-tocopherol reached a maximum during ripening, but ascorbate and glutathione content declined after the fullslip maturity stage.

In the nonclimacteric potato, ethylene application stimulated respiration and increased H_2O_2 levels (Chin and Frenkel, 1976; Warm and Laties, 1982). Studies on associated antioxidant enzyme activities are not available. The potato is interesting, however, as it is one of the few crops used to investigate the relationships between antioxidant status and susceptibility to calcium-related disorders. The contents of glutathione, and activities of DHAR, GR, POX, and SOD, were higher in the genotype that was resistant to a calcium deficiency disorder, internal rust spot, than in a more susceptible genotype, even at similar flesh calcium concentrations (Monk and Davies, 1989). No effects on ascorbate and α-tocopherol contents or activities of ascorbate free-radical reductase and CAT were detected. Although Monktalbot et al. (1991) found a negative correlation between percentage of tubers with calcium deficiency-related tissue necrosis and SOD activity in ten genotypes, calcium deficiency did not appear to induce specific SOD isoforms.

Evidence for the involvement of AOS in senescence of many other systems has been shown, e.g., petals (Droillard et al., 1987; Bartoli et al., 1996; Panavas and Rubinstein, 1998), maize ear shoot silks (Acevedo and Scandalios, 1996), seeds (Bailly et al., 1996), and stress-induced senescence of pea root nodules (Escuredo et al., 1996; Gogorcena et al., 1995). The literature is most extensive, however, for senescing leaf systems, which also involve complex, highly regulated degradative events including lipid oxidation (Thompson et al., 1998; Quirino et al., 2000). In senescing leaf systems, lipid peroxidation is typically associated with declining CAT and SOD activities (Patra et al., 1978; Dhindsa et al., 1981; Pastori and Trippi, 1993; Ranwala and Miller, 2000). Also, application of H_2O_2 or substrates that generate H_2O_2 accelerated senescence of excised leaves (Parida et al., 1978) and fruits (Brennan and Frenkel, 1977). The ability of leaves to respond to oxidative stress by induction of mRNA transcripts and/or activities of SOD may decline with age (Casano et al., 1994). Buchanan-Wollaston and Ainsworth (1997) demonstrated increased expression of a CAT gene during late senescence of *Brassica napus*.

AOS metabolism in senescencing cotyledons and leaves has been investigated using postharvest treatments that delay senescence. Calcium treatment of cucumber cotyledons was associated with lower H_2O_2 levels (Ferguson et al., 1983). In a study of the effects of ethylene, which promoted senescence, and lowered O_2 and elevated CO_2, which delayed senescence in spinach leaves, Hodges and Forney (2000) found strong evidence for relationships between measurements of senescence, AOS production, and associated enzymes. However, the relationships were often complex, highlighting the interactions of different antioxidants, and the enzymes involved in their interplay. In summary, a wide range of metabolite levels and activities of enzymes involved in AOS metabolism has been detected during ripening and senescence of different fruit and vegetative systems. Differences in observed relationships may be related to the use of diverse species and the nature of the ripening/senescence process.

Few studies have attempted to identify differences in AOS metabolism in relation to different rates of senescence and ripening among fruit and vegetable cultivars after harvest. However, existing studies sometimes support the view that these factors are related. Du and Bramlage (1994) found that total SOD activity and relative activities of Cu/ZnSOD, FeSOD, and MnSOD varied and changed greatly during senescence among three apple cultivars, but activities were not consistently related to differences in storability. A study of senescence in the stay-green mutant of *Festuca pratensis* in which senescing leaves maintain chlorophyll did not support a view that senescent leaves of the mutant and wild type were subjected to enhanced oxidative stress (Kingston-Smith et al., 1997). During senescence of both the mutant and the wild type, activities of APX decreased, while those of GR and MDHR remained unchanged, and those of SOD and DHAR increased. Similar H_2O_2 levels were found in senescent leaves of both genotypes, although reduced ascorbate levels were higher in those of the mutant compared with the wild type. However, the study was carried out using an actively photosynthesizing plant system, rather than a harvested product.

In contrast, Lacan and Baccou (1998) correlated the short storage life of one muskmelon cultivar with faster lipid peroxidation, compared with a long-life cultivar that had higher activities of CAT and isoforms of SOD. In broccoli, a cultivar that had more stable chlorophyll contents during senescence had higher SOD and POX activities and lower CAT activity than a faster yellowing cultivar (Toivonen and Sweeney, 1998). Of several methodologies to indicate oxidative status, only one incorporated total reducing capacity, correlated with the storage potential of different herbs (Philosoph-Hadas et al., 1994; Meir et al., 1995). It was concluded that different components of the senescence syndrome may be controlled by different genes and, therefore, may be susceptible to independent manipulation (Philosoph-Hadas et al., 1994). Further studies on the cultivar differences in rates of ripening and senescence in relation to AOS metabolism are warranted, including those designed to determine whether changes in enzymes associated with AOS metabolism are due to altered biosynthesis, activation, and/or degradation. However, better understanding of the processes will probably require utilization of genetic modification technology.

CHILLING INJURIES OF FRUITS AND VEGETABLES

Chilling Injury and Oxidative Metabolism

Chilling in sensitive fruits and vegetables is characterized by several visible symptoms and physiological dysfunctions (Raison and Orr, 1990; Saltveit and Morris, 1990). Chilling sensitivity can vary widely within species, including apples (Bramlage and Meir, 1990), avocados and bananas (Paull, 1990), citrus (Sala, 1998), cucumbers (Cabrera et al., 1992; Hakim et al., 1999), peaches (Brovelli et al., 1998), tomatoes (King and Ludford, 1983), and watermelons (Rise et al., 1990). Some of these studies include chilling susceptibility as a storability criterion rather than involve a study of chilling-injury-associated metabolism per se.

An increasing amount of evidence suggests that oxidative stress is an early response of fruits and vegetables to chilling. Metabolic processes involved in lipid peroxidation were stimulated in cucumber fruit (Hariyadi and Parkin, 1991), and lowered SOD and CAT activities and elevated H_2O_2 levels were associated with heart browning of 'Chili' and 'Yali' pears when exposed to low temperature (Ju et al., 1994). In potatoes, lipid peroxidation and ascorbate utilization was greater at $3°C$ than at $9°C$, and reduced ascorbate was associated with higher APX activity, but not ascorbate free-radical reductase, DHAR, or GR activities (Dipierro and DeLeonardis, 1997). In a similar study with storage of potatoes at $1°C$, $5°C$, and $20°C$, Mizuno et al. (1998) found that decreased ascorbic acid contents in the early part of storage at low temperatures were associated with higher APX activity but lower CAT activity. Postharvest applications of the antioxidants sodium benzoate or ethoxyquin decreased chilling severity in cucumber and sweet pepper (Wang and Baker, 1979), and paclobutrazol increased endogenous antioxidant levels and reduced chilling injury in green bell pepper fruits (Lurie et al., 1994).

Studies on AOS metabolism during chilling of cultivars within species are rare. Sala (1998) compared activities of CAT, SOD, APX, and GR of two chilling-susceptible and two chilling-resistant mandarin fruit cultivars. SOD activity increased similarly in both cultivar types during cold storage, but CAT, APX, and GR activities were highest in the tolerant cultivars. Although no complete statistical analyses were provided to quantify differences between the types, these results suggest that oxidative stress is involved in chilling injury development of mandarin fruit, and that chilling-resistant cultivars may have a more efficient antioxidant system than those that are chilling susceptible. Ascorbate contents of two potato cultivars decreased more rapidly at $1°C$ than at $20°C$, but to a similar extent (Kawakami et al., 2000). APX activity increased in 'Kitaakari' during the 15-week cold storage period, while that of 'Danshaku' increased to a maximum at nine weeks and returned to basal levels. SOD activity increased in 'Kitaakari', but not 'Danshaku', and CAT activity was not affected by cultivar. Unfortunately, no description of relative storability of the cultivars was provided. In a study that did not incorporate measurements of AOS, Hakim et al. (1999) used several cucumber introduction lines, F_1 hybrids of crosses between chilling-sensitive and chilling-resistant lines, and two commercial cultivars of cucumber to study chilling injury in relation to several physiological functions. Correlations among pitting, decay, weight loss, the chlorophyll fluorescence F_v/F_m ratio, respiration, electrolytic leakage, and pyruvate contents were low. Hakim et al. (1999) concluded that a continuum of sensitivity of each process to low temperature existed, and whether a cultivar was deemed to be chilling sensitive or chilling resistant depended on which particular physiological or biochemical process was evaluated. Similar results have been obtained for antioxidant concentrations, associated enzyme activities, and carbohydrate levels in differentially chilling-sensitive inbred maize lines (Hodges et al., 1997a,b). The cucumber system of Hakim et al. (1999) might be a useful postharvest system to test effects on AOS metabolites and enzymes during chilling in different cultivars.

Preharvest manipulations of crops that reduce chilling injury also influence antioxidant metabolites and activities of antioxidant enzymes. Conditioning zucchini squash by keeping them at 15°C for two days prior to storage at 5°C resulted in higher activities of CAT and GR, and higher ratios of reduced to oxidized forms of glutathione than in the control fruit (Wang, 1994, 1995). Preconditioned squash also had higher activities of APX, DHAR, and ascorbate free-radical reductase, and greater ascorbic acid content with a slower rate of decline under chilling conditions, than did control fruit (Wang, 1996). Sala and Lafuente (1999) found that heat treatment (37°C for three days) of mandarin fruit generally resulted in greater activities of CAT, SOD, and APX, and reduced GR activity, compared with nonconditioned fruit that were more susceptible to chilling injury. These studies, however, did not include measurements of AOS levels.

Superficial Scald

Superficial scald, a physiological storage disorder of some apple and pear cultivars, is probably the most well-characterized disorder involving oxidative processes. Although it may be a chilling-related disorder (Bramlage and Meir, 1990; Watkins et al., 1995), it is considered separately in this chapter because of its commercial importance.

Scald susceptibility of fruit is influenced greatly by cultivar (Wilkinson and Fidler, 1973), although occurrence of the disorder is also affected by many preharvest factors (Wilkinson and Fidler, 1973; Emongor et al., 1994). These factors include maturity and climate: fruit harvested early and/or grown in warmer, drier environments tend to be more susceptible to scald development than fruit harvested late and/or grown in colder environments. Nevertheless, cultivars such as 'Gala', 'Empire', and 'Idared' tend to be scald resistant, while others such as 'Cortland', 'Delicious', 'McIntosh', 'Granny Smith', and 'Law Rome' tend to be scald susceptible. Resistance of the former over a wide range of climactic conditions suggests that scald may be genetically determined. The genetics of scald susceptibility have been studied using a seedling population that resulted from a cross between 'Rome Beauty', a desert apple that is highly susceptible to scald, and 'White Angel', a crabapple that is resistant to scald (Weeden, 1993). These seedlings are hybrids of the same age and grown under similar environmental conditions. A total of 56 trees yielded fruit without any evidence of scald and 25 trees with a significant incidence of scald; this 2:1 ratio suggests that more than one gene for scald resistance was segregating in the population.

The generally accepted hypothesis of scald development is that the sesquiterpene α-farnesene is oxidized in the fruit peel, producing several conjugated trienol (CT) species including the triene hydroperoxide radical (Huelin and Coggiola, 1970a,b; Anet, 1972; Rowan et al., 1995; Whitaker et al., 1997). The precise oxidative events that occur during scald development are not well understood. However, free radicals produced by oxidation of α-farnesene may result in disruption, discoloration, and death of surface cells (Huelin and Coggiola, 1970a,b; Du and Bramlage, 1993). Another oxidation product of α-farnesene, 6-methyl-5-hepten-2-one (MHO) has also been implicated in scald development (Mir et al., 1999; Whitaker and Saftner, 2000), but relationships between MHO and scald can be poor (Rupasinghe et al., 2000).

The inhibitory role of antioxidant compounds, such as diphenylamine (Huelin and Coggiola, 1968; Lurie et al., 1989; Rupasinghe et al., 1998, 2000; Whitaker, 2000), ascorbyl palmitate and n-propyl gallate (Bauchot and John, 1996), and low oxygen (Rupasinghe et al., 2000; Whitaker et al., 1997; Whitaker, 2000), on the oxidation of α-farnesene supports the notion that antioxidant metabolites and/or enzymes may influence the resistance to scald.

Most studies on the involvement of α-farnesene and CTs in scald development are focused on scald-susceptible cultivars, differences in scald incidence generally being manipulated by factors such as harvest maturity (Anet, 1972; Meir and Bramlage, 1988; Du and Bramlage, 1993; Watkins et al., 1993). Nevertheless, an early study on scald development showed that accumulation of α-farnesene was earlier and greater in the scald-susceptible 'Granny Smith' than in scald-resistant 'Crofton' (Huelin and Coggiola, 1968). Subsequent studies with 'Granny Smith', 'Delicious', and 'Gala' also found lower α-farnesene and CT concentrations in the scald-resistant 'Gala' than the other cultivars (Whitaker et al., 1997), but some important exceptions have been noted. Anet (1974) found at least one anomalous 'Granny Smith' sample that had high α-farnesene production and CT accumulation but no scald development. Rupasinghe et al. (2000) could not detect a correlation between α-farnesene synthase, α-farnesene content, and

CTs in skin tissues with scald susceptibility or resistance in 11 apples tested. In an analysis of fruit from the 'White Angel' × 'Rome Beauty' crosses described previously, Whitaker et al. (2000) found that production of α-farnesene and CTs was not always closely related with scald susceptibility. Some scald-susceptible selections accumulated very few CTs, while some scald-resistant selections accumulated many CTs, comparable with other scald-susceptible lines.

Antioxidative defenses against AOS are likely to be a function of antioxidant concentrations and activities of antioxidative enzymes. Meigh and Filmer (1969) suggested that the poor correlation between α-farnesene and scald development reflected differences in defense systems. As for the role of α-farnesene and CTs in scald development, the majority of the available literature on antioxidant systems is centered on scald-susceptible cultivars (Meir and Bramlage, 1988; Gallerani et al., 1990; Barden and Bramlage, 1994a,b), rather than comparisons among cultivars of different susceptibility. Studies in which differences in scald have been obtained by modifying harvest date in these cultivars largely support a view that high endogenous antioxidant activities contribute to resistance against scald development, but no substantial losses of antioxidant activity or critical individual antioxidants have been identified.

Anet (1974) separated antioxidants (most unidentified) from different apple cultivars using thin-layer chromatography and suggested that scald did not occur if antioxidant concentrations remained adequate to prevent or sufficiently limit the extent of α-farnesene oxidation. Ju and Bramlage (1999) could not detect any consistent differences in lipid-soluble antioxidant activity between scald-resistant and -susceptible cultivars, although scald incidences of fruit used in the study was not assessed. A relationship between free phenolics in cuticle plus lipid-soluble antioxidant activity in the peel and scald resistance of cultivars (Ju and Bramlage, 1999) warrants further research.

Du and Bramlage (1994, 1995) found that SOD (total or metalloenzymes), CAT, POD, and PPO activities changed during storage, but the activities apparently were not related to expected scald susceptibilities in 'Cortland', 'Delicious', and 'Empire'. Accumulations of thiobarbituric acid-reactive substances or peroxides were also unrelated to cultivar differences. However, scald incidence, even in 'Cortland', was less than 23 percent in those studies, making comparisons between scalded and nonscalded cultivars inconclusive. In contrast, Fernández-Trujillo and Watkins (unpublished) found that POX activities of the highly scald-susceptible 'Law Rome' cultivar (98 percent scald incidence) were several times lower than those of the scald-resistant 'Idared' cultivar (0 percent scald incidence) at harvest and during storage.

Rao et al. (1998) found that lower activities of CAT and POX susceptibility in 'Rome Beauty' × 'White Angel' selections were associated with higher H_2O_2 pools and lipid peroxidation in the tissue, and were correlated with scald. Collectively, the available data indicate that genetically determined resistance to scald might be enhanced by modifying the ability of apples to metabolize AOS. However, over the whole population of selections, relationships between scald susceptibility and any single enzyme activity, e.g., POX, appear weaker (Fernández-Trujillo and Watkins, unpublished). In contrast to the earlier study of Rao et al. (1998), fruit maturity differed greatly among selections and may have accounted for the poorer relationships detected between factors.

The degree to which differences in AOS metabolism among cultivars and selections relate to other metabolic changes associated with chilling injury development is not known. Studies on AOS are difficult because simultaneous increases in antioxidative defense systems may be required to increase stress tolerance (Foyer et al., 1994; Allen et al., 1997). Moreover, in studies with tomato plants, the difference between chilling-tolerant and -susceptible species was their tendency to form free radicals at low temperature, rather than the ability to scavenge them (Walker et al., 1991).

GENETIC APPROACHES: ARE THEY FEASIBLE?

As described earlier, several studies have documented a relationship between progressive loss in antioxidant enzyme activities and tissue senescence and susceptibility to postharvest disorders in several fruits and vegetables, but no reports to date have demonstrated that modifying AOS scavenging capacity of plants minimizes postharvest losses. However, considering the fact that modifying AOS scavenging ability in whole plants has resulted in increased tolerance to various stress factors that act by generating oxidative stress, it is likely that prudent use of genetic engineering technology can help us design horticultural crops resistant to postharvest disorders. Watkins and Pritts (2001) concluded that no breakthroughs in understanding differences in tissue tolerances to postharvest stresses have been obtained by comparing cultivars. Conventional approaches, such as developing inbred lines for comparing similar genotypes, are not feasible in highly heterozygous fruit crops that exhibit inbreeding depression. Because of the large number of genes and corresponding enzyme systems apparently involved in postharvest responses, it may be more productive to manipulate genes within a cultivar using molecular genetics in order to better understand their role. Several examples, available for other plant systems, might be useful in designing crops resistant to postharvest disorders.

Modification of SOD Activity

The enzyme SOD is a metalloprotein that catalyzes the dismutation of $\cdot O_2^-$ to H_2O_2 and molecular oxygen, the initial step in Asada-Halliwell pathway (Figure 10.1). SOD enzymes are classified according to their metal cofactor and their subcellular localization. Cu/ZnSOD localized in both chloroplast and cytosol, and the MnSOD localized in the mitochondria are the most predominant SOD isoforms. Many plant species also contain a chloroplast-localized FeSOD isoform. The isoforms found in the organelles and cytosol differ widely in their biochemical properties and in the degree to which they are inhibited by H_2O_2 and cyanide (Bowler et al., 1992; Scandalios, 1997). To test the relationship between higher SOD activities and oxidative stress tolerance, different SOD genes have been expressed in transgenic plants; results vary (Allen, 1995). For example, overexpression of Cu/ZnSOD, MnSOD, and FeSOD genes provided tolerance to oxidative or environmental stress factors including chilling temperatures in various plant species (Van Camp et al., 1996; reviewed in Allen et al., 1997; Van Breusegem et al., 1999; McKersie et al., 2000). In contrast, studies by Pitcher et al. (1991) and Payton et al. (1997) found no significant improvements in plant tolerance to oxidative stress with up-regulation of SOD. This disparity was later attributed to the differences in cellular target and stress factors, and the fact that modification of the expression of one enzyme in a complex detoxification system would not confer greater advantage.

Modification of H_2O_2 Metabolizing Enzymes

APX, which is localized in the chloroplast, peroxisomes, glyoxysomes, mitochondria, and cytosol, is the primary H_2O_2 scavenging enzyme and has high substrate specificity for ascorbate (Asada, 1992). Similar to SOD, overexpression of APX in different plant systems has yielded contradictory results. Increased APX activity either in the cytosol or chloroplast resulted in greater tolerance to oxidants such as methyl viologen (MV) (Allen et al., 1997; Pitcher et al., 1994) but not to O_3 (Torsethaugen et al., 1997). However, down-regulation of APX increased sensitivity of tobacco plants to oxidative stress generated by O_3, a phytotoxic air pollutant (Orvar and Ellis, 1997). More recently, overexpression of an *Arabidopsis* peroxisomal APX3 conferred protection against oxidative stress generated by aminotriazole but not by MV (Wang et al., 1999). These studies clearly suggest that protection provided by APX is dependent on the specific localization of the isoform and the site of action of oxidative stresses.

Regeneration of oxidized ascorbate and reduced glutathione is a critical component of AOS scavenging system. Among various enzymes that influence the redox state of the ascorbate-glutathione cycle, GR is present in the chloroplast, cytosol, and mitochondria and has received wide attention (Broadbent et al., 1995). Overexpression of either the bacterial or plant GR gene was shown to maintain the reduction state of ascorbate pools more effectively and reduce plant sensitivity to oxidative stress compared with control plants (Aono et al., 1991, 1993; Broadbent et al., 1995; Foyer et al., 1995). Several redox cycles (reduction-oxidation) that scavenge AOS predominantly involving GR and APX were proposed to occur in plant cells. Glutathione, in either the reduced (GSH) or oxidized (GSSG) form, is a key component of antioxidant defenses in most aerobic organisms, including plants (Foyer et al., 1997). Interestingly, however, increased GSH biosynthetic capacity in the chloroplast of transformed tobacco resulted in greatly enhanced oxidative stress, and the phenotype was associated with foliar pools of both GSH and γ-glutamylcysteine (the immediate precursor to GSH) being in a more oxidized state (Creissen et al., 1999). It was concluded that oxidative damage was caused by a failure of the redox-sensing process in the chloroplast.

GSTs include a heterogenous family of enzymes that catalyze the conjugation of GSH to electrophilic sites of a variety of compounds such as lipid peroxides (Coles and Ketterer, 1990). Many GSTs in plants are multifunctional and several have been found to possess GPX activity (Roxas et al., 1997). Overexpression of GST/GPX in tobacco stimulated seedling growth and increased GSSG levels in response to low temperatures (Roxas et al., 1997), supporting the role of GSTs and GPX in oxidative stress responses. CAT is another major enzyme that rapidly removes H_2O_2 from plant cells. Overexpression of a CAT isoform reduced plant sensitivity to oxidative stress factors such as drought, MV, and high-intensity light conditions (Shikanai et al., 1998; Miyagawa et al., 2000). On the other hand, antisense expression of CAT constitutively increased susceptibility of transgenic plants to oxidative stress and chilling injury suggesting that CAT plays an important role in AOS metabolism (Takahashi et al., 1997; Chamnongpol et al., 1998; Kerdnaimongkol and Woodson, 1999). In addition to these antioxidant enzymes, metabolites known to possess antioxidant properties were also shown to influence plant tolerance to oxidative stress, e.g., a mutation with decreased ascorbic acid biosynthesis increased plant sensitivity to oxidative stress (Conklin et al., 1996).

Lowering AOS Production

The abundance of O_2 and the highly energetic electron transfer reactions associated with thylakoid membranes make chloroplasts a major source of AOS production in photosynthetic tissues of plants. However, most postharvest disorders develop during cold storage, usually in environmental conditions without photosynthetic activity. Therefore, the source of AOS production implicated in developing postharvest physiological disorders may be extrachloroplastic and located in the mitochondria (Purvis, 1997).

In some fruits and vegetables, the onset of ripening and/or senescence is accompanied by an increase in respiration called the respiratory climacteric (Biale and Young, 1981). There is ample evidence that the mitochondria of fruits remain functional and capable of conserving energy well into a post-climacteric phase, when other cellular components are exhibiting various signs of deterioration (Romani, 1987). Mitochondria of plant cells are highly sensitive to low temperature and several metabolic changes associated with chilling stress have been identified (Vanlerberghe and McIntosh, 1992; Prasad et al., 1994); the primary being inhibition of Cyt pathway of electron transport (Purvis and Shewfelt, 1993). During this process electron leakage from impaired electron transport chain reduces molecular O_2 to $\cdot O_2^-$, which is dismutated to H_2O_2, increasing cellular oxidative stress (Purvis et al., 1995). Since the alternative pathway bypasses complex III and the last two phosphorylation sites of the electron transport chain, the alternative oxidase (AOX) was postulated to influence the $\cdot O_2^-$ production (Purvis and Shewfelt, 1993; Vanlerberghe and McIntosh, 1997). Supporting this hypothesis, recent studies by Maxwell et al. (1999) using cultured tobacco cells have demonstrated that overexpression of AOX resulted in lower AOS production compared with the wild type.

CONCLUSION AND FUTURE RESEARCH

The postharvest period represents a time of stress in the life of a harvested product. During this time, natural senescence is potentially accelerated by removal of the product from the parent plant, but may be delayed by imposition of a number of postharvest treatments, primarily low temperature and modified atmospheres. Although relatively scarce for postharvest systems, both physiological and genetic studies clearly indicate that antioxidant enzymes are an important component of AOS metabolism in plant cells. Based on the examples of transgenic approaches provided in this chapter, it is likely that modification of antioxidant enzymes could affect senescence and provide additional tolerance to various AOS-induced disorders that are observed in harvested products. However, one must be cautious in interpreting the results obtained from transgenic plants and extrapolating the results toward postharvest disorders because the protective role is strongly dependent on the plant species, the subcellular localization of transgene, and the type of the stress.

AOS metabolism is dependent on a complex detoxification pathway; thus modification of a single component may not always provide maximal effect. It is apparent that approaches to lower AOS production, in conjunction with increased potential to scavenge AOS, may efficiently minimize AOS disorders in biological systems. The documented role of AOS in influencing oxidative stress can be utilized to engineer mitochondrial metabolism and minimize AOS production. At the same time, overexpression of an H_2O_2-scavenging enzyme in parallel with SOD might reduce the risk of unbalancing the $\cdot O_2^-/H_2O_2$ ratio and increase plant resistance to oxidative stress (Hérouart, Bowler, et al., 1993; Gressel and Galun, 1994; Rennenberg and Polle, 1994). This view is supported in studies by Aono et al. (1995) which demonstrated that plants overexpressing both SOD and GR are more tolerant to oxidative stress compared to plants expressing either SOD or GR alone. Thus, by utilizing a combination of different approaches, it appears feasible to design crops resistant to various physiological disorders that develop during postharvest periods.

Genetic modification of biological organisms is fascinating and interesting, but a cautionary approach should be taken before concluding that genetic modification of plants to withstand AOS alone may increase plant resistance to postharvest disorders. Most studies on transgenic plants are performed under laboratory conditions for relatively short periods, and their relevance under field conditions or during long-term storage needs to be ascertained. Our present ability to increase shelf life and minimize post-harvest disorders is largely due to our capacity to store horticultural produce at low temperatures and under modified atmospheres. Therefore, a combination of advances in the storage technology and genetic modification of horticultural produce may be required to maximize postharvest storability.

REFERENCES

Acevedo, A. and J.G. Scandalios (1996). Antioxidant gene (Cat/Sod) expression during the process of accelerated senescence in silks of the maize ear shoot. *Plant Physiology and Biochemistry* 34: 539-545.

Allen, R.D. (1995). Dissection of oxidative stress tolerance using transgenic plants. *Plant Physiology* 107: 1049-1054.

Allen, R.D., R.P. Webb, and S.A. Schale (1997). Use of transgenic plants to study antioxidant defenses. *Free Radicals in Biological Medicine* 23: 473-479.

Anet, E.F.L.J. (1972). Superficial scald: A functional disorder in stored apples. IX. Effect of maturity and ventilation. *Journal of the Science of Food and Agriculture* 23: 763-769.

Anet, E.F.L.J. (1974). Superficial scald: A functional disorder in stored apples. XI. Apple antioxidants. *Journal of the Science of Food and Agriculture* 25: 299-304.

Aono, M., A. Kubo, H. Saji, T. Natori, K. Tanaka, and N. Kondo (1991). Resistance to active oxygen toxicity of transgenic *Nicotiana tabacum* that expresses the gene for glutathione reductase from *Escherichia coli*. *Plant and Cell Physiology* 32: 691-697.

Aono, M., A. Kubo, H. Saji, K. Tanaka, and N. Kondo (1993). Enhanced tolerance to photooxidative stress of transgenic *Nicotiana tabacum* with high chloroplastic glutathione reductase activity. *Plant and Cell Physiology* 34: 129-135.

Aono, M., H. Saji, A. Sakamoto, K. Tanaka, N. Kondo, and K. Tanaka (1995). Paraquat tolerance of transgenic *Nicotiana tabacum* with enhanced activities of glutathione reductase and superoxide dismutase. *Plant and Cell Physiology* 36: 1687-1691.

Asada, K. (1992). Ascorbate peroxidase—a hydrogen peroxide scavenging enzyme in plants. *Physiologia Plantarum* 85: 235-241.

Bailly, C., A. Benamar, F. Corbineau, and D. Come (1996). Changes in the malondialdehyde content and in superoxide dismutase, catalase and glutathione reductase activities in sunflower seeds as related to deterioration during accelerated aging. *Physiologia Plantarum* 97: 104-110.

Baker, J.E. (1976). Superoxide dismutase in ripening fruits. *Plant Physiology* 58: 644-647.

Barden, C.L. and W.J. Bramlage (1994a). Accumulation of antioxidants in apple peel as related to postharvest factors and superficial scald susceptibility of the fruit. *Journal of the American Society for Horticultural Science* 119: 264-269.

Barden, C.L. and W.J. Bramlage (1994b). Relationships of antioxidants in apple peel to changes in α-farnesene and conjugated trienes during storage and to superficial scald development after storage. *Postharvest Biology and Technology* 4: 23-33.

Bartoli, C.G., M. Simontacchi, E. Montaldi, and S. Puntarulo (1996). Oxidative stress, antioxidant capacity and ethylene production during aging of cut carnation *(Dianthus caryophyllus)* petals. *Journal of Experimental Botany* 297: 595-601.

Bauchot, A.D. and P. John (1996). Scald development and the levels of alpha-farnesene and conjugated triene hydroperoxides in apple peel after treatment with sucrose ester-based coatings in combination with food-approved antioxidants. *Postharvest Biology and Technology* 7: 41-49.

Biale, J.B. and R.E. Young (1981). Respiration and ripening in fruit—retrospect and prospect. In *Recent Advances in Biochemistry of Fruits and Vegetables*, J. Friend and M.J.C. Rhodes (Eds.). New York: Academic Press, pp. 1-39.

Bowler, C., M. Van Montagu, and D. Inzé (1992). Superoxide dismutase and stress tolerance. *Annual Review of Plant Physiology and Molecular Biology* 43: 83-116.

Bramlage, W.J. (1993). Interactions of orchard factors and mineral nutrition on quality of pome fruit. *Acta Horticulturae* 326: 15-28.

Bramlage, W.J. and S. Meir (1990). Chilling injury of crops of temperate origin. In *Chilling Injury of Horticultural Crops*, C.Y. Wang (Ed.). Boca Raton, FL: CRC Press, pp. 37-49.

Brennan, T. and C. Frenkel (1977). Involvement of hydrogen peroxide in the regulation of senescence in pear. *Plant Physiology* 59: 411-416.

Brennan, T., A. Rychter, and C. Frenkel (1979). Activity of enzymes involved in the turnover of hydrogen peroxide during fruit senescence. *Botanical Gazette* 140: 384-388.

Broadbent, P., G.P. Creissen, B. Kular, A.R. Wellburn, and P. Mullineaux (1995). Oxidative stress responses in transgenic tobacco containing altered levels of glutathione reductase activity. *The Plant Journal* 8: 247-255.

Brovelli, E.A., J.K. Brecht, W.B. Sherman, and C.A. Sims (1998). Anatomical and physiological responses of melting- and nonmelting-peaches to postharvest chilling. *Journal of the American Society for Horticultural Science* 123: 668-674.

Buchanan-Wollaston, V. (1997). The molecular biology of leaf senescence. *Journal of Experimental Botany* 48: 181-199.

Buchanan-Wollaston, V. and C. Ainsworth (1997). Leaf senescence in *Brassica napus:* Cloning of senescence related genes by subtractive hybridization. *Plant Molecular Biology* 33: 821-834.

Cabrera, R.M., M.E. Saltveit, and K. Owens (1992). Cucumbers differ in their response to chilling temperatures. *Journal of the American Society for Horticultural Science* 117: 802-807.

Casano, L.M., M. Martin, and B. Sabater (1994). Sensitivity of superoxide dismutase transcript levels and activities to oxidative stress is lower in mature-senescent than young barley leaves. *Plant Physiology* 106: 1033-1039.

Chamnongpol, S., H. Willekens, W. Moeder, C. Langebartels, H. Sandermann, M. Van Montagu, M.D. Inze, and W. Van Camp (1998). Defense activation and enhanced pathogen tolerance induced by H_2O_2 in transgenic tobacco. *Proceedings of the National Academy of Science, USA* 95: 5818-5823.

Chin, C.-K. and C. Frenkel (1976). Influence of ethylene and oxygen on respiration and peroxide formation in potato tubers. *Nature* 264: 60.

Coles, B. and B. Ketterer (1990). The role of glutathione and glutathione-s-transferases in chemical carcinogenesis. *CRC Critical Reviews in Biochemistry* 25: 47-70.

Conklin, P.L., E.H. Williams, and R.L. Last (1996). Environmental stress sensitivity of an ascorbic acid-deficient *Arabidopsis* mutant. *Proceedings of the National Academy of Science USA* 93: 9970-9974.

Conner, P.J. and T.J. Ng (1999). Changes in lipid peroxidation and antioxidant status in ripening melon (*Cucumis melo* L.) fruit. *Report of the Cucurbit Genetics Cooperative* 22: 24-27.

Creissen, G.P., E.A. Edwards, and P.M. Mullineaux (1994). Glutathione reductase and ascorbate peroxidase. In *Causes of Photo-Oxidative Stress and Amelioration of Defense Systems in Plants,* C.H. Foyer and P.M. Mullineaux (Eds.). Boca Raton, FL: CRC Press, pp. 343-364.

Creissen, G., J. Firmin, M. Fryer, B. Kular, N. Leyland, H. Reynolds, G. Pastori, F. Wellburn, N. Baker, A. Wellburn, and P. Mullineaux (1999). Elevated glutathione biosynthetic capacity in the chloroplasts of transgenic tobacco plants paradoxically causes increased oxidative stress. *The Plant Cell* 11: 1277-1291.

Dhindsa, R.S., P. Plumb-Dhindsa, and T.A. Thorpe (1981). Leaf senescence: Correlated with increased levels of membrane permeability and lipid peroxidation, and decreased levels of superoxide dismutase and catalase. *Journal of Experimental Botany* 32: 93-101.

Dipierro, S. and S. DeLeonardis (1997). The ascorbate system and lipid peroxidation in stored potato (*Solanum tuberosum* L.) tubers. *Journal of Experimental Botany* 48: 779-783.

Droillard, M.J., A. Paulin, and J.C. Massot (1987). Free radical production, catalase, and superoxide dismutase activities and membrane integrity during senescence of petals of cut carnations *(Dianthus caryophyllus)*. *Physiologia Plantarum* 71: 197-202.

Du, Z. and W.J. Bramlage (1993). A modified hypothesis on the role of conjugated trienes in superficial scald development on stored apples. *Journal of the American Society for Horticultural Science* 118: 807-813.

Du, Z. and W.J. Bramlage (1994). Superoxide dismutase activities in senescing apple fruit (*Malus domestica* Borkh.). *Journal of Food Science* 59: 581-584.

Du, Z. and W.J. Bramlage (1995). Peroxidative activity of apple peel in relation to development of post-storage disorders. *HortScience* 30: 205-209.

Elstner, E.F. (1991). Mechanism of oxygen activation in different compartments of plant cells. In *Active Oxygen/Oxidative Stress and Plant Metabolism,* E.J. Pell and K.L. Steffen (Eds.). Rockville, MD: American Society of Plant Physiologists, pp. 13-25.

Emongor, V.E., D.P. Murr, and E.C. Lougheed (1994). Preharvest factors that predispose apples to superficial scald. *Postharvest Biology and Technology* 4: 289-300.

Escuredo, P.R., F.R. Minchin, Y. Gogorcena, I. Iturbe-Ormaetxe, R.V. Klucas, and M. Becana (1996). Involvement of activated oxygen in nitrate-induced senescence of pea root nodules. *Plant Physiology* 110: 1187-1195.

Ferguson, I.B., C.B. Watkins, and J.E. Harman (1983). Inhibition by calcium of senescence of detached cucumber cotyledons: Effect on ethylene and hydroperoxide production. *Plant Physiology* 71: 182-186.

Foyer, C.H., P. Descourvieres, and K.J. Kunert (1994). Protection against oxygen radicals: An important defense mechanism studied in transgenic plants. *Plant Cell and Environment* 17: 507-523.

Foyer, C.H., H. Lopez-Delgado, J.F. Dat, and I.M. Scott (1997). Hydrogen peroxide- and glutathione-associated mechanisms of acclimatory stress tolerance and signaling. *Physiologia Plantarum* 100: 241-254.

Foyer, C.H., N. Souriau, S. Perret, M. Lelandais, K.J. Kunert, C. Pruvost, and L. Jouanin (1995). Overexpression of glutathione reductase but not glutathione synthetase leads to increase in antioxidant capacity and resistance to photoinhibition. *Plant Physiology* 109: 1047-1057.

Frenkel, C. and M. Eskin (1977). Ethylene evolution as related to changes in hydroperoxides in ripening tomato fruit. *HortScience* 12: 552-553.

Gallerani, G., G.C. Pretella, and RA. Budini (1990). The distribution and role of natural antioxidant substances in apple fruit affected by superficial scald. *Advances in Horticultural Science* 4: 144-146.

Gan, S. and R. Amasino (1995). Inhibition of leaf senescence by autoregulated production of cytokinin. *Science* 270: 1986-1988.

Gogorcena, Y., I. Iturbe-Ormaetxe, P.R. Escuredo, and M. Becana (1995). Antioxidant defenses against activated oxygen in pea nodules subjected to water stress. *Plant Physiology* 108: 753-759.

Green, P.J. (1994). The ribonucleases of higher plants. *Annual Review of Plant Physiology and Plant Molecular Biology* 45: 421-445.

Gressel, J. and E. Galun (1994). Genetic controls of photooxidative tolerance. In *Causes of Photo-Oxidative Stress and Amelioration of Plant Defense Systems,* C.H. Foyer and P.M. Mullineaux (Eds.). Boca Raton, FL: CRC Press, pp. 237-274.

Hakim, A., A.C. Purvis, and B.G. Mullinix (1999). Differences in chilling sensitivity of cucumber varieties depends on storage temperature and the physiological dysfunction evaluated. *Postharvest Biology and Technology* 17: 97-104.

Halliwell, B. and J.M.C. Gutteridge (1989). *Free Radicals in Biology And Medicine.* Oxford, United Kingdom: Clarendon Press.

Hariyadi, P. and K.L. Parkin (1991). Chilling-induced oxidative stress in cucumber fruits. *Postharvest Biology and Technology* 1: 33-45.

Hérouart, D., C. Bowler, H. Willekens, W. Van Camp, L. Slooten, M. Van Montagu, and D. Inzé (1993). Genetic engineering of oxidative stress resistance in higher plants. *Philosophical Transactions of the Royal Society of London-Biological Science* 342: 235-240.

Hérouart, D., M. Van Montagu, and D. Inzé (1993). Redox activated expression of the cytosolic Cu/Zn-superoxide dismutase gene in *Nicotiana*. *Proceedings of the National Academy of Science USA* 90: 3108-3112.

Hess, J.G. (1993). Vitamin E and alpha-tocopherol. In *Antioxidants in Higher Plants,* R.G. Alscher and J.G. Hess (Eds.). Boca Raton, FL: CRC Press, pp. 111-134.

Hodges, D.M., C.J. Andrews, D.A. Johnson, and R.I. Hamilton (1997a). Antioxidant enzyme and compound responses to chilling stress and their combining abilities in differentially sensitive maize hybrids. *Crop Science* 37: 857-863.

Hodges, D.M., C.J. Andrews, D.A. Johnson, and R.I. Hamilton (1997b). Antioxidant enzyme responses to chilling stress in differentially sensitive inbred maize lines. *Journal of Experimental Botany* 48: 1105-1113.

Hodges, D.M. and C.F. Forney (2000). The effects of ethylene, depressed oxygen and elevated carbon dioxide of antioxidant profiles of senescing spinach leaves. *Journal of Experimental Botany* 51: 645-655.

Holland, D., G. Ben-Hayyin, Z. Faltin, L. Camoin, A.D. Strosberg, and Y. Eshdat (1993). Molecular characterization of salt-stress-associated protein in citrus: Protein and cDNA sequence homology to mammalian glutathione peroxidase. *Plant Molecular Biology* 21: 923-927.

Huelin, F.E. and I.M. Coggiola (1968). Superficial scald, a functional disorder of stored apples. IV. Effect of variety, maturity, oil wraps and diphenylamine on the concentrations of α-farnesene in the fruit. *Journal of the Science of Food and Agriculture* 19: 297-301.

Huelin, F.E. and I.M. Coggiola (1970a). Superficial scald, a functional disorder of stored apples. V. Oxidation of α-farnesene and its inhibition by diphenylamine. *Journal of the Science of Food and Agriculture* 21: 44-48.

Huelin, F.E. and I.M. Coggiola (1970b). Superficial scald, a functional disorder of stored apples. VII. Effect of applied α-farnesene, temperature and diphenylamine on scald and the concentration and oxidation of α-farnesene in the fruit. *Journal of the Science of Food and Agriculture* 21: 584-589.

Jabs, T. (1999). Reactive oxygen intermediates as mediators of programmed cell death in plants and animals. *Biochemical Pharmacology* 57: 231-245.

John, I., R. Drake, A. Farrell, W. Cooper, P. Lee, P. Horton, and D. Grierson (1995). Delayed leaf senescence in ethylene deficient ACC-oxidase antisense tomato plants—molecular and physiological analysis. *Plant Journal* 7: 483-490.

Ju, Z. and W.J. Bramlage (1999). Phenolics and lipid-soluble antioxidants in fruit cuticle of apples and their antioxidant activities in model systems. *Postharvest Biology and Technology* 16: 107-118.

Ju, Z., Y. Yuan, C. Liou, C. S. Zhan, and S. Xin (1994). Effects of low temperature on H_2O_2 and heart browning of Chili and Yali pear (*Pyrus bretscheideri*, R.). *Scientia Agricultura Sinica* 27: 77-81.

Kawakami, S., M. Mizuno, and H. Tsuchida (2000). Comparison of antioxidant enzyme activities between *Solanum tuberosum* L. cultivars Danshaku and Kitaakari during low-temperature storage. *Journal of Agricultural and Food Chemistry* 48: 2117-2121.

Kays, S.J. (1997). *Postharvest Physiology of Perishable Plant Products*. Athens, GA: Exon Press.

Kerdnaimongkol, K. and W.R. Woodson (1999). Inhibition of catalase by antisense RNA increases susceptibility to oxidative stress and chilling injury in transgenic tomato plants. *Journal of the American Society for Horticultural Science* 124: 330-336.

King, M.M. and P.M. Ludford (1983). Chilling injury and electrolyte leakage in fruit of different tomato cultivars. *Journal of the American Society for Horticultural Science* 108: 74-77.

Kingston-Smith, A.H., H. Thomas, and C.H. Foyer (1997). Chlorophyll *a* fluorescence, enzyme and antioxidant analyses provide evidence for the operation of alternative electron sinks during leaf senescence in a stay-green mutant of *Festuca pratensis*. *Plant, Cell and Environment* 20: 1323-1337.

Lacan, D. and J.C. Baccou (1998). High levels of antioxidant enzymes correlate with delayed senescence in nonnetted muskmelon fruits. *Planta* 204: 377-382.

Lurie, S., J. Klein, and R. Ben-Arie (1989). Physiological changes in diphenylamine-treated 'Granny Smith' apples. *Israel Journal of Botany* 38: 199-207.

Lurie, S., R. Ronen, Z. Lipsker, and B. Aloni (1994). Effects of paclobutrazol and chilling temperatures on lipids, antioxidants and ATPase activity of plasma membrane isolated from green bell pepper fruits. *Physiologia Plantarum* 91: 593-598.

Maxwell, D.P., Y. Wang, and L. McIntosh (1999). The alternative oxidase lowers mitochondrial reactive oxygen production in plant cells. *Proceedings of the National Academy of Science USA* 96: 8271-8276.

McKersie, B.D. (1991). The role of oxygen free radicals and desiccation stress in plants. In *Active Oxygen/Oxidative Stress and Plant Metabolism*, E.J. Pell and

K.L. Steffen (Eds.). Rockville, MD: American Society of Plant Physiologists, pp. 107-118.

McKersie, B.D., J. Murnaghan, K.S. Jones, and S.R. Bowley (2000). Iron-superoxide dismutase expression in transgenic alfalfa increases winter survival without a detectable increase in photosynthetic oxidative stress tolerance. *Plant Physiology* 122: 1427-1437.

Meigh, D.F. and A.A.E. Filmer (1969). Natural skin coating of the apple and its influence on scald in storage. III. α–farnesene. *Journal of the Science of Food and Agriculture* 18: 307-313.

Meir, S. and W.J. Bramlage (1988). Antioxidant activity in Cortland apple peel and susceptibility to superficial scald after storage. *Journal of the American Society for Horticultural Science* 113: 412-418.

Meir, S., K. Kanner, B. Akiri, and S. Philosoph-Hadas (1995). Determination and involvement of aqueous reducing compounds in oxidative defense systems of various senescing leaves. *Journal of Agricultural and Food Chemistry* 43: 1813-1819.

Mir, N., R. Perez, and R.M. Beaudry (1999). A poststorage burst of 6-methyl-5-hepten-2-one (MHO) may be related to superficial scald in 'Cortland' apples. *Journal of the American Society for Horticultural Science* 124: 173-176.

Miyagawa, Y., M. Tamoi, and S. Shigeoka (2000). Evaluation of defense system in chloroplasts to photooxidative stress caused by paraquat using transgenic tobacco plants expressing catalase from *Escherichia coli*. *Plant and Cell Physiology* 41: 311-320.

Mizuno, M., M. Kamei, and H. Tsuchida (1998). Ascorbate peroxidase and catalase cooperate for protection against hydrogen peroxide generated in potato tubers during low-temperature storage. *Biochemistry and Molecular Biology International* 44: 717-726.

Monk, L.S. and H.V. Davies (1989). Antioxidant status of the potato tuber and Ca^{2+} deficiency as a physiological stress. *Physiologia Plantarum* 75: 411-416.

Monktalbot, L.S., H.V. Davies, M. Macaulay, and B.P. Forster (1991). Superoxide dismutase and susceptibility of potato (*Solanum tuberosum* L.) tubers to calcium related disorders. *Journal of Plant Physiology* 137: 499-501.

Nam, H.G. (1997). The molecular genetic analysis of leaf senescence. *Current Opinion in Biotechnology* 8: 200-207.

Orvar, B.L. and B.E. Ellis (1997). Transgenic tobacco plants expressing antisense RNA for cytosolic ascorbate peroxidase show increased susceptibility to ozone injury. *The Plant Journal* 11: 1297-1306.

Panavas, T. and B. Rubinstein (1998). Oxidative events during programmed cell death of daylily (*Hemerocallis* hybrid) petals. *Plant Science* 133: 125-138.

Parida, R.K., M. Kar, and D. Mishra (1978). Enhancement of senescence in excised rice leaves by hydrogen peroxide. *Canadian Journal of Botany* 56: 2937-2941.

Parkin, K.L., A. Marangoni, R.L. Jackman, R.Y. Yada, and D.W. Stanley (1989). Chilling injury: A review of possible mechanisms. *Journal of Food Biochemistry* 13: 127-153.

Pastori, G.M. and V.S. Trippi (1993). Antioxidative protection in a drought-resistant maize strain during leaf senescence. *Physiologia Plantarum* 87: 227-231.

Patra, H.K., M. Kar, and D. Mishra (1978). Catalase activity in leaves and cotyledons during plant development and senescence. *Biochemie und Physiologie der Pflanzen* 172: 385-390.

Payton, P., R.D. Allen, N. Trolinder, and A.S. Holaday (1997). Overexpression of chloroplast-targeted Mn superoxide dismutase in cotton (*Gossypium hirsutum* L., cv. Coker 312) does not alter the reduction of photosynthesis after short exposures to low temperature and high light intensity. *Photosynthetic Research* 52: 233-244.

Philosoph-Hadas, S., S. Meir, and N. Aharoni (1994). Oxidative defense systems in leaves of three edible herb species in relation to their senescence rate. *Journal of Agricultural and Food Chemistry* 42: 2376-2381.

Picton, S., J.E. Gray, and D. Grierson (1995). The manipulation and modification of tomato fruit ripening by expression of antisense RNA in transgenic plants. *Euphytica* 85: 193-202.

Pitcher, L.H., E. Brennan, A. Hurley, P. Dunsmuir, J.M. Tepp, and B.A. Zilinskas. (1991). Overproduction of copper/zinc-superoxide dismutase does not confer ozone tolerance in transgenic tobacco. *Plant Physiology* 97: 452-455.

Pitcher, L.II., P. Repetti, and B.A. Zilinskas (1994). Overproduction of ascorbate peroxidase protects transgenic tobacco against oxidative stress (abstract no. 623). *Plant Physiology* 105: S-116.

Prasad, T.K., M.D. Anderson, and C.R. Stewart (1994). Acclimation, hydrogen peroxide, and abscisic acid protect mitochondria against irreversible chilling injury in maize seedlings. *Plant Physiology* 105: 619-627.

Purvis, A.C. (1997). The role of adaptive enzymes in carbohydrate oxidation by stressed and senescing plant tissues. *HortScience* 32: 1165-1168.

Purvis, A.C. and R.L. Shewfelt (1993). Does the alternative pathway ameliorate chilling injury in senescent plant tissues? *Physiologia Plantarum* 88: 712-718.

Purvis, A.C., R.L. Shewfelt, and J.W. Gegogeine (1995). Superoxide production by mitochondria isolated from green bell pepper fruit. *Physiologia Plantarum* 94: 743-749.

Quirino, B.F., Y-S. Noh, E. Himelblau, and R.M. Amasino (2000). Molecular aspects of leaf senescence. *Trends in Plant Science* 5: 278-282.

Rabinowitch, H.D. and D. Sklan (1980). Superoxide dismutase: A possible protective agent against sunscald in tomatoes (*Lycopersicum esculentum* Mill.). *Planta* 148: 162-167.

Rabinowitch, H.D. and D. Sklan (1981). Superoxide dismutase activity in ripening cucumber and pepper fruit. *Physiologia Plantarum* 52: 380-384.

Raison, J.K. and G.R. Orr (1990). Proposals for a better understanding of the molecular basis of chilling injury. In *Chilling Injury of Horticultural Crops,* C.Y. Wang (Ed.). Boca Raton, FL: CRC Press, pp. 145-164.

Ranwala, A.P. and W.B. Miller (2000). Preventative mechanisms of gibberellin$_{4+7}$ and light on low-temperature-induced leaf senescence in *Lilium* cv. Stargazer. *Postharvest Biology and Technology* 19: 85-92.

Rao, M.V., J.R. Koch, and K.R. Davis (2000). Ozone: A tool for probing programmed cell death in plants. *Plant Molecular Biology* 44: 346-358.

Rao, M.V., C.B. Watkins, S.K. Brown, and N.F. Weeden (1998). Active oxygen species metabolism in superficial scald resistant and susceptible 'White Angel' × 'Rome Beauty' apple selections. *Journal of the American Society for Horticultural Science* 123: 299-304.

Rennenberg, H. and A. Polle (1994). Protection from oxidative stress in transgenic plants. *Biochemical Society Transactions* 22: 936-940.

Rise, L.A., J.K. Brecht, S.A. Sargent, S.J. Locascio, J.M. Crall, G.W. Elmstrom, and D.N. Maynard (1990). Storage characteristics of small watermelon cultivars. *Journal of the American Society for Horticultural Science* 115: 440-443.

Romani, R.J. (1987). Senescence and homeostasis in postharvest research. *HortScience* 22: 865-868.

Rowan, D.D., J.M. Allen, S. Fielder, J.A. Spicer, and M.A. Brimble (1995). Identification of conjugated triene oxidation products of α-farnesene in apple skin. *Journal of Agricultural and Food Chemistry* 123: 2040-2045.

Roxas, V.P., R.K. Smith, E.R. Allen, and R.D. Allen (1997). Overexpression of glutathione-S-transferase/glutathione peroxidase enhances the growth of transgenic tobacco seedlings during stress. *Nature Biotechnology* 15: 988-991.

Rupasinghe, H.P.V., G. Paliyath, and D.P. Murr (1998). Biosynthesis of α-farnesene and its relation to superficial scald development in 'Delicious' apples. *Journal of the American Society for Horticultural Science* 123: 882-886.

Rupasinghe, H.P.V., G. Paliyath, and D.P. Murr (2000). Sesquiterpene α-farnesene synthase: Partial purification, characterization, and activity in relation to superficial scald development in apples. *Journal of the American Society for Horticultural Science* 125: 111-119.

Sala, J.M. (1998). Involvement of oxidative stress in chilling injury in cold-stored mandarin fruits. *Postharvest Biology and Technology* 13: 255-261.

Sala, J.M. and M.T. Lafuente (1999). Catalase in the heat-induced chilling tolerance of cold-stored hybrid Fortune mandarin fruit. *Journal of Agricultural and Food Chemistry* 47: 2410-2414.

Saltveit, M.E. and L.L. Morris (1990). Overview on chilling injury of horticultural crops. In *Chilling Injury of Horticultural Crops,* C.Y. Wang (Ed.). Boca Raton, FL: CRC Press, pp. 3-15.

Scandalios, J.G. (1997). Molecular genetics of superoxide dismutases in plants. In *Oxidative Stress and the Molecular Biology of Antioxidant Defenses,* J.G. Scandalios (Ed.). New York: Cold Spring Harbor Laboratory Press, pp. 527-568.

Skikanai, T., T. Takeda, H. Yamauchi, S. Sano, K.I. Tomizawa, A. Yokota, and S. Shigeoka (1998). Inhibition of ascorbate peroxidase under oxidative stress in tobacco having bacterial catalase in chloroplasts. *FEBS Letters* 428: 47-51.

Takahashi, H., Z. Chen, H. Du, Y. Liu, and D.F. Klessig (1997). Development of necrosis and activation of disease resistance in transgenic tobacco plants with severely reduced catalase activities. *The Plant Journal* 11: 993-1005.

Thompson, J.E., C.D. Froese, E. Madey, M.D. Smith, and Y. Hong (1998). Lipid metabolism during plant senescence. *Progress in Lipid Research* 37: 119-141.

Toivonen, P.M.A. and M. Sweeney (1998). Differences in chlorophyll loss at 13°C for two broccoli (*Brassica oleraea* L.) cultivars associated with antioxidant enzyme activities. *Journal of Agricultural and Food Chemistry* 46: 20-24.

Torsethaugen, G., L.H. Pitcher, B.A. Zilinskas, and E.J. Pell (1997). Overproduction of ascorbate peroxidase in the tobacco chloroplast does not provide protection against ozone. *Plant Physiology* 114: 529-537.

Van Breusegem, F., L. Slooten, J.M. Stassart, T. Moens, J. Botterman, M. Van Montagu, and D. Inzé (1999). Overproduction of *Arabidopsis thaliana* FeSOD confers oxidative stress tolerance to transgenic maize. *Plant Cell Physiology* 40: 515-523.

Van Camp, W., K. Capiau, M. Van Montagu, D. Inzé, and L. Slooten (1996). Enhancement of oxidative stress tolerance in transgenic tobacco plants overproducing Fe-superoxide dismutase in chloroplasts. *Plant Physiology* 112: 1703-1714.

Vanlerberghe, G.C. and L. McIntosh (1992). Lower growth temperature increases alternative capacity and alternative oxidase protein in tobacco. *Plant Physiology* 100: 115-119.

Vanlerberghe, G.C. and L. McIntosh (1997). Alternative oxidase: From gene to function. *Annual Review of Plant Physiology and Plant Molecular Biology* 48: 703-734.

Walker, M.A., B.D. McKersie, and K.P. Pauls (1991). Effects of chilling injury on the biochemical and functional properties of thylakoid membranes. *Plant Physiology* 97: 663-669.

Wang, C.Y. (1993). Approaches to reduce chilling injury of fruits and vegetables. *Horticultural Reviews* 15: 63-95.

Wang, C.Y. (1994). Effect of temperature preconditioning on catalase, peroxidase, and superoxide dismutase in chilled zucchini squash. *Postharvest Biology and Technology* 5: 67-76.

Wang, C.Y. (1995). Temperature preconditioning affects glutathione content and glutathione reductase activity in chilled zucchini squash. *Journal of Plant Physiology* 145: 148-152.

Wang, C.Y. (1996). Temperature preconditioning affects ascorbate antioxidant system in chilled zucchini squash. *Postharvest Biology and Technology* 8: 29-36.

Wang, C.Y. and J.E. Baker (1979). Effects of two free radical scavengers and intermittent warming on chilling injury and polar lipid composition of cucumber and sweet pepper fruits. *Plant Cell Physiology* 20: 243-251.

Wang, J., H. Zhang, and R.D. Allen (1999). Overexpression of an *Arabidopsis* peroxisomal ascorbate peroxidase gene in tobacco increases protection against oxidative stress. *Plant Cell Physiology* 40: 725-732.

Warm, E. and G.C. Laties (1982). Quantification of hydrogen peroxide in plant extracts by the chemiluminescence reaction with luminol. *Phytochemistry* 21: 827-831

Watkins, C.B., C.L. Barden, and W.J. Bramlage (1993). Relationships between alpha-farnesene, ethylene production and superficial scald development of apples. *Acta Horticulturae* 343: 155-160.

Watkins, C.B., W.J. Bramlage, and B.A. Cregoe (1995). Superficial scald of Granny Smith apples is expressed as a typical chilling injury. *Journal of the American Society for Horticultural Science* 120: 88-94.

Watkins, C.B. and M.P. Pritts (2001). The influence of cultivar on postharvest performance of fruits and vegetables. *Acta Horticulturae* 553: 59-63.

Weeden, N.F. (1993). Genetic control of apple storage scald. *New York Fruit Quarterly* Winter: 12-13.

Whitaker, B.D. (2000). DPA treatment alters α-farnesene metabolism in peel of 'Empire' apples stored in air or 1.5 percent O_2 atmosphere. *Postharvest Biology and Technology* 18: 91-97.

Whitaker, B.D., J.F. Nock, and C.B. Watkins (2000). Peel tissue α-farnesene and conjugated trienol concentrations during storage of 'White Angel' × 'Rome Beauty' hybrid apple selections susceptible and resistant to superficial scald. *Postharvest Biology and Technology* 20: 231-241.

Whitaker, B.D. and R.A. Saftner (2000). Temperature-dependent autoxidation of conjugated trienols from apple peel yields 6-methyl-5-hepten-2-one, a volatile implicated in induction of scald. *Journal of Agricultural and Food Chemistry* 48: 2040-2043.

Whitaker, B.D., T. Solomos, and D.J. Harrison (1997). Quantification of α-farnesene and its conjugated trienol oxidation products from apple peel by C_{18}-HPLC with UV detection. *Journal of Agricultural and Food Chemistry* 45: 760-765.

Wilkinson, B.G. and J.C. Fidler (1973). Physiological disorders. In *The Biology of Apple and Pear Storage,* J.C. Fidler, B.G. Wilkinson, K.L. Edney, and R.O. Sharples (Eds.). Slough, England: Commonwealth Agricultural Bureau, pp. 67-131.

Chapter 11

Postharvest Treatments to Control Oxidative Stress in Fruits and Vegetables

Peter M. A. Toivonen

INTRODUCTION

Oxidative stress has been associated with the development of postharvest physiological disorders such as superficial scald, tissue browning, and chilling injury (Shewfelt and Purvis, 1995). It has also been associated with loss in nutritive value in regards to antioxidant vitamins such as ascorbic acid, α-tocopherol, and carotenoids (Leshem, 1988; Alscher et al., 1997; Pérez et al., 1999). Approaches to modulate or control oxidative stress in plant tissues can therefore be very important to improving shelf life and quality retention during postharvest handling of fruits and vegetables.

Treatments to control oxidative stress-induced injury in fruit and vegetable tissues generally involve two approaches: (1) the use of dips or coatings to directly prevent oxidative reactions, and (2) postharvest treatments (such as low or high temperatures, atmospheric treatments, and growth regulators) to enhance endogenous resistance or tolerance to oxidative stress.

In relation to the first approach, there has been extensive work done using antioxidant chemicals in dips or infusions to control oxidative browning in fruits and vegetables. Much of this work has been thoroughly reviewed previously (McEvily et al., 1992; Sapers, 1993; Friedman, 1996), but recent developments in this area merit discussion. There have also been advances using edible coatings, containing additives which provide improved control of oxidative reactions, so these are also discussed.

In relation to the second approach, enhancement of endogenous resistance within the tissue, there have been many new strategies using various types of treatments. Temperature treatments, growth regulator applications, atmospheric treatments, nitric oxide fumigation, and ethanol fumigation have all shown some potential to reduce oxidative injury in fruit and/or vegetable tissues. Hence, these treatments are discussed in detail.

CONTROL OF OXIDATIVE INJURY USING
ANTIOXIDANT DIPS AND EDIBLE COATINGS

Antioxidant Dips

Antioxidants and/or inhibitors of polyphenol oxidase (PPO) have been used to control enzymatically mediated oxidative browning (or discoloration) in fresh-cut fruit and vegetable products (Sapers, 1993). Although some of this browning can be attributed to direct cell damage from the cutting process, most of the browning occurs after secondary injury-induced processes cause membrane breakdown in tissues that are proximal to the cut surfaces (Rolle and Chism, 1987). Single component dips have generally not been found as effective as dips containing two or more antioxidants and/or PPO enzyme inhibitors (Toivonen, 1992; Sapers, 1993; Kim et al., 1997). Methods involving the use of antioxidant dips in combination with modified atmosphere packaging have been tested, and these combination treatments seem to provide better control of tissue browning than either treatment used separately (Chen et al., 1990a; Gil et al., 1998; Buta et al., 1999).

Numerous synthetic compounds have been identified which are very effective for controlling oxidative browning reactions, including butylated hydroxytoluene, sulfites, ethoxyquin, and diphenylamine in many fruits and vegetables (Chen et al., 1990b; Blanpied, 1993; Sapers, 1993; Whitaker, 2000). The use of synthetic chemicals for controlling browning is becoming less acceptable to the consumer (Lu and Toivonen, 2000), and the use of synthetic materials such as sulfites, ethoxyquin, and diphenylamine is either limited by regulatory bodies or potentially facing a complete ban (Lau, 1993; Buta et al., 1999). In response to consumer concerns, efforts have been focused on natural materials such as 4-hexylresorcinol, cinnamic acid, benzoic acid, ascorbic acid, and citric acid (Sapers, 1993), but even some of these may not be accepted by regulatory agencies. The first three compounds (4-hexylresorcinol, cinnamic acid, benzoic acid) are PPO inhibitors and are often used together with ascorbic acid, a water-soluble antioxidant, which prevents quinone accumulation through reduction of quinones back to their corresponding native phenolics (Sapers, 1993). The natural dip components are not as universally effective as sulfites, therefore, one of the limitations with these new dip formulations is that they may be useful only for specific fruit and vegetable products (Sapers, 1993). In addition, antioxidant components must readily reach the site of oxidative stress in order to be effective. Ascorbic acid is not lipid soluble, and so is very ineffective in the control of superficial scald (Lotz et al., 1997). Ascorbyl palmitate is more effective in suppressing superficial scald because it is lipid soluble and can

better penetrate the cuticular wax of the apple skin and control oxidation reactions of α-farnesene in the peel.

One area that has been poorly explored is the use of dips to control membrane deterioration. Membrane deterioration is a consequence of senescence-related peroxidation of membrane lipids (Thompson et al., 1987) and
is associated with tissue browning. The main strategy currently employed
with dip treatments is to prevent the oxidation of polyphenols by PPO,
which is a secondary oxidation process. Polyphenols and PPO spatially sep-
arate in intact cells, with the polyphenols being localized in the vacuole and
PPO in the plastids and mitochondria (Murata et al., 1997). Therefore, oxidative membrane injury is the primary process that allows the intermixing of the
normally spatially separated enzyme (PPO) and oxidizable substrates (polyphenols), which leads to browning. Calcium dips have been implicated in enhancing membrane stability, slowing senescence, and enhancing the retention
of membrane integrity (Legge et al., 1982; Poovaiah, 1988; Picchioni et al.,
1995). Calcium dips have been shown to delay membrane deterioration in
cabbage (*Brassica oleracea* L. var. *Capitata*) (Chéour et al., 1992) and apples (*Malus domestica* Borkh) (Picchioni et al., 1995). However, the level of
benefit has not been as promising as the use of other treatments in controlling
membrane injuries that lead to tissue deterioration (Whitaker et al., 1997).

Edible Coatings

Edible coatings have shown some promise for use in fresh and minimally
processed products. Coatings, unlike dips, provide physical protection to
the fruit or vegetable surface. These coatings are semipermeable and therefore control gas exchange and water loss for the treated fruit or vegetable
products (Baldwin et al., 1995). Antioxidants can also be incorporated into
them, further controlling tissue degradation and discoloration.

Several edible coating materials have been shown to control oxidative injuries via their effect on gas exchange/water loss properties of the tissue surface. How modified atmospheres can reduce oxidative injury and how water
loss can lead to oxidative injury in tissues has been discussed previously
(Toivonen, 2003). Cellulose-based edible coatings were shown to reduce
the loss of the lipid-soluble antioxidant carotenes in carrots (*Daucus carota*
L.) (Li and Barth, 1998). Several different edible-coating mixtures have
been shown to be effective in controlling browning in apple slices. Pennisi
(1992) used a mixture of chitosan and lauric acid, while Avena-Bustillos
and Krochta (1993) used casein and one of several lipid materials, and
Wong et al. (1994) found an even more complex mixture of alginic acid, casein, and lipids to be useful to control oxidative browning in apple slices.
Oxidative browning of shredded cabbage was significantly reduced with a

complex of sucrose fatty acids (Sakane et al., 1990). In all these cases, the effects of these coatings have been attributed to either reduced oxygen levels within the tissue or reduced water loss from the tissue. Although edible coatings have been shown to significantly modify the internal atmospheres (Baldwin et al., 1995) and control water losses (Kester and Fennema, 1986) of fruit and vegetable products, it is difficult to distinguish the effects due to atmospheric modification versus effects due to control of water loss. The literature in this area is still limited, and more work in evaluating new coating materials and mixtures should be pursued, especially for those which improve gas and water permeability characteristics.

The incorporation of antioxidants into coating materials or mixtures have been shown to control some oxidative injuries (Bauchot et al., 1995). Addition of ascorbyl palmitate to a sucrose ester-based coating mixture resulted in significant control of scald in 'Granny Smith' apples which had been stored for four months. However, the mechanism of control was not clear because the response could not be directly associated with the oxidation of α-farnesene to conjugated triene hydroperoxides (Bauchot and John, 1996). In addition, the coating was found not to be as effective as an antioxidant dip. In another case, a coating containing ascorbic acid and calcium disodium ethylenediamine tetraacetic acid was shown to reduce enzymatic browning in mushrooms (Nisperos-Carriedo et al., 1991). Limited evidence suggests that the incorporation of antioxidants and other compounds in edible coatings has some promise for the control of oxidative browning in cut fruit and vegetable products. It is clear that more work needs to be done in this area of research.

POSTHARVEST TREATMENTS TO CONTROL OXIDANT INJURY IN FRUITS AND VEGETABLES

Conditioning Plant Tissues Against Oxidative Injury

Induction of cross-tolerance or cross-resistance to a stress by using another type of sublethal (nondamaging) stress has been well demonstrated and has been reported to involve the enhancement of antioxidant enzyme activities in plant tissue (Bowler et al., 1992). Recently, the process termed *general adaptation syndrome* (GAS) (see Figure 11.1) has been coined for the induction of cross-resistance to stress in environmental plant research (Leshem and Kuiper, 1996). In essence, GAS refers to the hypothesis that plant responses to most stresses have common mechanisms which alleviate or prevent oxidative injury at the cellular level. Resistance to one stress (e.g., drought) can be enhanced by treating a plant with another type of stress (e.g., heat) at a sublethal level. This results either through the development of resistance or

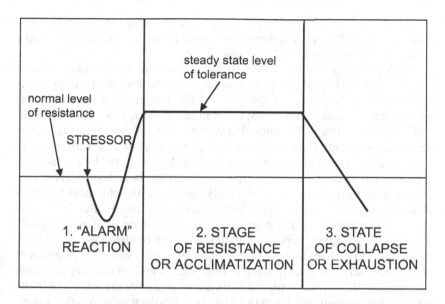

FIGURE 11.1. Triphasic kinetics of the general adaptation syndrome (GAS). (1) Alarm reaction: this is the somatic response of the first exposure to a localized or general stressor and initially lowers basal state of resistance. If the stressor is sufficiently strong (e.g., extremes of temperature, salinity, etc.), death may result. (2) Stage of resistance or acclimatization: if continued exposure to the stressor is compatible with adaptation, resistance mechanisms of either a syntoxic (coexistence with stress) or catatoxic (stressor removal) nature ensues. The cellular manifestations characteristic of the alarm reaction have virtually disappeared or are strictly held in abeyance and resistance rises above normal. (3) State of collapse or exhaustion: following long-term exposure to the stressor, adaptation capability is eventually exhausted. Signs of the alarm reaction reappear and the organ or the whole organism senesces and dies. (From Leshem and Kuiper [1996], used with kind permission from Kluwer Academic Publishers.)

through an acclimatization process by the plant tissues. Leshem and Kuiper (1996) suggest this can be accomplished via modification of syntoxic and/or catatoxic factors in the plant tissues. Syntoxic factors are identified as ABA, jasmonates, cytokinins, flavonoids, etc. which allow the plant to cope with the stress (i.e., a passive tolerance mechanism, or peaceful coexistence with the stressor). Catatoxic factors are identified as relating directly to antioxidant protection systems and include superoxide dismutase, α-tocopherol, ascorbic acid, and glutathione. Catatoxic factors allow the tissues to reduce or eliminate the oxidizing stress agents (i.e., actively attacking the stressor). The level of benefit from these processes may not always be large enough to provide a cross-tolerance or cross-resistance sufficient to completely protect against oxidative damage.

Much is known about oxidative stress response in plant tissues, and several models have been developed to explain the complex relationships of oxygen free radicals, protectant systems, and the occurrence of membrane injuries (Shewfelt and Purvis, 1995; Leshem and Kuiper, 1996; Alscher et al., 1997; Dat et al., 1998). The antioxidant protectant system in plant tissues is very complex, and efficient operation requires a balance of various enzymatic and nonenzymatic components. This is demonstrated by the fact that application of ascorbic acid in a dip or infusion has been shown to provide only transient protection which declines rapidly as ascorbic acid becomes oxidized (Vamos-Vigyazo, 1981). In the intact plant tissues, ascorbic acid is only one component of a system that involves balance of ascorbate pools with enzyme-coupled processes which regenerate oxidized ascorbate back to the reduced (active) form (Wang, 1996; Dat et al., 1998). The addition of exogenous ascorbate does nothing to improve the endogenous regeneration mechanisms in the fruit or vegetable tissue, so that once the ascorbate in the dip solution is consumed, the antioxidant benefits disappear. Another example is that enhanced superoxide dismutase activity will not lead to enhanced resistance to oxidative stress if it is not balanced with increases in the hydrogen peroxide scavenging systems (Sen Gupta et al., 1993). Hydrogen peroxide is a product of superoxide dismutation, therefore, increased superoxide dismutase activity, without a parallel increase in hydrogen peroxide scavenging capability in tissues, can lead to accumulations of hydrogen peroxide in the tissues; this results in peroxidation-mediated oxidative injury (Sen Gupta et al., 1993; Toivonen and Sweeney, 1997). Genetic transformations and mild stress treatments that increase superoxide dismutase activity in plant tissues most often also induce increases in hydrogen peroxide scavenging capability, thereby ensuring an increase in oxidative stress resistance (Sen Gupta et al., 1993; Wang, 1995a).

Temperature Treatments

Heat treatments for fruits and vegetables have been studied extensively and thoroughly reviewed in the past decade (Lurie, 1998). Treatments with hot water or air have been shown to reduce several disorders in fruits and vegetables including superficial scald and chilling injury. The impact of heat treatments on oxidative stress response and how this relates to postharvest disorders, however, has not been reviewed.

Heat treatments have been shown to reduce injury-associated disorders such as irradiation-induced pitting of citrus peel (Abdellaoui et al., 1995;

Miller and McDonald, 1998), chilling injury (Lafuente et al., 1990; Lurie and Klein, 1991; Saltveit, 1991; Wang, 1994; Lange and Cameron, 1997; Laamim et al., 1998; Porat et al., 2000), superficial scald in apples (Lotz et al., 1997), and yellowing in broccoli (Kazami et al., 1991; Forney, 1995; Tian et al., 1997). Many of these injuries have been correlated with oxidative damage to membranes.

Heat treatments have been shown to affect antioxidant systems in the tissues of fruits and vegetables. Hot water treatments enhance ascorbic acid levels in broccoli (*Brassica oleracea* L., var *Italica*) (Kazami et al., 1991) and clementine oranges *(Citrus reticulata)* (Abdellaoui et al., 1995). Hot-water treatment of broccoli also enhances the activity of superoxide dismutase and catalase (Toivonen, unpublished data), suggesting that effects could be attributed through an imposition of mild stress by the treatment. Edreva et al. (1998) have shown that sublethal heat treatments (i.e., those not resulting in direct tissue injury) enhance antioxidant levels as well as polyamine levels in plant tissues, again suggesting that the treatments impose a mild stress on the tissue. Hot-air treatment resulted in reduced membrane leakage of cucumbers (*Cucumis sativa* L.) when they were subsequently placed in chilling temperatures (Lafuente et al., 1990; Saltveit, 1991). Heat shock proteins have also been implicated in the response to heat treatments, yet their role in the response is not clear (Lafuente et al., 1990; Lurie and Klein, 1991).

Nondamaging heat treatments have been shown to have two effects over time: a short-term perturbation followed by a long-term stabilization of membranes (Whitaker et al., 1997). This fits well with the model developed by Leshem and Kuiper (1996), i.e., a sublethal stress results in an injury response (i.e., perturbation) which is subsequently followed by physiological stabilization and increased stress resistance and/or adaptation to oxidative stress. This increased resistance and/or adaptation can be attributed to changes in oxidative stress protection systems, resulting in increases in superoxide dismutase, catalase, and peroxidase activities and polyamine accumulations. Therefore, the general adaptation syndrome (see Figure 11.1), as described by Leshem and Kuiper (1996) appears to be validated by the responses to heat treatments reported by Whitaker et al. (1997).

Low-temperature conditioning is another approach that has been developed to protect sensitive products, such as zucchini squash (*Cucurbita pepo* L.), against oxidative injury. This procedure involves conditioning of chilling-sensitive products at temperatures that are below optimal yet still above the cell damage threshold. Zucchini fruit held at 15°C develops resistance against chilling-induced degradation of membrane glycerolipids (Wang et al., 1992). It was established that superoxide dismutase and catalase activities remained higher in pretreated zucchini squash as compared with nontreated fruit, which were subsequently placed into chilling temperatures (Wang, 1995a). The glutathione antioxidant-protection system (Wang, 1995b) and the ascorbate antioxidant system were also at higher capacities in conditioned fruits (Wang, 1996). Polyamine levels are enhanced with the conditioning treatment (Wang, 1994). The increase in the antioxidant enzyme activities suggests an enhancement of oxygen-radical quenching capacity in the tissues, while increases in polyamine levels are associated with increase in tolerance to oxidative stress in plant tissues as discussed by Leshem and Kuiper (1996). This work with temperature conditioning in zucchini squash shows that stress adaption treatment can result in a wide range of increases in various antioxidant protectant system components in plant tissues.

The impact of temperature conditioning for chilling-insensitive fruits and vegetables has been also reported to reduce oxidatively induced physiological disorders. In these cases, a low storage temperature apparently induces a stress resistance that may be expressed only when the product is subsequently placed into higher shelf-temperature conditions or other stressful handling conditions. Toivonen et al. (1993) found that holding carrots at 0°C for six days resulted in resistance to surface browning when placed into simulated shelf temperatures (13°C). Low-temperature storage for two weeks prior to application of controlled atmospheres has been found to reduce a flesh browning disorder in 'Braeburn' apples (Elgar et al., 1998). Two weeks of low-temperature storage of 'Braeburn' apples reduces the total peroxide content of the flesh, and this correlates with lowered superoxide dismutase activity and higher levels of lipid antioxidants (carotenoids) in the tissue (Toivonen, unpublished data), which may partially explain the increase in resistance to browning disorder in subsequent controlled atmosphere storage. Yellowing in broccoli at shelf temperatures has been linked to lipid peroxidation reactions (Zhuang et al., 1995), while enzymatic enhancement of enzymes such as superoxide dismutase, catalase, and/or peroxidase are expected to reduce lipid peroxidation of membranes (Thompson et al., 1987). Short-term low-temperature storage of broccoli has increased the resistance to yellowing when placed under a simulated shelf temperature of 13°C (Toivonen, 1996). This resistance was associated with increases in antioxidant enzyme activities during storage at 1°C (Toivonen, unpublished data). However, longer terms of low temperature storage (five and ten weeks) may actually increase yellowing in broccoli placed into a 20°C shelf temperature (Pogson and Morris, 1997). This suggests that the benefits obtained with low-temperature conditioning are transient and that longer durations of low-temperature exposure leads to oxidative injury in broccoli. Other reports have shown that benefits of low-temperature conditioning are associated with relatively short treatment durations (Wang, 1993). The transient nature for benefit in response to low-temperature stress fits the GAS model (Figure

11.1). If the duration of treatment is extended, the protectant systems will eventually be exhausted, and at that point tissue injury ensues.

Another approach to controlling chilling injury has been to apply a regime of intermittent warming during low-temperature storage (Ben-Arie et al., 1970; Cohen et al., 1983; Cabrera and Saltveit, 1990; Chessa and Shirra, 1992; Alwan and Watkins, 1999; Dai et al., 1999; Schirra and Cohen, 1999). Intermittent warming has been shown to ameliorate membrane changes or injury associated with chilling expressed as surface pitting (Wang and Baker, 1979; Cabrera and Saltveit, 1990; Dai et al., 1999; Schirra and Cohen, 1999). The development of pitting is associated with oxidative injury (Wang and Baker, 1979), and there is some correlative evidence that intermittent warming results in maintenance of antioxidant components of the cell such as glutathione and ascorbic acid (Dai et al., 1999). Intermittent warming has been shown to reduce apple scald in several cultivars, and this has been primarily attributed to mechanisms which reduce accumulation of toxic products via volatilization or through metabolic conversion in the peel tissues (Alwan and Watkins, 1999). Volatilization of toxic products is known to be important in reducing apple scald because both hypobaric storage (Wang and Dilley, 2000a) and ventilation (Toivonen, 1997) are effective in reducing scald. However, metabolic conversion the toxic products or stress adaptation responses of apple tissues in response to intermittent warming cannot be ruled out. Information is limited regarding how intermittent warming mediates the reduction in chilling injury in apples or other fruit. This area of research probably merits further investigation, since reliable use of this approach will require a better understanding of the mechanism involved.

Atmospheric Treatments

Ultralow-oxygen (ULO) storage involves storing a product at 0.7 percent or lower oxygen levels. The application of ULO controlled atmosphere storage has been used commercially for successful control of superficial scald in apples in certain growing areas (Lau, 1993; Lau et al., 1998) and in pears (*Pyrus communis* L.) (Chen and Varga, 1993). This procedure has also been shown to reduce membrane leakage and chilling injury in avocado (*Persea americana* Mill.) fruit (Pesis et al., 1993). With ULO there is a risk of anaerobic off-flavor development, so another technique has been developed, called dynamic ultralow-oxygen (coined as dynamic control system or DCS) storage, in which oxygen levels are adjusted such that the ethanol in the storage air space is kept below 1 ppm (Schouten et al., 1997). Ultralow oxygen levels in minimally processed products have been shown to maintain endogenous ascorbic acid and reduce oxidative browning (Izumi et al.,

1996; Gil et al., 1998). The use of ultralow oxygen to control oxidative browning in packaged lettuce (*Lactuca sativa* L.) salads is a current commercial practice (López-Gálvez et al., 1997). However, in some cases, quality problems can emerge as a consequence of the accumulation of fermentation products in the fruit and vegetable tissues (Ke et al., 1991; Gil et al., 1998; Lu and Toivonen, 2000).

A technique called initial low oxygen stress (ILOS) also has been shown to be effective in controlling scald development of several apple cultivars (Lurie and Klein, 1993; Truter et al., 1993; Van der merwe et al., 1997; Wang and Dilley, 2000b). ILOS involves holding the apples at 0.5 percent O_2 or less for up to two weeks prior to conventional controlled atmosphere storage. It is likely that ILOS induces an anaerobic metabolism, which results in accumulation of ethanol, thereby controlling scald as demonstrated by Ghahramani and Scott (1998b). ILOS treatments may not be as effective as heat treatments in controlling superficial scald (Lurie and Klein, 1993).

Although low oxygen treatment effects may be mediated by ethanol accumulation (see following discussion regarding ethanol treatments), there is some evidence that low oxygen treatments can enhance antioxidant enzyme systems in plant tissues (Monk et al., 1987). Very little study has been done on the mechanisms of ULO treatment on reduction of scald or other oxidative browning disorders in fruits and vegetables, except those regarding the involvement of ethanol accumulation (Ghahramani and Scott, 1998a). Future work should be initiated regarding the effects of ultralow oxygen treatments on antioxidant protection systems in tissues, as it may be that the effects of ULO are not completely attributable to ethanol accumulation.

The application of high oxygen treatments to prevent oxidative injury is a relatively new concept. High levels of oxygen (100 percent) for up to 12 days at 1°C have been shown to reduce membrane leakage and cut surface browning of slices made from treated apples (Lu and Toivonen, 2000). High oxygen treatment of apples for 12 days at 1°C reduced superoxide dismutase activity and increased lipid-soluble antioxidant capacity in the apple tissue, and this was correlated with reduction of total peroxides (Lu and Toivonen, unpublished data). A combination of high oxygen and high carbon dioxide has also been shown to reduce oxidative injury in carrot tissues (Amanatidou et al., 2000). Caution regarding this approach is required for two reasons: (1) high levels of oxygen require careful handling to avoid fire and/or explosion risks, and (2) if oxygen treatment is extended for longer durations, the plant tissues can become damaged (Solomos et al., 1997; Lu and Toivonen, 2000). This second caution underlies the importance of the level and duration of stress in determining whether the treatment imposed is sublethal (not causing measurable membrane injury) or lethal (causing measurable membrane injury) in nature. Therefore, if superatmospheric oxygen treatments are investigated in the future, the oxygen levels and treatment duration relationships in ascertaining tissue tolerance limits need to be carefully studied.

Ozone, another approach to application of controlled oxidative stress, was found to improve fruit quality and increase ascorbic acid levels in strawberries (Pérez et al., 1999). Sublethal doses of ozone induce an enhanced level of antioxidant enzyme activities in whole plants (Rao et al., 1996). This limited information on ozone response suggests that future work should be pursued to evaluate the potential for ozone treatment to enhance the antioxidant protection systems in plant tissues.

Growth Regulators

There are few examples of growth regulators and their effects on induction of stress resistance or tolerance to oxidative injury in fruit and vegetable tissues. Methyl jasmonate application reduces chilling injury in avocado, grapefruit (*Citrus x paradisi* Macfod.), and bell pepper (*Capsicum annuum* L.) (Meir et al., 1996). Polyamine levels are also enhanced by methyl jasmonate, suggesting that it is acting as a stress signal in the plant tissue (Saniewski, 1995). Polyamine accumulation is also known to increase tolerance to oxidative stress in plant tissues (Leshem and Kuiper, 1996). However, methyl jasmonate may also enhance both the rate of senescence of leafy tissues and chlorophyll loss in apples (Saniewski, 1995). The multitude of positive and negative effects in different plant tissues found with methyl jasmonate application (Sembdner and Parthier, 1993; Saniewski, 1995) may limit the usefulness of this regulator in controlling oxidative injury. Further work is required to better understand the responses to methyl jasmonate that mediate resistance to oxidative stress.

Cytokinin applications are also known to control yellowing in broccoli (Rushing, 1990; Clarke et al., 1994; Downs et al., 1997). Since yellowing of broccoli is known to be strongly associated with lipid peroxidation (Zhuang et al., 1995), it is probable that cytokinin applications control reactions leading to lipid peroxidation. The explanation may lie in the fact that cytokinins are known to inhibit the generation of oxygen free radicals via inhibition of xanthine oxidase and also via direct scavenging of oxygen free radicals (Leshem, 1988).

Other growth regulators have been shown to reduce chilling injury. Abscisic acid applications reduced chilling injury in grapefruit (Kawada et al., 1979) and zucchini squash (Wang, 1991), while ethylene treatments reduced chilling injury in honeydew melons (*Cucumis melo* L., Inodorus Group) (Lipton and Aharoni, 1979). Polyamine applications also reduced chilling-induced brown core in 'McIntosh' apples (Kramer et al., 1991) and pitting in zucchini squash (Kramer and Wang, 1989). Diazocyclopenta-diene, an inhibitor of ethylene action, has also been shown to reduce super-ficial scald of apples, presumably via a direct effect on delaying ripening of the fruit (Gong and Tian, 1998).

It appears that most of the growth regulators associated with stress re-sponse have some effect on controlling oxidative stress-induced injuries in fruits and vegetables. Leshem and Kuiper (1996) indicate that growth regu-lators operate in a syntoxic manner (as agents to enhance coping with stress) in regard to preventing oxidative stress-induced injuries. The only exception to this may be the direct antioxidant capability that cytokinins possess (Leshem, 1988). Future development of the use of regulators to control oxi-dative stress response requires basic knowledge of all the effects of these compounds on physiology and ripening, as well as on the induction of oxi-dative protective systems.

Nitric Oxide

Nitric oxide has recently been identified as a senescence-delaying agent, acting to reduce ethylene production (Leshem and Wills, 1998) thereby controlling rates of ripening in climacteric fruit as well as deterioration of flowers and vegetables (Leshem et al., 1998). It has also been shown to de-crease superoxide generation (Caro and Puntarulo, 1998). One of the modes of action for reducing superoxide generation is believed to be through the inhibition of cytochrome P_{450} activity, which is a key component of the oxi-datively coupled electron transport chain. The inhibition of oxidative me-tabolism is known to induce protective systems in plant tissues (Shewfelt and Purvis, 1995). Therefore, the role of nitric oxide may be as an oxidative stress syntoxic agent (i.e., agent to enhance coping with the stress) as op-posed to a catatoxic agent (i.e., scavenger of oxygen radicals) (Leshem and Kuiper, 1996). However some caution must be maintained regarding nitric oxide, as higher levels can result in membrane injury in chloroplasts (Lesh-em et al., 1997). This is because nitric oxide itself is a free radical and can also form peroxynitrite in the presence of superoxide anions. Both of these radicals can injure membrane-bound proteins in the photosynthetic centers of the chloroplast (Leshem et al., 1997).

Ethanol Treatment

Ethanol vapors have been shown to control superficial scald in 'Granny Smith' apples (Ghahramani and Scott, 1998a), yellowing in broccoli (Cor-cuff et al., 1996), and chilling-induced membrane injury in potatoes (*Sola-num tuberosum* L.) (Frenkel et al., 1995). The use of low oxygen to induce ethanol accumulation results in a similar response as application of exogenous ethanol application (Ghahramani and Scott, 1998b) and has been discussed previously in this chapter. The mode of action of ethanol in controlling scald is associated with reduced accumulation of conjugated trienes, which are intermediates that lead to scald development (Ghahramani and Scott, 1998a,b). Because yellowing is known to be related to lipid peroxidation (Zhuang et al., 1995), it is presumed that the ethanol is acting to control the level of lipid peroxidation in broccoli tissue. Ethanol has been shown to have effects on membrane fluidity (Toivonen, 1997), and may thereby mediate the effects observed for potatoes since membrane fluidity is important for maintaining membrane function (Legge et al., 1982). However, little evidence at the present time excludes the possibility that the response to ethanol may involve a more complex set of biochemical changes in plant tissue. More work is required to better understand the mode of action and limitations for the use of ethanol to modify oxidative stress resistance.

Combined Treatments

The previous discussion has focused on single treatments to control oxidative stress-induced injury. However, an increasing number of reports suggest that combined treatments may give superior results. The application of hot-water treatments in combination with modified atmosphere packaging has been shown to reduce chilling injury in peppers better than either approach separately (González-Aguilar et al., 1997). Gil et al. (1998) found that the application of an ascorbic acid dip along with low oxygen resulted in the best control of cut-surface browning of 'Fuji' apple slices. Chen et al. (1990a) found that a combination approach using both controlled atmospheres and ethoxyquin allowed a reduction in the dose of ethoxyquin required to achieve effective control of scald in 'd'Anjou' pears. Curry and Sugar (1993) used a combination of heat treatments with ethoxyquin, which resulted in a reduction of the amount of ethoxyquin required for good scald control in 'd'Anjou' pears. Ascorbyl palmitate emulsion application in addition to proper CA storage conditions can eliminate scald in 'Granny Smith' apples (Lotz et al., 1997). Wang (1994) found that treating zucchini squash with a combination of heat and low-temperature conditioning gave better control of chilling injury than either treatment used alone. Although it is evident that combination treatments can improve the control of oxidative stress-induced injury, this approach has been studied only superficially to date. Although the literature regarding the use of combination treatments is limited, there appears to be some promise in using this approach because synergistic reduction of oxidative injury can result.

FUTURE DIRECTIONS

The reduction of oxidatively induced injuries can be achieved by various means. The direct approach is to prevent the oxidative reactions altogether by using antioxidants or enzyme inhibitors. Synthetic antioxidants such as sulfites and diphenylamine are very effective; however, there is considerable pressure to move to the use of natural antioxidants for fruits and vegetables or to indirect, nonchemical approaches. If natural antioxidants are to be effective, it is clear that they will need to be used in combination with other compounds and/or treatments. The nonchemical approaches of ULO, heat, and low-temperature conditioning show the greatest promise for use as components of a combined treatment strategy.

The general adaptation syndrome hypothesis (Leshem and Kuiper, 1996) infers that indirect, nonchemical approaches operate via similar mechanisms to increase resistance to oxidative stress. This might possibly explain why so many different treatments can reduce certain oxidatively induced postharvest disorders. However, this model cannot explain why the combination of two different nondirect treatments result in synergistic effects on oxidative stress resistance. The work with heat treatments, low-temperature conditioning, and the combination of both treatments in zucchini squash suggests that there are differential effects of these treatments on various components of the endogenous antioxidant protection systems. This underlies the complexity of the antioxidative stress protection/coping mechanisms in plant tissues. Work is required to better elucidate the complete set of responses of the antioxidant protectant systems to the different indirect treatments. Therefore, future research should include a complete analysis of the effects of individual treatments on the modification of the complete set of protectant/coping systems in various commodities. Once better understanding of the biochemical responses is documented, the conditions required for reliable induction of antioxidant protection using combined treatments can be better predicted. It must be kept in mind that the application of mild stresses to enhance resistance systems in the tissues can, at some point in time or at some stress level, lead to oxidative injury (Figure 11.1). Therefore, studies will be required to evaluate when a combined treatment stress level or duration becomes excessive. The species, cultivar, developmental stage, and age of tissue will likely be very important determinants for sensitivity to oxygen radical stress and therefore should also be considered in such studies. Such information will be invaluable in the development of these treatments for practical commercial use.

REFERENCES

Abdellaoui, S., M. Lacroix, M. Jobin, C. Boubekri, and M. Gagnon (1995). Effets de l'irradiation gamma avec et sans traitement a l'eau chaude sur les propetietes physicochimiques, la teneur en vitamine C et les qualites organoleptiques des clementines. *Sciences des Aliments* 15: 217-235.

Alscher, R.G., J.L. Donahue, and C.L. Cramer (1997). Reactive oxygen species and antioxidants: Relationships in green cells. *Physiologia Plantarum* 100: 224-233.

Alwan, T.F. and C.B. Watkins (1999). Intermittent warming effects on superficial scald development of 'Cortland', 'Delicious' and 'Law Rome' apple fruit. *Postharvest Biology and Technology* 16: 203-212.

Amanatidou, A., R.A. Slump, L.G.M. Gorris, and E.J. Smid (2000). High oxygen and high carbon dioxide modified atmospheres for shelf-life extension of minimally processed carrots. *Journal of Food Science* 65: 61-66.

Avena-Bustillos, R.A. and J.M. Krochta (1993). Water vapor permeability of caseinate-based edible films as affected by pH, calcium crosslinking and lipid content. *Journal of Food Science* 58: 904-907.

Baldwin, E.A., M.O. Nisperos-Carreido, and R.A. Baker (1995). Edible coatings for lightly processed fruits and vegetables. *HortScience* 30:35-38.

Bauchot, A.D. and P. John (1996). Scald development and the levels of α-farnesene and conjugated triene hydroperoxides in apple peel after treatment with sucrose ester-based coatings in combination with food-approved antioxidants. *Postharvest Biology and Technology* 7: 41-49.

Bauchot, A.D., P. John, Y. Soria, and I. Recasens (1995). Sucrose ester-based coatings formulated with food-compatible antioxidants in the prevention of superficial scald in stored apples. *Journal of the American Society for Horticultural Science* 120: 491-496.

Ben-Arie, R., S. Lavee, and S. Guelfat-Reicht (1970). Control of wooly breakdown of Elberta peaches in cold storage by intermittent exposure to room temperature. *Proceedings of the American Society for Horticultural Science* 95: 801-802.

Blanpied, G.D. (1993). Effect of repeated postharvest applications of butylated hydroxytoluene (BHT) on storage scald of apples. *Proceedings of the 6th International Controlled Atmosphere Research Conference.* Ithaca, NY: Northeast Regional Agricultural Engineering Service, Cornell University. pp. 466-469.

Bowler, C., M. Van Montagu, and D. Inzé (1992). Superoxide dismutase and stress tolerance. *Annual Review of Plant Physiology and Plant Molecular Biology* 43: 83-116.

Buta, J.G., H.E. Moline, D.W. Spaulding, and C.Y. Wang (1999). Extending storage life of fresh-cut apples using natural products and their derivatives. *Journal of Agricultural and Food Chemistry* 47: 1-6.

Cabrera, R.M. and M.E. Saltveit Jr. (1990). Physiological response to chilling temperatures of intermittently warmed cucumber fruit. *Journal of the American Society for Horticultural Science* 115: 256-261.

Caro, A. and S. Puntarulo (1998). Nitric oxide decreases superoxide anion generation by microsomes from soybean embryonic axes. *Physiologia Plantarum* 104: 357-364.

Chen, P.M. and D.M. Varga (1993). Efficacy of step-wise low O_2, low O_2, and initial high CO_2 treatments on the control of superficial scald of 'd'Anjou' pears. *Proceedings 6th International Controlled Atmosphere Research Conference.* Ithaca, NY: Northeast Regional Agricultural Engineering Service, Cornell University. pp. 453-465.

Chen, P.M., D.M. Varga, E.A. Mielke, T.J. Facteau, and S.R. Drake (1990a). Control of superficial scald on d'Anjou pears by ethoxyquin: Effect of ethoxyquin concentration, time and method of application, and a combined effect with controlled atmosphere storage. *Journal of Food Science* 55: 167-170.

Chen, P.M., D.M. Varga, E.A. Mielke, T.J. Facteau, and S.R. Drake (1990b). Control of superficial scald on d'Anjou pears by ethoxyquin: Oxidation of α-farnesene and its inhibition. *Journal of Food Science* 55: 171-175, 180.

Chéour, F., J. Arul, J. Makhlouf, and C. Willemot (1992). Delay of membrane lipid degradation by calcium treatment during cabbage leaf senescence. *Plant Physiology* 100: 1656-1660.

Chessa, I. and M. Shirra (1992). Prickly pear cv. "Gialla": Intermittent and constant refrigeration trials. *Acta Horticulturae* 296: 129-137.

Clarke, S.F., P.E. Jameson, and C.G. Downs (1994). The influence of 6-benzyl-aminopurine on postharvest senescence of floral tissues of broccoli (*Brassica oleracea* var Italica). *Plant Growth Regulation* 14: 21-27.

Cohen, E., M. Shuali, and Y. Shalom (1983). Effect of intermittent warming on the reduction of chilling injury of Villa Franca lemon fruits stored at low temperature. *Journal of Horticultural Science* 58: 593-598.

Corcuff, R., J. Arul, F. Hamza, F. Castaigne, and J. Makhlouf (1996). Storage of broccoli florets in ethanol vapor enriched atmospheres. *Postharvest Biology and Technology* 7: 219-229.

Curry, E. and D. Sugar (1993). Reducing scald in 'Red d'Anjou' pears with hot water and reduced rates of ethoxyquin. *Acta Horticulturae* 367: 426-431.

Dai, H.F., Z.L. Ji, and Z.Q. Zhang (1999). The effect of intermittent warming on chilling injury and metabolism of glutathione, ascorbic acid of mango fruit. *Journal of South China Agricultural University* 20: 51-54. [English summary]

Dat, J.F., C.H. Foyer, and I.M. Scott (1998). Changes in salicylic acid and antioxidants during induced thermotolerance in mustard seedlings. *Plant Physiology* 118: 1455-1461.

Downs, C.G., S.D. Somerfield, and M.C. Davey (1997). Cytokinin treatment delays senescence but not sucrose loss in harvested broccoli. *Postharvest Biology and Technology* 11: 93-100.

Edreva, A., I. Yordanov, R. Kardjieva, and E. Gesheva (1998). Heat shock responses of bean plants: Involvement of free radical, antioxidants and free radical/active oxygen scavenging systems. *Biologia Plantarum* 41: 185-191.

Elgar, H.J., D.M. Burmeister, and C.B. Watkins (1998). Storage and handling effects on a CO_2-related internal browning disorder of 'Braeburn' apples. *HortScience* 33: 719-722.

Forney, C.F. (1995). Hot-water dips extend the shelf life of fresh broccoli. *HortScience* 30: 1054-1057.

Frenkel, C., A. Enez, and M.R. Henninger (1995). Ethanol-induced cold-tolerance in chilling-sensitive crops. *Proceedings of the International Society of Agricultural Engineers.* St. Joseph, MI: American Society of Agricultural Engineers. pp. 512-521.

Friedman, M. (1996). Food browning and its prevention: An overview. *Journal of Agricultural and Food Chemistry* 44: 631-653.

Ghahramani, F. and K.J. Scott (1998a). The action of ethanol in controlling superficial scald of apples. *Australian Journal of Agricultural Research* 49: 199-205.

Ghahramani, F. and K.J. Scott (1998b). Oxygen stress of 'Granny Smith' apples in relation to superficial scald, ethanol, α-farnesene, and conjugated tricncs. *Australian Journal of Agricultural Research* 49: 207-210.

Gil, M.I., J.R. Gorny, and A.A. Kader (1998). Responses of 'Fuji' apple slices to ascorbic acid treatment and low-oxygen atmospheres. *HortScience* 33: 305-309.

Gong, Y. and M.S. Tian (1998). Inhibitory effect of diazocyclopentadiene on the development of superficial scald in Granny Smith apple. *Plant Growth Regulation* 26: 117-121.

González-Aguilar, G.A., R. Cruz, M. Granados, and R. Báez (1997). Hot water dips and film packaging extend the shelf life of bell peppers. *Proceedings of the 7th International Controlled Atmosphere Research Conference* 4: 66-72.

Izumi, H., A.E. Watada, and W. Douglas (1996). Low O_2 atmospheres affect storage quality of zucchini squash slices treated with calcium. *Journal of Food Science* 61: 317-321.

Kawada, K., T.A. Wheaton, A.C. Purvis, and W. Grierson (1979). Levels of growth regulators and reducing sugars of 'Marsh' grapefruit peel as related to seasonal resistance to chilling injury. *HortScience* 14: 446.

Kazami, D., T. Sato, H. Nakagawa, and N. Ogura (1991). Effect of pre-storage hot water dipping of broccoli heads on shelf and quality during storage. *Nippin Ngeikagaku Kaishi* 65: 19-26. [English summary]

Ke, D., L. Rodriguez-Sinobas, and A.A. Kader (1991). Physiology and prediction of fruit tolerance to low-oxygen atmosphere. *Journal of the American Society for Horticultural Science* 116: 253-260.

Kester, J.J. and O.R. Fennema (1986). Edible films and coatings: A review. *Food Technology* 40: 47-59.

Kim, H.-S., K.-Y. Lee, H.-G. Lee, O. Han, and U.-J. Chang (1997). Studies in the extension of the shelf-life of kochujang during storage. *Journal of the Korean Society of Food Science and Nutrition* 26: 595-600. [English summary]

Kramer, G.F. and C.Y. Wang (1989). Correlation of reduced chilling injury with increased spermine and spermidine levels in zucchini squash. *Physiologia Plantarum* 76: 479-484.

Kramer, G.F., C.Y. Wang, and W.C. Conway (1991). Inhibition of softening by polyamine application in 'Golden Delicious' and 'McIntosh' apples. *Journal of the American Society for Horticultural Science* 116: 813-817.

Laamim, M., Z. Lapsker, E. Fallik, A. Ait-Oubahou, and S. Lurie (1998). Treatments to reduce chilling in harvested cucumbers. *Advances in Horticultural Science* 12: 175-178.

Lafuente, M.T., A. Belver, M.G. Guye, and M.E. Saltveit Jr. (1990). Effect of temperature conditioning on chilling injury of cucumber cotyledons. Possible role of abscisic acid and heat shock proteins. *Plant Physiology* 95: 443-449.

Lange, D.L. and A.C. Cameron (1997). Pre- and postharvest temperature conditioning of greenhouse-grownsweet basil. *HortScience* 32: 114-116.

Lau, O.L. (1993). Prediction and nonchemical control of 'Delicious' apple scald. *Proceedings of the 6th International Controlled Atmosphere Research Conference.* Ithaca, NY: Northeast Regional Agricultural Engineering Service, Cornell University. pp. 435-446.

Lau, O.L., C.L. Barden, S.M. Blankenship, P.M. Chen, E.A. Curry, J.R. DeEll, L. Lehman-Salada, E.J. Mitcham, R.K. Prange, and C.B. Watkins (1998). A North American cooperative survey of 'Starkrimson Delicious' apple responses to 0.7 percent O_2 storage on superficial scald and other disorders. *Postharvest Biology and Technology* 13: 10-26.

Legge, R.L., J.E. Thompson, J.E. Baker, and M. Lieberman (1982). The effect of calcium on the fluidity and phase properties of microsomal membranes isolated from postclimacteric golden delicious apples. *Plant and Cell Physiology* 23: 161-169.

Leshem, Y. (1988). Plant senescence processes and free radicals. *Free Radical Biology and Medicine* 5: 39-49.

Leshem, Y.Y., E. Haramaty, D. Iluz, Z. Malik, Y. Sofer, L. Roitman, and Y. Leshem (1997). Effect of stress nitric oxide (NO): Interaction between chlorophyll fluorescence, galactolipid fluidity and lipoxygenase activity. *Plant Physiology and Biochemistry* 35: 573-579.

Leshem, Y.Y. and P.J.C. Kuiper (1996). Is there a GAS (general adaptation syndrome) response to various types of environmental stress? *Biologia Plantarum* 38: 1-18.

Leshem, Y.Y. and R.B.H. Wills (1998). Harnessing senescence delaying gases nitric oxide and nitrous oxide: A novel approach to postharvest control of fresh horticultural produce. *Biologia Plantarum* 41: 1-10.

Leshem, Y.Y., R.B.H. Wills, and V.V.-V. Ku (1998). Evidence for the function of the free radical gas—nitric oxide (NO˙)—as an endogenous maturation and senescence regulating factor in higher plants. *Plant Physiology and Biochemistry* 36: 825-833.

Li, P. and M.M. Barth (1998). Impact of edible coatings on nutritional and physiological changes in lightly-processed carrots. *Postharvest Biology and Technology* 14: 51-60.

Lipton, T.J. and Y. Aharoni (1979). Chilling injury and ripening of 'Honey Dew' muskmelons stored at 2.5 or 5°C after ethylene treatment at 20°C. *Journal of the American Society for Horticultural Science* 104: 327-330.

López-Gálvez, G., G. Peiser, X. Nice, and M. Cantwell (1997). Quality changes in packaged salad products during storage. *Zeitschrift für Lebensmittel-Untersuchung und-Forschung* 205: 64-72.

Lotz, E., F.J. Barnard, and J.C. Combrink (1997). Evaluation of alternative treatments for control of superficial scald in apples. *Deciduous Fruit Grower* 47: 443, 445-449.

Lu, C. and P.M.A. Toivonen (2000). Effect of 1 and 100 kPa O_2 atmospheric pretreatments of whole 'Spartan' apples on subsequent quality and shelf life of slices stored in modified atmosphere packages. *Postharvest Biology and Technology* 18: 99-107.

Lurie, S. (1998). Postharvest heat treatments. *Postharvest Biology and Technology* 14: 257-269.

Lurie, S. and J.D. Klein (1991). Acquisition of low-temperature tolerance in tomatoes by exposure to high-temperature stress. *Journal of the American Society of Horticultural Science* 116: 1007-1012.

Lurie, S. and J.D. Klein (1993). Prestorage heat and anaerobic treatments to control apple scald. *Proceedings of the 6th International Controlled Atmosphere Research Conference*. Ithaca, NY: Northeast Regional Agricultural Engineering Service, Cornell University. pp. 447-452.

McEvily, A.J., R. Iyengar, and W.S. Otwell (1992). Inhibition of enzymatic browning in foods and beverages. *Critical Reviews in Food Science and Nutrition* 32: 253-273.

Meir, S., S. Philosoph-Hadas, S. Lurie, S. Droby, M. Akerman, G. Zauberman, P. Shapiro, L. Cohen, and Y. Fuchs (1996). Reduction of chilling injury in stored avocado, grapefruit, and bell pepper by methyl jasmonate. *Canadian Journal of Botany* 74: 870-874.

Miller, W.R. and R.E. McDonald (1998). Short-term heat conditioning of grapefruit to alleviate irradiation injury. *HortScience* 33: 1224-1227.

Monk, L.S., K.V. Fagerstedt, and R.M.M. Crawford (1987). Superoxide dismutase as an anaerobic polypeptide. A key factor in recovery from oxygen deprivation in *Iris pseudacorus*? *Plant Physiology* 85: 1016-1020.

Murata, M., M. Tsurutani, S. Hagiwara, and S. Homma (1997). Subcellular location of polyphenol oxidase in apples. *Bioscience, Biotechnology, and Biochemistry* 61: 1495-1499.

Nisperos-Carreido, M.O., E.A. Baldwin, and P.E. Shaw (1991). Development of an edible coating for extending postharvest life of selected fruits and vegetables. *Proceedings of the Florida State Horticulture Society* 104: 122-125.

Pennisi, E. (1992). Sealed in edible film. *Science News* 141: 12.

Pérez, A.G., C. Sanz, J.J. Ríos, R. Olías, and J.M. Olías (1999). Effects of ozone treatment on postharvest strawberry quality. *Journal of Agricultural and Food Chemistry* 47: 1652-1656.

Pesis, E., R. Marinansky, G. Zauberman, and Y. Fuchs (1993). Reduction of chilling injury symptoms of stored avocado fruit by prestorage treatment with high nitrogen atmosphere. *Acta Horticulturae* 343: 251-255.

Picchioni, G.A., A.E. Watada, W.S. Conway, B.D. Whitaker, and C.E. Sams (1995). Phospholipid, galactolipid, and steryl lipid composition of apple fruit cortical tissue following postharvest $CaCl_2$ infiltration. *Phytochemistry* 39: 763-769.

Pogson, B.J. and S.C. Morris (1997). Consequences of cool storage of broccoli on physiological and biochemical changes and subsequent senescence at 20°C. *Journal of the American Society for Horticultural Science* 122: 553-558.

Poovaiah, B.W. (1988). Molecular aspects of calcium action in plants. *HortScience* 23: 267-271.

Porat, R., D. Pavoncello, J. Peretz, S. Ben-Yehoshua, and S. Lurie (2000). Effects of various heat treatments on the induction of cold tolerance and on the postharvest qualities of 'Star Ruby' grapefruit. *Postharvest Biology and Technology* 18: 159-165.

Rao, M.V., G. Paliyath, and D.P. Ormrod (1996). Ultraviolet-B- and ozone-induced biochemical changes in antioxidant enzymes of *Arabidopsis thaliana*. *Plant Physiology* 110: 125-136.

Rolle, R.S. and G.W. Chism (1987). Physiological consequences of minimally processed fruits and vegetables. *Journal of Food Quality* 10: 157-177.

Rushing, J. (1990). Cytokinins affect respiration, ethylene production, and chlorophyll retention of packaged broccoli florets. *HortScience* 25: 88-90.

Sakane, Y., N. Arita, S. Shimokana, H. Ito, and Y. Osajima (1990). Storage of shredded cabbage in plastic films using ethylene-acetaldehyde or sucrose fatty acid esters. *Nippon Shokuhin Kogyo Gakkaishi* 37: 281-286. [English Summary]

Saltveit Jr., M.E. (1991). Prior temperature exposure affects subsequent chilling sensitivity. *Physiologia Plantarum* 82: 529-536.

Saniewski, M. (1995). Methyl jasmonate in relation to ethylene production and other physiological processes in selected horticultural crops. *Acta Horticulturae* 394: 85-98.

Sapers, G.M. (1993). Browning of foods: Control by sulfites, antioxidants, and other means. *Food Technology* 47(10): 75-84.

Schirra, M. and E. Cohen (1999). Long-term storage of 'Olinda' oranges under chilling and intermittent warming temperatures. *Postharvest Biology and Technology* 16: 63-69.

Schouten, S.P., R.K. Prange, J. Verschoor, T.R. Lammers, and J. Ossterhaven (1997). Improvement of quality of Elstar apples by dynamic control of ULO conditions. *Proceedings of the 7th International Controlled Atmosphere Research Conference.* University of California Davis 2:71-78.

Sembdner, G. and B. Parthier (1993). The biochemistry and the physiological and molecular actions of jasmonates. *Annual Review of Plant Physiology and Plant Molecular Biology* 44: 569-589.

Sen Gupta, A., R.P. Webb, A.S. Holaday, and R.D. Allen (1993). Overexpression of superoxide dismutase protects plants from oxidative stress. *Plant Physiology* 102: 1067-1073.

Shewfelt, R.L. and A.C. Purvis (1995). Toward a comprehensive model for lipid peroxidation in plant tissue disorders. *HortScience* 30: 213-218.

Solomos, T., B. Whitaker, and C. Lu (1997). Deleterious effects of pure oxygen on 'Gala' and 'Granny Smith' apples. *HortScience* 32: 458.

Thompson, J.E., R.L. Legge, and R.F. Barber (1987). The role of free radicals in senescence and wounding. *New Phytologist* 105: 317-344.

Tian, M.S., T. Islam, D.G. Stevenson, and D.E. Irving (1997). Color, ethylene production, respiration, and compositional changes in broccoli dipped in hot water. *Journal of the American Society for Horticultural Science* 122: 112-116.

Toivonen, P.M.A. (1992). The reduction of browning in parsnips. *Journal of Horticultural Science* 67: 547-551.

Toivonen, P.M.A. (1996). The effects of storage temperature, storage duration, hydro-cooling and micro-perforated wrap on shelf life of broccoli (*Brassica oleracea* L., Italica Group). *Postharvest Biology and Technology* 10: 59-65.

Toivonen, P.M.A. (1997). Non-ethylene, non-respiratory vocatives in harvested fruits and vegetables: Their occurrence, biological activity and control. *Postharvest Biology and Technology* 12: 109-125.

Toivonen, P.M.A. (2003). Effects of storage conditions and postharvest procedures on oxidative stress in fruits and vegetables. In *Postharvest Oxidative Stress in Horticultural Crops*, D.M. Hodges (Ed.). pp. 69-70. Binghamton, NY: Food Products Press.

Toivonen, P.M.A. and M. Sweeney (1997). Differences in chlorophyll loss at 13°C for two broccoli (*Brassica oleracea* L.) cultivars associated with antioxidant enzyme activities. *Journal of Agricultural and Food Chemistry* 46: 20-24.

Toivonen, P.M.A., M.R. Upadhyaya, and M.M. Gaye (1993). Low temperature preconditioning to improve the shelf life of fresh market carrots. *Acta Horticulturae* 343: 339-340.

Truter, A.B., J.C. Combrink, and F.J. Calitz (1993). Control of superficial scald of apples by ultra-low and stress levels of oxygen as an alternative to diphenylamine. *Proceedings of the 6th International Controlled Atmosphere Research Conference*. Ithaca, NY: Northeast Regional Agricultural Engineering Service, Cornell University. pp. 471-480.

Vamos-Vigyazo, L. (1981). Polyphenol oxidase and peroxidase in fruits and vegetables. *Critical Reviews in Food Science and Nutrition* 15: 49-127.

Van der merwe, J.R., J.C. Combrink, A.B. Truter, and F.J. Calitz (1997). Effect of initial low oxygen stress treatment and CA storage at increased carbon dioxide levels on post-storage quality of South African-grown 'Granny Smith' and 'Toped' apples. *Proceedings of the 7th International Controlled Atmosphere Research Conference*. University of California Davis 2: 79-84.

Wang, C.Y. (1991). Effect of abscisic acid on chilling injury of zucchini squash. *Journal of Plant Growth Regulation* 10: 101-105.

Wang, C.Y. (1993). Approaches to reduce chilling injury of fruits and vegetables. *Horticultural Reviews* 15: 63-95.

Wang, C.Y. (1994). Combined treatment of heat shock and low temperature conditioning reduces chilling in zucchini squash. *Postharvest Biology and Technology* 4: 65-73.

Wang, C.Y. (1995a). Effect of temperature preconditioning on catalase, peroxidase, and superoxide dismutase in chilled zucchini squash. *Postharvest Biology and Technology* 5: 67-76.

Wang, C.Y. (1995b). Temperature preconditioning affects glutathione content and glutathione reductase activity in chilled zucchini squash. *Journal of Plant Physiology* 145: 148-152.

Wang, C.Y. (1996). Temperature preconditioning affects ascorbate antioxidant system in chilled zucchini squash. *Postharvest Biology and Technology* 8: 29-36.

Wang, C.Y. and J.E. Baker (1979). Effects of two free radical scavengers and intermittent warming on chilling injury and polar lipid composition of cucumber and sweet pepper fruits. *Plant and Cell Physiology* 20: 243-251.

Wang, C.Y., G.F. Kramer, B.D. Whitaker, and W.R. Lusby (1992). Temperature preconditioning increases tolerance to chilling injury and alters lipid composition in zucchini squash. *Journal of Plant Physiology* 140: 229-235.

Wang, Z. and D.R. Dilley (2000a). Hypobaric storage removes scald-related vocatives during the low temperature induction of superficial scald of apples. *Postharvest Biology and Technology* 18: 191-199.

Wang, Z. and D.R. Dilley (2000b). Initial low oxygen stress controls superficial scald of apples. *Postharvest Biology and Technology* 18: 201-213.

Whitaker, B.D. (2000). DPA treatment alters α-farnesene metabolism in peel of 'Empire' apples stored in air or 1.5 percent O_2 atmosphere. *Postharvest Biology and Technology* 18: 91-97.

Whitaker, B.D., J.D. Klein, W.S. Conway, and C.E. Sams (1997). Influence of prestorage heat and calcium treatments on lipid metabolism in 'Golden Delicious' apples. *Phytochemistry* 45: 465-472.

Wong, D.W.S., S.J. Tillin, J.S. Hudson, and A.E. Pavlath (1994). Gas exchange in cut apples with bilayer coatings. *Journal of Agricultural and Food Chemistry* 42: 2278-2285.

Zhuang, H., D.F. Hildebrand, and M.M. Barth (1995). Senescence of broccoli buds is related to changes in lipid peroxidation. *Journal of Agricultural and Food Chemistry* 43: 2585-2591.

Index

Printed in the United States
by Baker & Taylor Publisher Services